Two-dimensional Crossing and Product Cubic Systems, Vol. II

Albert C. J. Luo

Two-dimensional Crossing and Product Cubic Systems, Vol. II

Crossing-linear and Self-quadratic
Product Vector Field

 Springer

Albert C. J. Luo
Department of Mechanical & Mechatronics Engineering
Southern Illinois University Edwardsville
Edwardsville, IL, USA

ISBN 978-3-031-57099-5   ISBN 978-3-031-57100-8   (eBook)
https://doi.org/10.1007/978-3-031-57100-8

This Springer imprint is published by the registered company Springer Nature Switzerland AG
The registered company address is: Gewerbestrasse 11, 6330 Cham, Switzerland

If disposing of this product, please recycle the paper.

茫茫人海遇，他乡逢知己，兄长同袍衣，华山论剑事。

——朝俊【顾克勤兄退休】癸卯秋

Preface

In this book, consider crossing and product cubic systems with a crossing-linear and self-quadratic product vector field. Nonlinear dynamics and singularity for such crossing and product cubic systems are presented. The parabola-saddles and third-order saddles and centers are discussed as the appearing bifurcations. The switching bifurcations are based on up-down hyperbolic upper-to-lower saddles on the up-down and down-up infinite-equilibriums, parabola-saddles on the inflection-saddle infinite-equilibriums, hyperbolic and circular upper-to-lower saddles on parabola-saddle infinite equilibriums, and parabola-saddles on the inflection-source and sink infinite-equilibriums. The materials in this book are scattered in five chapters.

In Chap. 1, nonlinear dynamics and singularity of a crossing and product cubic system with a crossing-linear and self-quadratic product vector field are discussed, and the corresponding switching dynamics are discussed through the infinite-equilibriums. A theory for nonlinear dynamics behaviors of such a cubic system is presented through a theorem. Such a cubic system possesses parabola-saddle, third-order saddles and centers, and hyperbolic singular flows for appearing bifurcations. The switching bifurcations are up-down hyperbolic upper-to-lower saddles, parabola-saddles on the inflection-saddle infinite-equilibriums, parabola-hyperbolic and circular upper-to-lower saddles, and parabola-saddle on the inflection-source and sink infinite-equilibriums.

In Chap. 2, the parabola-saddle, third-order centers and saddles, and hyperbolic singular flows in the crossing and product cubic systems are presented, and the corresponding switching dynamics are discussed through infinite-equilibriums in such cubic systems. There are three switching bifurcations on the infinite-equilibriums. The up-down hyperbolic upper-to-lower saddles are the switching bifurcations of hyperbolic singular flows with saddle and center. The parabola-saddles are the switching bifurcations of parabola-saddles with hyperbolic-to-hyperbolic-secant flows. The hyperbolic and circular upper-to-lower saddles are the switching bifurcations of a third-order saddle and a hyperbolic-to-hyperbolic-secant flow with a third-order center and a hyperbolic-secant-to hyperbolic flow.

In Chap. 3, series of centers and saddles with hyperbolic singular flows in the crossing and product cubic systems are presented, and the corresponding switching dynamics are discussed through infinite-equilibriums. The hyperbolic singular flows are the appearing bifurcations of hyperbolic and hyperbolic-secant flows. The up-down hyperbolic upper-to-lower saddles on the infinite-equilibriums are the switching bifurcations of parabola-saddle with saddle and center. The same directional saddles and centers are discovered in such cubic systems.

In Chap. 4, parabola-saddle and hyperbolic singular flows in the crossing and product cubic systems are presented, and the corresponding switching dynamics are discussed through infinite-equilibriums. The parabola-saddles on the inflection-source and sink infinite-equilibriums are the switching bifurcations of a saddle and a hyperbolic-secant flow with a center and hyperbolic flow. The parabola-hyperbolic sink-to-source on the parabola-sink and source infinite-equilibriums are the switching bifurcations of a parabola-saddle and a hyperbolic flow with another parabola-saddle with hyperbolic-secant flow. The third-order parabola-saddles are the switching bifurcations of a third-order saddle and a hyperbolic-secant flow with third-order center and hyperbolic flow.

In Chap. 5, simple equilibriums and hyperbolic flows forming a series in the crossing and product cubic systems are presented, and the corresponding switching dynamics are discussed through the inflection-source and sink infinite-equilibriums. The parabola-saddles on single and double inflection-source and sink infinite-equilibriums are discussed. The parabola-saddles are the switching bifurcations of a saddle and a hyperbolic-secant flow with a center and hyperbolic flow.

Finally, the author hopes the materials presented herein can provide a better understanding of cubic nonlinear systems in science and engineering.

Edwardsville, IL, USA   Albert C. J. Luo

Contents

Chapter 1
Crossing and Product-Cubic Systems

In this chapter, the nonlinear dynamics and singularity of a crossing and product-cubic system with a crossing-linear and self-quadratic product vector field are discussed, and the corresponding switching dynamics are discussed through the infinite equilibriums. A theory for the nonlinear dynamic behaviors of such a cubic system is presented through a theorem. Such a cubic system possesses parabola-saddles, third-order saddles and centers. Parabola-saddles are the appearing bifurcations of saddles and centers. Third-order saddles are the appearing bifurcations from a saddle to saddle, center, and saddle. A third-order center is the appearing bifurcation from a center to center, saddle, and center. Up-down hyperbolic upper-to-lower saddles are the switching bifurcations of hyperbolic singular flows with a saddle and center. Parabola-saddles are the switching bifurcations of parabola-saddles with hyperbolic-to-hyperbolic-secant flows. Parabola-hyperbolic upper-to-lower saddles are the switching bifurcations of a third-order saddle and a hyperbolic-to-hyperbolic-secant flow. Parabola-circular upper-to-lower saddles are the switching bifurcations of a third-order center and a hyperbolic-secant-to-hyperbolic flow. Parabola-saddles on the inflection source and sink infinite equilibriums are the switching bifurcations of a saddle and a hyperbolic-secant flow with a center and hyperbolic flow.

1.1 Crossing-Linear and Self-Quadratic Product Vector Fields

In this section, a crossing and product cubic system with a crossing-linear and self-quadratic product vector field will be discussed. The corresponding dynamical behaviors will be presented through the following theorem.

Theorem 1.1 *Consider a two-dimensional, cubic dynamical system with a product-cubic vector field and a crossing-cubic vector field as*

© The Author(s), under exclusive license to Springer Nature Switzerland AG 2025
A. C. J. Luo, *Two-dimensional Crossing and Product Cubic Systems, Vol. II,*
https://doi.org/10.1007/978-3-031-57100-8_1

$$\dot{x}_{j_2} = a_{j_2 j_2 0}(x_{j_1} - b_{j_2 j_1 1})(x_{j_1}^2 + B_{j_2 j_1} x_{j_1} + C_{j_2 j_1}),$$
$$\dot{x}_{j_1} = a_{j_1 j_1 0}(x_{j_2} - a_{j_1 j_2 1})(x_{j_1}^2 + B_{j_1 j_1} x_{j_1} + C_{j_1 j_1}),$$

(1.1)

with

$$\Delta_{j_1 j_1} = B_{j_1 j_1}^2 - 4C_{j_1 j_1}, \; \Delta_{j_2 j_1} = B_{j_2 j_1}^2 - 4C_{j_2 j_1}.$$

(1.2)

(i) For $\Delta_{j_1 j_2} < 0$ and $\Delta_{j_2 j_1} < 0$, the standard form is

$$\dot{x}_{j_2} = a_{j_2 j_2 0}(x_{j_1} - a_{j_2 j_1 1})\left[(x_{j_1} - a_{j_2 j_1})^2 + b_{j_2 j_1}\right],$$
$$\dot{x}_{j_1} = a_{j_1 j_1 0}(x_{j_2} - a_{j_1 j_2 1})\left[(x_{j_1} - a_{j_1 j_1})^2 + b_{j_1 j_1}\right]$$

(1.3)

where

$$a_{j_1 j_1} = -\frac{1}{2}B_{j_1 j_1}, b_{j_1 j_1} = \frac{1}{4}(-\Delta_{j_1 j_1}),$$
$$a_{j_2 j_1 1} = b_{j_2 j_1 1}, a_{j_2 j_1} = -\frac{1}{2}B_{j_2 j_1}, b_{j_2 j_1} = \frac{1}{4}(-\Delta_{j_2 j_1}).$$

(1.4)

The first integral manifold is

$$\frac{1}{2}\left[(x_{j_1} - a_{j_2 j_1 1})^2 - (x_{j_1 0} - a_{j_2 j_1 1})^2\right]$$
$$+ 2(a_{j_1 j_1} - a_{j_2 j_1})(x_{j_1} - x_{j_1 0}) + \left\{(a_{j_1 j_1} - a_{j_2 j_1})(a_{j_1 j_1} - a_{j_2 j_1 1})\right.$$
$$+ \frac{1}{2}\left[(a_{j_1 j_1} - a_{j_2 j_1})^2 + b_{j_2 j_1} - b_{j_1 j_1}\right]\right\} \ln \frac{|(x_{j_1} - a_{j_1 j_1})^2 + b_{j_1 j_1}|}{|(x_{j_1 0} - a_{j_1 j_1})^2 + b_{j_1 j_1}|}$$
$$+ \left\{(a_{j_1 j_1} - a_{j_2 j_1 1})\left[(a_{j_1 j_1} - a_{j_2 j_1})^2 + b_{j_2 j_1} - b_{j_1 j_1}\right]\right.$$
$$\left. - 2b_{j_1 j_1}(a_{j_1 j_1} - a_{j_2 j_1})\right\}\frac{1}{\sqrt{b_{j_1 j_1}}}(\arctan \frac{x_{j_1} - a_{j_1 j_1}}{\sqrt{b_{j_1 j_1}}} - \arctan \frac{x_{j_1 0} - a_{j_1 j_1}}{\sqrt{b_{j_1 j_1}}})$$
$$= \frac{1}{2}\frac{a_{j_1 j_1 0}}{a_{j_2 j_2 0}}\left[(x_{j_2} - a_{j_1 j_2 1})^2 - (x_{j_2 0} - a_{j_1 j_2 1})^2\right].$$

(1.5)

The equilibrium of $(x_{j_2}^*, x_{j_1}^*) = (a_{j_1 j_2 1}, a_{j_2 j_1 1})$ has the following properties:

- *For $a_{j_1 j_1 0} > 0$ and $a_{j_2 j_2 0} > 0$,*

$$(a_{j_1 j_2 1}, a_{j_2 j_1 1}) = \underbrace{(\text{UP}_+, \text{UP}_+)}_{\text{positive saddle}}.$$

(1.6)

The equilibrium of $(x_{j_2}^*, x_{j_1}^*) = (a_{j_1 j_2 1}, a_{j_2 j_1 1})$ is a (UP$_+$,UP$_+$)-positive saddle.

- *For $a_{j_1 j_1 0} < 0$ and $a_{j_2 j_2 0} > 0$,*

$$(a_{j_1 j_2 1}, a_{j_2 j_1 1}) = \underbrace{(\mathrm{DP}_+, \mathrm{DP}_-)}_{\text{CCW center}}. \tag{1.7}$$

The equilibrium of $(x_{j_2}^*, x_{j_1}^*) = (a_{j_1 j_2 1}, a_{j_2 j_1 1})$ is a $(\mathrm{DP}_+, \mathrm{DP}_-)$-counter-clockwise center.

- *For $a_{j_1 j_1 0} > 0$ and $a_{j_2 j_2 0} < 0$,*

$$(a_{j_1 j_2 1}, a_{j_2 j_1 1}) = \underbrace{(\mathrm{DP}_-, \mathrm{DP}_+)}_{\text{CW center}}. \tag{1.8}$$

The equilibrium of $(x_{j_2}^*, x_{j_1}^*) = (a_{j_1 j_2 1}, a_{j_2 j_1 1})$ is a $(\mathrm{DP}_-, \mathrm{DP}_+)$-clockwise center.

- *For $a_{j_1 j_1 0} < 0$ and $a_{j_2 j_2 0} < 0$,*

$$(a_{j_1 j_2 1}, a_{j_2 j_1 1}) = \underbrace{(\mathrm{UP}_-, \mathrm{UP}_-)}_{\text{negative saddle}}. \tag{1.9}$$

The equilibrium of $(x_{j_2}^*, x_{j_1}^*) = (a_{j_1 j_2 1}, a_{j_2 j_1 1})$ is a $(\mathrm{UP}_-, \mathrm{UP}_-)$-negative saddle.

(ii) For $\Delta_{j_1 j_1} < 0$ and $\Delta_{j_2 j_1} = 0$, the standard form is

$$
\begin{aligned}
\dot{x}_{j_2} &= a_{j_2 j_2 0}(x_{j_1} - a_{j_2 j_1 s_1})(x_{j_1} - a_{j_2 j_1 s_2})^2, \\
\dot{x}_{j_1} &= a_{j_1 j_1 0}(x_{j_2} - a_{j_1 j_2 1})\left[(x_{j_1} - a_{j_1 j_1})^2 + b_{j_1 j_1}\right]
\end{aligned}
\tag{1.10}
$$

where

$$
\begin{aligned}
a_{j_1 j_1} &= -\frac{1}{2} B_{j_1 j_1}, \quad b_{j_1 j_1} = \frac{1}{4}(-\Delta_{j_1 j_1}), \\
a_{j_2 j_1 s_1} &= b_{j_2 j_1 1}, \quad a_{j_2 j_1 s_2} = -\frac{1}{2} B_{j_2 j_1}.
\end{aligned}
\tag{1.11}
$$

(ii₁) The first integral manifold is

$$
\begin{aligned}
&\frac{1}{2}\left[(x_{j_1} - a_{j_2 j_1 s_1})^2 - (x_{j_1 0} - a_{j_2 j_1 s_1})^2\right] \\
&+ 2(a_{j_1 j_1} - a_{j_2 j_1 s_1})(x_{j_1} - x_{j_1 0}) + \{(a_{j_1 j_1} - a_{j_2 j_1 s_2})(a_{j_1 j_1} - a_{j_2 j_1 s_1}) \\
&+ \frac{1}{2}\left[(a_{j_1 j_1} - a_{j_2 j_1 s_2})^2 - b_{j_1 j_1}\right]\} \ln \frac{|(x_{j_1} - a_{j_1 j_1})^2 + b_{j_1 j_1}|}{|(x_{j_1 0} - a_{j_1 j_1})^2 + b_{j_1 j_1}|} \\
&+ \{(a_{j_1 j_1} - a_{j_2 j_1 s_1})\left[(a_{j_1 j_1} - a_{j_2 j_1 s_2})^2 - b_{j_1 j_1}\right] \\
&- 2b_{j_1 j_1}(a_{j_1 j_1} - a_{j_2 j_1 s_1})\} \frac{1}{\sqrt{b_{j_1 j_1}}}(\arctan \frac{x_{j_1} - a_{j_1 j_1}}{\sqrt{b_{j_1 j_1}}} - \arctan \frac{x_{j_1 0} - a_{j_1 j_1}}{\sqrt{b_{j_1 j_1}}}) \\
&= \frac{1}{2}\frac{a_{j_1 j_1 0}}{a_{j_2 j_2 0}}\left[(x_{j_2} - a_{j_1 j_2 1})^2 - (x_{j_2 0} - a_{j_1 j_2 1})^2\right].
\end{aligned}
\tag{1.12}
$$

(ii$_{1a}$) The equilibrium of $(x_{j_2}^*, x_{j_1}^*) = (a_{j_1 j_2 1}, a_{j_2 j_1 s_1})$ has the following properties:

- *For $a_{j_1 j_1 0} > 0$ and $a_{j_2 j_2 0}(a_{j_2 j_1 s_1} - a_{j_2 j_1 s_2})^2 > 0$,*

$$(a_{j_1 j_2 1}, a_{j_2 j_1 s_1}) = \underbrace{(\mathrm{UP}_+, \mathrm{UP}_+)}_{\text{positive saddle}}. \tag{1.13}$$

  The equilibrium of $(x_{j_2}^*, x_{j_1}^*) = (a_{j_1 j_2 1}, a_{j_2 j_1 s_1})$ is a (UP$_+$,UP$_+$)-positive saddle.
- *For $a_{j_1 j_1 0} < 0$ and $a_{j_2 j_2 0}(a_{j_2 j_1 s_1} - a_{j_2 j_1 s_2})^2 > 0$,*

$$(a_{j_1 j_2 1}, a_{j_2 j_1 s_1}) = \underbrace{(\mathrm{DP}_+, \mathrm{DP}_-)}_{\text{CCW center}}. \tag{1.14}$$

  The equilibrium of $(x_{j_2}^*, x_{j_1}^*) = (a_{j_1 j_2 1}, a_{j_2 j_1 s_1})$ is a (DP$_+$,DP$_-$)-counter-clockwise center.
- *For $a_{j_1 j_1 0} > 0$ and $a_{j_2 j_2 0}(a_{j_2 j_1 s_1} - a_{j_2 j_1 s_2})^2 < 0$,*

$$(a_{j_1 j_2 1}, a_{j_2 j_1 s_1}) = \underbrace{(\mathrm{DP}_-, \mathrm{DP}_+)}_{\text{CW center}}. \tag{1.15}$$

  The equilibrium of $(x_{j_2}^*, x_{j_1}^*) = (a_{j_1 j_2 1}, a_{j_2 j_1 s_1})$ is a (DP$_-$,DP$_+$)-clockwise center.
- *For $a_{j_1 j_1 0} < 0$ and $a_{j_2 j_2 0}(a_{j_2 j_1 s_1} - a_{j_2 j_1 s_2})^2 < 0$,*

$$(a_{j_1 j_2 1}, a_{j_2 j_1 s_1}) = \underbrace{(\mathrm{UP}_-, \mathrm{UP}_-)}_{\text{negative saddle}}. \tag{1.16}$$

  The equilibrium of $(x_{j_2}^*, x_{j_1}^*) = (a_{j_1 j_2 1}, a_{j_2 j_1 s_1})$ is a (UP$_-$,UP$_-$)-negative saddle.

(ii$_{1b}$) The equilibrium of $(x_{j_2}^*, x_{j_1}^*) = (a_{j_1 j_2 1}, a_{j_2 j_1 s_2})$ has the following properties:

- *For $a_{j_1 j_1 0} > 0$ and $a_{j_2 j_2 0}(a_{j_2 j_1 s_2} - a_{j_2 j_1 s_1}) > 0$,*

$$(a_{j_1 j_2 1}, a_{j_2 j_1 s_2}) = \underbrace{(\mathrm{UP}, \mathrm{US})}_{\text{up-parabola upper-saddle}}. \tag{1.17}$$

  The equilibrium of $(x_{j_2}^*, x_{j_1}^*) = (a_{j_1 j_2 1}, a_{j_2 j_1 s_2})$ is a (UP,US)-up-parabola upper-saddle.
- *For $a_{j_1 j_1 0} < 0$ and $a_{j_2 j_2 0}(a_{j_2 j_1 s_2} - a_{j_2 j_1 s_1}) > 0$,*

$$(a_{j_1 j_2 1}, a_{j_2 j_1 s_2}) = \underbrace{(\mathrm{DP}, \mathrm{US})}_{\text{down-parabola upper-saddle}}. \tag{1.18}$$

  The equilibrium of $(x_{j_2}^*, x_{j_1}^*) = (a_{j_1 j_2 1}, a_{j_2 j_1 s_2})$ is a (DP,US)-down-parabola upper-saddle.

- *For $a_{j_1j_10} > 0$ and $a_{j_2j_20}(a_{j_2j_1s_2} - a_{j_2j_1s_1}) < 0$,*

$$(a_{j_1j_21}, a_{j_2j_1s_2}) = \underbrace{(\text{DP}, \text{LS})}_{\text{down-parabola lower-saddle}} . \tag{1.19}$$

The equilibrium of $(x_{j_2}^*, x_{j_1}^*) = (a_{j_1j_21}, a_{j_2j_1s_2})$ is a (DP,LS)-down-parabola lower-saddle.

- *For $a_{j_1j_10} < 0$ and $a_{j_2j_20}(a_{j_2j_1s_2} - a_{j_2j_1s_1}) < 0$,*

$$(a_{j_1j_21}, a_{j_2j_1s_2}) = \underbrace{(\text{UP}, \text{LS})}_{\text{up-parabola lower-saddle}} . \tag{1.20}$$

The equilibrium of $(x_{j_2}^*, x_{j_1}^*) = (a_{j_1j_21}, a_{j_2j_1s_2})$ is a (UP,LS)-up-parabola lower-saddle.

(ii_2) *For $a_{j_2j_1s_1} = a_{j_2j_1s_2} = a_{j_2j_11}$,*

$$\begin{aligned}\dot{x}_{j_2} &= a_{j_2j_20}(x_{j_1} - a_{j_2j_11})^3, \\ \dot{x}_{j_1} &= a_{j_1j_10}(x_{j_2} - a_{j_1j_21})\left[(x_{j_1} - a_{j_1j_1})^2 + b_{j_1j_1}\right].\end{aligned} \tag{1.21}$$

The first integral manifold is

$$\begin{aligned}
&\frac{1}{2}\left[(x_{j_1} - a_{j_2j_11})^2 - (x_{j_10} - a_{j_2j_11})^2\right] + 2(a_{j_1j_1} - a_{j_2j_11})(x_{j_1} - x_{j_10}) \\
&+ \frac{1}{2}\left[3(a_{j_1j_1} - a_{j_2j_11})^2 - b_{j_1j_1}\right] \ln \frac{|(x_{j_1} - a_{j_1j_1})^2 + b_{j_1j_1}|}{|(x_{j_10} - a_{j_1j_1})^2 + b_{j_1j_1}|} \\
&+ (a_{j_1j_1} - a_{j_2j_11})\left[(a_{j_1j_1} - a_{j_2j_11})^2 - 3b_{j_1j_1}\right] \frac{1}{\sqrt{b_{j_1j_1}}}(\arctan \frac{x_{j_1} - a_{j_1j_1}}{\sqrt{b_{j_1j_1}}} - \arctan \frac{x_{j_10} - a_{j_1j_1}}{\sqrt{b_{j_1j_1}}}) \\
&= \frac{1}{2}\frac{a_{j_1j_10}}{a_{j_2j_20}}\left[(x_{j_2} - a_{j_1j_21})^2 - (x_{j_20} - a_{j_1j_21})^2\right].
\end{aligned} \tag{1.22}$$

The equilibrium of $(x_{j_2}^*, x_{j_1}^*) = (a_{j_1j_21}, a_{j_2j_1s_2})$ has the following properties:

- *For $a_{j_1j_10} > 0$ and $a_{j_2j_20} > 0$,*

$$(a_{j_1j_21}, a_{j_2j_11}) = \underbrace{(\text{UP}_+, 3^{\text{rd}}\text{UP}_+)}_{\text{third-order positive saddle}} . \tag{1.23}$$

The equilibrium of $(x_{j_2}^*, x_{j_1}^*) = (a_{j_1j_21}, a_{j_2j_11})$ is a $(\text{UP}_+, 3^{\text{rd}}\text{UP}_+)$-third-order positive saddle.

- *For $a_{j_1 j_1 0} < 0$ and $a_{j_2 j_2 0} > 0$,*

$$(a_{j_1 j_2 1}, a_{j_2 j_1 1}) = \underbrace{(DP_+, 3^{rd}DP_-)}_{\text{third-order CCW center}} . \tag{1.24}$$

The equilibrium of $(x^*_{j_2}, x^*_{j_1}) = (a_{j_1 j_2 1}, a_{j_2 j_1 1})$ is a $(DP_+, 3^{rd}DP_-)$-third-order counter-clockwise center.

- *For $a_{j_1 j_1 0} > 0$ and $a_{j_2 j_2 0} < 0$,*

$$(a_{j_1 j_2 1}, a_{j_2 j_1 1}) = \underbrace{(DP_-, 3^{rd}DP_+)}_{\text{third-order CW center}} . \tag{1.25}$$

The equilibrium of $(x^*_{j_2}, x^*_{j_1}) = (a_{j_1 j_2 1}, a_{j_2 j_1 1})$ is a $(DP_-, 3^{rd}DP_+)$-third-order clockwise center.

- *For $a_{j_1 j_1 0} < 0$ and $a_{j_2 j_2 0} < 0$,*

$$(a_{j_1 j_2 1}, a_{j_2 j_1 1}) = \underbrace{(UP_-, 3^{rd}UP_-)}_{\text{third-order negative saddle}} . \tag{1.26}$$

The equilibrium of $(x^*_{j_2}, x^*_{j_1}) = (a_{j_1 j_2 1}, a_{j_2 j_1 1})$ is a $(UP_-, 3^{rd}UP_-)$-third-order negative saddle.

(iii) For $\Delta_{j_1 j_2} < 0$ and $\Delta_{j_2 j_1} > 0$, the standard form is

$$\begin{aligned}
\dot{x}_{j_2} &= a_{j_2 j_2 0}(x_{j_1} - a_{j_2 j_1 1})(x_{j_1} - a_{j_2 j_1 2})(x_{j_1} - a_{j_2 j_1 3}), \\
\dot{x}_{j_1} &= a_{j_1 j_1 0}(x_{j_2} - a_{j_1 j_2 1})\left[(x_{j_1} - a_{j_1 j_1})^2 + b_{j_1 j_1}\right]
\end{aligned} \tag{1.27}$$

where

$$\begin{aligned}
&a_{j_1 j_1} = -\frac{1}{2}B_{j_1 j_1}, b_{j_1 j_1} = \frac{1}{4}(-\Delta_{j_1 j_1}), \\
&b_{j_2 j_1 2}, b_{j_2 j_1 3} = -\frac{1}{2}(B_{j_2 j_1} \pm \sqrt{\Delta_{j_2 j_1}}), \\
&\{a_{j_2 j_1 1}, a_{j_2 j_1 2}, a_{j_2 j_1 3}\} = \text{sort}\{b_{j_2 j_1 1}, b_{j_2 j_1 2}, b_{j_2 j_1 2}\}, \\
&a_{j_2 j_1 s_1} < a_{j_2 j_1 s_2}, s_1, s_2 \in \{1, 2, 3\}, s_1 < s_2.
\end{aligned} \tag{1.28}$$

The first integral manifold is

$$\frac{1}{2}\left[(x_{j_1} - a_{j_2 j_1 s_1})^2 - (x_{j_1 0} - a_{j_2 j_1 s_1})^2\right] + (2a_{j_1 j_1} - a_{j_2 j_1 s_2} - a_{j_2 j_1 s_3})(x_{j_1} - x_{j_1 0})$$

$$+\frac{1}{2}\left\{(2a_{j_1 j_1} - a_{j_2 j_1 s_2} - a_{j_2 j_1 s_3})(a_{j_1 j_1} - a_{j_2 j_1 s_1})\right.$$

$$+\left.\left[(a_{j_1 j_1} - a_{j_2 j_1 s_2})(a_{j_1 j_1} - a_{j_2 j_1 s_3}) - b_{j_1 j_1}\right]\right\} \ln \frac{\mid (x_{j_1} - a_{j_1 j_1})^2 + b_{j_1 j_1}\mid}{\mid (x_{j_1 0} - a_{j_1 j_1})^2 + b_{j_1 j_1}\mid}$$

$$+\left\{(a_{j_1 j_1} - a_{j_2 j_1 s_1})\left[(a_{j_1 j_1} - a_{j_2 j_1 s_2})(a_{j_1 j_1} - a_{j_2 j_1 s_3}) - b_{j_1 j_1}\right]\right.$$

$$\left. - b_{j_1 j_1}(2a_{j_1 j_1} - a_{j_2 j_1 s_2} - a_{j_2 j_1 s_3})\right\} \frac{1}{\sqrt{b_{j_1 j_1}}}(\arctan \frac{x_{j_1} - a_{j_1 j_1}}{\sqrt{b_{j_1 j_1}}} - \arctan \frac{x_{j_1 0} - a_{j_1 j_1}}{\sqrt{b_{j_1 j_1}}})$$

$$= \frac{1}{2}\frac{a_{j_1 j_1 0}}{a_{j_2 j_2 0}}\left[(x_{j_2} - a_{j_1 j_2 1})^2 - (x_{j_2 0} - a_{j_1 j_2 1})^2\right].$$

$$(1.29)$$

The equilibrium of $(x_{j_2}^*, x_{j_1}^*) = (a_{j_1 j_2 1}, a_{j_2 j_1 s_1})$ $(s_1 = 1, 2, 3)$ *has the following properties:*

- *For* $a_{j_1 j_1 0} > 0$ *and* $a_{j_2 j_2 0}\prod_{s_2 = 1, s_2 \neq s_1}^3 (a_{j_2 j_1 s_1} - a_{j_2 j_1 s_2}) > 0$,

$$(a_{j_1 j_2 1}, a_{j_2 j_1 s_1}) = \underbrace{(\text{UP}_+, \text{UP}_+)}_{\text{positive saddle}}. \qquad (1.30)$$

  The equilibrium of $(x_{j_2}^*, x_{j_1}^*) = (a_{j_1 j_2 1}, a_{j_2 j_1 s_1})$ *is a* $(\text{UP}_+, \text{UP}_+)$*-positive saddle.*

- *For* $a_{j_1 j_1 0} < 0$ *and* $a_{j_2 j_2 0}\prod_{s_2 = 1, s_2 \neq s_1}^3 (a_{j_2 j_1 s_1} - a_{j_2 j_1 s_2}) > 0$,

$$(a_{j_1 j_2 1}, a_{j_2 j_1 1}) = \underbrace{(\text{DP}_+, \text{DP}_-)}_{\text{CCW center}}. \qquad (1.31)$$

  The equilibrium of $(x_{j_2}^*, x_{j_1}^*) = (a_{j_1 j_2 1}, a_{j_2 j_1 s_1})$ *is a* $(\text{DP}_+, \text{DP}_-)$*-counter-clockwise center.*

- *For* $a_{j_1 j_1 0} > 0$ *and* $a_{j_2 j_2 0}\prod_{s_2 = 1, s_2 \neq s_1}^3 (a_{j_2 j_1 s_1} - a_{j_2 j_1 s_2}) < 0$,

$$(a_{j_1 j_2 1}, a_{j_2 j_1 1}) = \underbrace{(\text{DP}_-, \text{DP}_+)}_{\text{CW center}}. \qquad (1.32)$$

  The equilibrium of $(x_{j_2}^*, x_{j_1}^*) = (a_{j_1 j_2 1}, a_{j_2 j_1 s_1})$ *is a* $(\text{DP}_-, \text{DP}_+)$*-clockwise center.*

- *For* $a_{j_1 j_1 0} < 0$ *and* $a_{j_2 j_2 0}\prod_{s_2 = 1, s_2 \neq s_1}^3 (a_{j_2 j_1 s_1} - a_{j_2 j_1 s_2}) < 0$,

$$(a_{j_1 j_2 1}, a_{j_2 j_1 1}) = \underbrace{(\text{UP}_-, \text{UP}_-)}_{\text{negative saddle}}. \qquad (1.33)$$

  The equilibrium of $(x_{j_2}^*, x_{j_1}^*) = (a_{j_1 j_2 1}, a_{j_2 j_1 s_1})$ *is a* $(\text{UP}_-, \text{UP}_-)$*-negative saddle.*

(iv) For $\Delta_{j_1 j_2} = 0$ and $\Delta_{j_2 j_1} < 0$, the standard form is

$$
\begin{aligned}
\dot{x}_{j_2} &= a_{j_2 j_2 0}(x_{j_1} - a_{j_2 j_1 1})\big[(x_{j_1} - a_{j_2 j_1})^2 + b_{j_2 j_1}\big], \\
\dot{x}_{j_1} &= a_{j_1 j_1 0}(x_{j_2} - a_{j_1 j_2 1})(x_{j_1} - a_{j_1 j_1 1})^2
\end{aligned}
\tag{1.34}
$$

where

$$
\begin{aligned}
a_{j_1 j_1 1} &= -\frac{1}{2} B_{j_1 j_1}, a_{j_2 j_1 1} = b_{j_2 j_1 1}, \\
a_{j_2 j_1} &= -\frac{1}{2} B_{j_2 j_1}, b_{j_2 j_1} = \frac{1}{4}(-\Delta_{j_2 j_1}).
\end{aligned}
\tag{1.35}
$$

(iv₁) The first integral manifold for $a_{j_2 j_1 1} \neq a_{j_1 j_1 1}$ is

$$
\begin{aligned}
&\frac{1}{2}\big[(x_{j_1} - a_{j_2 j_1 1})^2 - (x_{j_1 0} - a_{j_2 j_1 1})^2\big] \\
&+ 2(a_{j_1 j_1 1} - a_{j_2 j_1})(x_{j_1} - x_{j_1 0}) + \Big\{ 2(a_{j_1 j_1 1} - a_{j_2 j_1})(a_{j_1 j_1 1} - a_{j_2 j_1 1}) \\
&+ \big[(a_{j_1 j_1 1} - a_{j_2 j_1})^2 + b_{j_2 j_1}\big]\Big\} \ln \frac{|x_{j_1} - a_{j_1 j_1 1}|}{|x_{j_1 0} - a_{j_1 j_1 1}|} \\
&- (a_{j_1 j_1 1} - a_{j_2 j_1 1})\big[(a_{j_1 j_1 1} - a_{j_2 j_1})^2 + b_{j_2 j_1}\big]\Big(\frac{1}{x_{j_1} - a_{j_1 j_1 1}} - \frac{1}{x_{j_1 0} - a_{j_1 j_1 1}}\Big) \\
&= \frac{1}{2}\frac{a_{j_1 j_1 0}}{a_{j_2 j_2 0}}\big[(x_{j_2} - a_{j_1 j_2 1})^2 - (x_{j_2 0} - a_{j_1 j_2 1})^2\big].
\end{aligned}
\tag{1.36}
$$

(iv₁ₐ) The equilibrium of $(x_{j_2}^*, x_{j_1}^*) = (a_{j_1 j_2 1}, a_{j_2 j_1 1})$ has the following properties:

- *For $a_{j_1 j_1 0}(a_{j_2 j_1 1} - a_{j_1 j_1 1})^2 > 0$ and $a_{j_2 j_2 0} > 0$,*

$$
(a_{j_1 j_2 1}, a_{j_2 j_1 1}) = \underbrace{(\mathrm{UP}_+, \mathrm{UP}_+)}_{\text{positive saddle}}.
\tag{1.37}
$$

  The equilibrium of $(x_{j_2}^*, x_{j_1}^*) = (a_{j_1 j_2 1}, a_{j_2 j_1 1})$ is a $(\mathrm{UP}_+, \mathrm{UP}_+)$-positive saddle.
- *For $a_{j_1 j_1 0}(a_{j_2 j_1 1} - a_{j_1 j_1 1})^2 < 0$ and $a_{j_2 j_2 0} > 0$,*

$$
(a_{j_1 j_2 1}, a_{j_2 j_1 1}) = \underbrace{(\mathrm{DP}_+, \mathrm{DP}_-)}_{\text{CCW center}}.
\tag{1.38}
$$

  The equilibrium of $(x_{j_2}^*, x_{j_1}^*) = (a_{j_1 j_2 1}, a_{j_2 j_1 1})$ is a $(\mathrm{DP}_+, \mathrm{DP}_-)$-counter-clockwise center.
- *For $a_{j_1 j_1 0}(a_{j_2 j_1 1} - a_{j_1 j_1 1})^2 > 0$ and $a_{j_2 j_2 0} < 0$,*

$$
(a_{j_1 j_2 1}, a_{j_2 j_1 1}) = \underbrace{(\mathrm{DP}_-, \mathrm{DP}_+)}_{\text{CW center}}.
\tag{1.39}
$$

  The equilibrium of $(x_{j_2}^*, x_{j_1}^*) = (a_{j_1 j_2 1}, a_{j_2 j_1 1})$ is a $(\mathrm{DP}_-, \mathrm{DP}_+)$-clockwise center.

- *For $a_{j_1 j_1 0}(a_{j_2 j_1 1} - a_{j_1 j_1 1})^2 < 0$ and $a_{j_2 j_2 0} < 0$,*

$$(a_{j_1 j_2 1}, a_{j_2 j_1 1}) = \underbrace{(\text{UP}_-, \text{UP}_-)}_{\text{negative saddle}}. \tag{1.40}$$

The equilibrium of $(x^*_{j_2}, x^*_{j_1}) = (a_{j_1 j_2 1}, a_{j_2 j_1 1})$ is a (UP$_-$,UP$_-$)-negative saddle.

(iv$_{1b}$) The equilibrium of $(x^*_{j_1}, x^*_{j_2}) = (a_{j_1 j_1 1}, a_{j_1 j_2 1})$ has the following properties:

- *For $a_{j_1 j_1 0} > 0$ and $a_{j_2 j_2 0}(a_{j_1 j_1 1} - a_{j_2 j_1 1}) > 0$,*

$$(a_{j_1 j_1 1}, a_{j_1 j_2 1}) = \underbrace{(\text{UP:UP}, \text{pF})}_{\text{hyperbolic-secant-to-hyperbolic flow } (+)}. \tag{1.41}$$

The equilibrium of $(x^*_{j_1}, x^*_{j_2}) = (a_{j_1 j_1 1}, a_{j_1 j_2 1})$ is a (UP:UP,pF)-positive hyperbolic-secant-to-hyperbolic flow.
- *For $a_{j_1 j_1 0} < 0$ and $a_{j_2 j_2 0}(a_{j_1 j_1 1} - a_{j_2 j_1 1}) > 0$,*

$$(a_{j_1 j_1 1}, a_{j_1 j_2 1}) = \underbrace{(\text{DP:DP}, \text{pF})}_{\text{hyperbolic-to-hyperbolic-secant flow } (+)}. \tag{1.42}$$

The equilibrium of $(x^*_{j_1}, x^*_{j_2}) = (a_{j_1 j_1 1}, a_{j_1 j_2 1})$ is a (DP:DP,pF)-positive hyperbolic-to-hyperbolic-secant flow.
- *For $a_{j_1 j_1 0} > 0$ and $a_{j_2 j_2 0}(a_{j_1 j_1 1} - a_{j_2 j_1 1}) < 0$,*

$$(a_{j_1 j_1 1}, a_{j_1 j_2 1}) = \underbrace{(\text{DP:DP}, \text{nF})}_{\text{hyperbolic-to-hyperbolic-secant flow } (-)}. \tag{1.43}$$

The equilibrium of $(x^*_{j_1}, x^*_{j_2}) = (a_{j_1 j_1 1}, a_{j_1 j_2 1})$ is a (DP:DP,nF)-negative hyperbolic-secant-to-hyperbolic flow.
- *For $a_{j_1 j_1 0} < 0$ and $a_{j_2 j_2 0}(a_{j_1 j_1 1} - a_{j_2 j_1 1}) < 0$,*

$$(a_{j_1 j_1 1}, a_{j_1 j_2 1}) = \underbrace{(\text{UP:UP}, \text{nF})}_{\text{hyperbolic-secant-to-hyperbolic flow } (-)}. \tag{1.44}$$

The equilibrium of $(x^*_{j_1}, x^*_{j_2}) = (a_{j_1 j_1 1}, a_{j_1 j_2 1})$ is a (UP:UP,nF)-negative hyperbolic-secant to hyperbolic flow.

(iv$_2$) The first integral manifold for $a_{j_2 j_1 1} = a_{j_1 j_1 1}$ is

$$\frac{1}{2}\left[(x_{j_1} - a_{j_1 j_1 1})^2 - (x_{j_1 0} - a_{j_1 j_1 1})^2\right]$$

$$+ 2(a_{j_1 j_1 1} - a_{j_2 j_1})(x_{j_1} - x_{j_1 0}) + \left[(a_{j_1 j_1 1} - a_{j_2 j_1})^2 + b_{j_2 j_1}\right] \ln \frac{|x_{j_1} - a_{j_1 j_1 1}|}{|x_{j_1 0} - a_{j_1 j_1 1}|} \tag{1.45}$$

$$= \frac{1}{2} \frac{a_{j_1 j_1 0}}{a_{j_2 j_2 0}}\left[(x_{j_2} - a_{j_1 j_2 1})^2 - (x_{j_2 0} - a_{j_1 j_2 1})^2\right].$$

(iv$_{2a}$) The infinite-equilibrium of $x_{j_1}^* = a_{j_2 j_1 1} = a_{j_1 j_1 1}$ with $\bar{x}_{j_2} \neq a_{j_1 j_2 1}$ has the following properties:

- *For $a_{j_1 j_1 0}(\bar{x}_{j_2} - a_{j_1 j_2 1}) > 0$ and $a_{j_2 j_2 0} > 0$,*

$$(a_{j_1 j_1 1}, \bar{x}_{j_2}) = \underbrace{(\text{US}, \text{DU})}_{\text{down-up upper-saddle}} . \tag{1.46}$$

The infinite-equilibrium of $x_{j_1}^* = a_{j_2 j_1 1} = a_{j_1 j_1 1}$ is a (US,DU)-down-up asymptotic upper-saddle.
- *For $a_{j_1 j_1 0}(\bar{x}_{j_2} - a_{j_1 j_2 1}) < 0$ and $a_{j_2 j_2 0} > 0$,*

$$(a_{j_1 j_1 1}, \bar{x}_{j_2}) = \underbrace{(\text{LS}, \text{UD})}_{\text{up-down lower-saddle}} . \tag{1.47}$$

The infinite-equilibrium of $x_{j_1}^* = a_{j_2 j_1 1} = a_{j_1 j_1 1}$ is an (LS,UD)-up-down asymptotic lower-saddle.
- *For $a_{j_1 j_1 0}(\bar{x}_{j_2} - a_{j_1 j_2 1}) > 0$ and $a_{j_2 j_2 0} < 0$,*

$$(a_{j_1 j_1 1}, \bar{x}_{j_2}) = \underbrace{(\text{US}, \text{UD})}_{\text{up-down upper-saddle}} . \tag{1.48}$$

The infinite-equilibrium of $x_{j_1}^* = a_{j_2 j_1 1} = a_{j_1 j_1 1}$ is a (US,UD)-up-down asymptotic upper-saddle.
- *For $a_{j_1 j_1 0}(\bar{x}_{j_2} - a_{j_1 j_2 1}) < 0$ and $a_{j_2 j_2 0} < 0$,*

$$(a_{j_1 j_1 1}, \bar{x}_{j_2}) = \underbrace{(\text{LS}, \text{DU})}_{\text{down-up lower-saddle}} . \tag{1.49}$$

The infinite-equilibrium of $x_{j_1}^* = a_{j_2 j_1 1} = a_{j_1 j_1 1}$ is an (LS,DU)-down-up asymptotic lower-saddle.

(iv$_{2b}$) The equilibrium of $(x_{j_2}^*, x_{j_1}^*) = (a_{j_1 j_2 1}, a_{j_2 j_1 1})$ with $a_{j_1 j_1 1} = a_{j_2 j_1 1}$ has the following properties:

- *For $a_{j_1 j_1 0} > 0$ and $a_{j_2 j_2 0} > 0$,*

$$(a_{j_1 j_2 1}, a_{j_2 j_1 1}) = \underbrace{(_{\text{UD}}\text{LS} : {}_{\text{DU}}\text{US}, \text{DP}_- : \text{UP}_+)}_{\text{hyperbolic lower-to-upper saddle}}. \tag{1.50}$$

The equilibrium of $(x_{j_2}^*, x_{j_1}^*) = (a_{j_1 j_2 1}, a_{j_2 j_1 1})$ for $a_{j_1 j_1 1} = a_{j_2 j_1 1}$ is a ($_{\text{UD}}$LS: $_{\text{DU}}$US, DP$_-$:UP$_+$)-hyperbolic lower-to-upper saddle.

- *For $a_{j_1j_10} < 0$ and $a_{j_2j_20} > 0$,*

$$(a_{j_1j_21}, a_{j_2j_11}) = \underbrace{(_{DU}US:{}_{UD}LS, UP_-:DP_+)}_{\text{hyperbolic-secant upper-to-lower saddle}}.\tag{1.51}$$

The equilibrium of $(x_{j_2}^*, x_{j_1}^*) = (a_{j_1j_21}, a_{j_2j_11})$ for $a_{j_1j_11} = a_{j_2j_11}$ is a $(_{DU}US:{}_{UD}LS, UP_-:DP_+)$-hyperbolic-secant upper-to-lower saddle.
- *For $a_{j_1j_10} > 0$ and $a_{j_2j_20} < 0$,*

$$(a_{j_1j_21}, a_{j_2j_11}) = \underbrace{(_{DU}LS:{}_{UD}US, UP_+:DP_-)}_{\text{hyperbolic-secant lower-to-upper saddle}}.\tag{1.52}$$

The equilibrium of $(x_{j_2}^*, x_{j_1}^*) = (a_{j_1j_21}, a_{j_2j_11})$ for $a_{j_1j_11} = a_{j_2j_11}$ is a $(_{DU}LS:{}_{UD}US, UP_+:DP_-)$-hyperbolic-secant lower-to-upper saddle.
- *For $a_{j_1j_10} < 0$ and $a_{j_2j_20} < 0$,*

$$(a_{j_1j_21}, a_{j_2j_11}) = \underbrace{(_{UD}US:{}_{DU}LS, DP_+:UP_-)}_{\text{hyperbolic upper-to-lower saddle}}.\tag{1.53}$$

The equilibrium of $(x_{j_2}^*, x_{j_1}^*) = (a_{j_1j_21}, a_{j_2j_11})$ for $a_{j_1j_11} = a_{j_2j_11}$ is a $(_{UD}US:{}_{DU}LS, DP_+:UP_-)$-hyperbolic upper-to-lower saddle.

(v) For $\Delta_{j_1j_2} = 0$ and $\Delta_{j_2j_1} = 0$, the standard form is

$$\begin{aligned}
\dot{x}_{j_2} &= a_{j_2j_20}(x_{j_1} - a_{j_2j_1s_1})(x_{j_1} - a_{j_2j_1s_2})^2,\\
\dot{x}_{j_1} &= a_{j_1j_10}(x_{j_2} - a_{j_1j_21})(x_{j_1} - a_{j_1j_11})^2
\end{aligned}\tag{1.54}$$

where

$$a_{j_1j_11} = -\frac{1}{2}B_{j_1j_1}, a_{j_2j_1s_1} = b_{j_2j_11}, a_{j_2j_1s_2} = -\frac{1}{2}B_{j_2j_1}.\tag{1.55}$$

(v_1) The first integral manifold for $a_{j_1j_11} \neq a_{j_2j_1s_1} \neq a_{j_2j_1s_2}$ is

$$\begin{aligned}
&\frac{1}{2}\left[(x_{j_1} - a_{j_2j_1s_1})^2 - (x_{j_10} - a_{j_2j_1s_1})^2\right]\\
&+2(a_{j_1j_11} - a_{j_2j_1s_2})(x_{j_1} - x_{j_10})\\
&+\left[(a_{j_1j_11} - a_{j_2j_1s_2})^2 + 2(a_{j_1j_11} - a_{j_2j_1s_1})(a_{j_1j_11} - a_{j_2j_1s_2})\right]\ln\frac{|x_{j_1} - a_{j_1j_11}|}{|x_{j_10} - a_{j_1j_11}|}\\
&-(a_{j_1j_11} - a_{j_2j_1s_1})(a_{j_1j_11} - a_{j_2j_1s_2})^2(\frac{1}{x_{j_1} - a_{j_1j_11}} - \frac{1}{x_{j_10} - a_{j_1j_11}})\\
&=\frac{1}{2}\frac{a_{j_1j_10}}{a_{j_2j_20}}\left[(x_{j_2} - a_{j_1j_21})^2 - (x_{j_20} - a_{j_1j_21})^2\right].
\end{aligned}\tag{1.56}$$

(v_{1a}) *The equilibrium of* $(x_{j_2}^*, x_{j_1}^*) = (a_{j_1 j_2 1}, a_{j_2 j_1 s_1})$ *has the following properties:*

- *For* $a_{j_1 j_1 0}(a_{j_2 j_1 s_1} - a_{j_1 j_1 1})^2 > 0$ *and* $a_{j_2 j_2 0}(a_{j_2 j_1 s_1} - a_{j_2 j_1 s_2})^2 > 0$,

$$(a_{j_1 j_2 1}, a_{j_2 j_1 s_1}) = \underbrace{(UP_+, UP_+)}_{\text{positive saddle}}. \tag{1.57}$$

The equilibrium of $(x_{j_2}^*, x_{j_1}^*) = (a_{j_1 j_2 1}, a_{j_2 j_1 s_1})$ *is a* (UP$_+$,UP$_+$)-*positive saddle.*
- *For* $a_{j_1 j_1 0}(a_{j_2 j_1 s_1} - a_{j_1 j_1 1})^2 < 0$ *and* $a_{j_2 j_2 0}(a_{j_2 j_1 s_1} - a_{j_2 j_1 s_2})^2 > 0$,

$$(a_{j_1 j_2 1}, a_{j_2 j_1 s_1}) = \underbrace{(DP_+, DP_-)}_{\text{CCW center}}. \tag{1.58}$$

The equilibrium of $(x_{j_2}^*, x_{j_1}^*) = (a_{j_1 j_2 1}, a_{j_2 j_1 s_1})$ *is a* (DP$_+$,DP$_-$)-*counter-clockwise center.*
- *For* $a_{j_1 j_1 0}(a_{j_2 j_1 s_1} - a_{j_1 j_1 1})^2 > 0$ *and* $a_{j_2 j_2 0}(a_{j_2 j_1 s_1} - a_{j_2 j_1 s_2})^2 < 0$,

$$(a_{j_1 j_2 1}, a_{j_2 j_1 s_1}) = \underbrace{(DP_-, DP_+)}_{\text{CW center}}. \tag{1.59}$$

The equilibrium of $(x_{j_2}^*, x_{j_1}^*) = (a_{j_1 j_2 1}, a_{j_2 j_1 s_1})$ *is a* (DP$_-$,DP$_+$)-*clockwise center.*
- *For* $a_{j_1 j_1 0}(a_{j_2 j_1 s_1} - a_{j_1 j_1 1})^2 < 0$ *and* $a_{j_2 j_2 0}(a_{j_2 j_1 s_1} - a_{j_2 j_1 s_2})^2 < 0$,

$$(a_{j_1 j_2 1}, a_{j_2 j_1 s_1}) = \underbrace{(UP_-, UP_-)}_{\text{negative saddle}}. \tag{1.60}$$

The equilibrium of $(x_{j_2}^*, x_{j_1}^*) = (a_{j_1 j_2 1}, a_{j_2 j_1 s_1})$ *is a* (UP$_-$,UP$_-$)-*negative saddle.*

(v_{1b}) *The equilibrium of* $(x_{j_2}^*, x_{j_1}^*) = (a_{j_1 j_2 1}, a_{j_2 j_1 s_2})$ *with* $a_{j_1 j_1 1} \neq a_{j_2 j_1 s_2}$ *has the following properties:*

- *For* $a_{j_1 j_1 0}(a_{j_2 j_1 s_2} - a_{j_1 j_1 1})^2 > 0$ *and* $a_{j_2 j_2 0}(a_{j_2 j_1 s_2} - a_{j_2 j_1 s_1}) > 0$,

$$(a_{j_1 j_2 1}, a_{j_2 j_1 s_2}) = \underbrace{(UP, US)}_{\text{up-parabola upper-saddle}}. \tag{1.61}$$

The equilibrium of $(x_{j_2}^*, x_{j_1}^*) = (a_{j_1 j_2 1}, a_{j_2 j_1 s_2})$ *is a* (UP,US)-*up-parabola upper-saddle.*
- *For* $a_{j_1 j_1 0}(a_{j_2 j_1 s_2} - a_{j_1 j_1 1})^2 < 0$ *and* $a_{j_2 j_2 0}(a_{j_2 j_1 s_2} - a_{j_2 j_1 s_1}) > 0$,

$$(a_{j_1 j_2 1}, a_{j_2 j_1 s_2}) = \underbrace{(DP, US)}_{\text{down-parabola upper-saddle}}. \tag{1.62}$$

The equilibrium of $(x_{j_2}^*, x_{j_1}^*) = (a_{j_1 j_2 1}, a_{j_2 j_1 s_2})$ *is a* (DP,US)-*down-parabola upper-saddle.*

- *For $a_{j_1j_10}(a_{j_2j_1s_2} - a_{j_1j_11})^2 > 0$ and $a_{j_2j_20}(a_{j_2j_1s_2} - a_{j_2j_1s_1}) < 0$,*

$$(a_{j_1j_21}, a_{j_2j_1s_2}) = \underbrace{(\text{DP}, \text{LS})}_{\text{down-parabola lower-saddle}}. \tag{1.63}$$

The equilibrium of $(x_{j_2}^*, x_{j_1}^*) = (a_{j_1j_21}, a_{j_2j_1s_2})$ is a (DP,LS)-down-parabola lower-saddle.

- *For $a_{j_1j_10}(a_{j_2j_1s_2} - a_{j_1j_11})^2 < 0$ and $a_{j_2j_20}(a_{j_2j_1s_2} - a_{j_2j_1s_1}) < 0$,*

$$(a_{j_1j_21}, a_{j_2j_1s_2}) = \underbrace{(\text{UP}, \text{LS})}_{\text{up-parabola lower-saddle}}. \tag{1.64}$$

The equilibrium of $(x_{j_2}^*, x_{j_1}^*) = (a_{j_1j_21}, a_{j_2j_1s_2})$ is a (UP,LS)-up-parabola lower-saddle.

– *The parabola-saddles are the appearing bifurcations of saddle and center.*

(v_{1c}) The equilibrium of $(x_{j_2}^*, x_{j_1}^*) = (a_{j_1j_21}, a_{j_2j_11})$ has the following properties:

- *For $a_{j_1j_10} > 0$ and $a_{j_2j_20}(a_{j_1j_11} - a_{j_2j_1s_1})(a_{j_1j_11} - a_{j_2j_1s_2})^2 > 0$,*

$$(a_{j_1j_11}, a_{j_1j_21}) = \underbrace{(\text{UP}:\text{UP}, \text{pF})}_{\text{hyperbolic-secant-to-hyperbolic flow } (+)}. \tag{1.65}$$

The equilibrium of $(x_{j_1}^*, x_{j_2}^*) = (a_{j_1j_11}, a_{j_1j_21})$ is a (UP:UP, pF)-positive hyperbolic-secant-to-hyperbolic flow.

- *For $a_{j_1j_10} < 0$ and $a_{j_2j_20}(a_{j_1j_11} - a_{j_2j_1s_1})(a_{j_1j_11} - a_{j_2j_1s_2})^2 > 0$,*

$$(a_{j_1j_11}, a_{j_1j_21}) = \underbrace{(\text{DP}:\text{DP}, \text{pF})}_{\text{hyperbolic-to-hyperbolic-secant flow } (+)}. \tag{1.66}$$

The equilibrium of $(x_{j_1}^*, x_{j_2}^*) = (a_{j_1j_11}, a_{j_1j_21})$ is a (DP:DP, pF)-positive hyperbolic-to-hyperbolic-secant flow.

- *For $a_{j_1j_10} > 0$ and $a_{j_2j_20}(a_{j_1j_11} - a_{j_2j_1s_1})(a_{j_1j_11} - a_{j_2j_1s_2})^2 < 0$,*

$$(a_{j_1j_11}, a_{j_1j_21}) = \underbrace{(\text{DP}:\text{DP}, \text{nF})}_{\text{hyperbolic-to-hyperbolic-secant flow } (-)}. \tag{1.67}$$

The equilibrium of $(x_{j_1}^*, x_{j_2}^*) = (a_{j_1j_11}, a_{j_1j_21})$ is a (DP:DP, nF)-negative hyperbolic-secant-to-hyperbolic flow.

- *For $a_{j_1j_10} < 0$ and $a_{j_2j_20}(a_{j_1j_11} - a_{j_2j_1s_1})(a_{j_1j_11} - a_{j_2j_1s_2})^2 < 0$,*

$$(a_{j_1j_11}, a_{j_1j_21}) = \underbrace{(\text{UP:UP, nF})}_{\text{hyperbolic-secant-to-hyperbolic flow } (-)} . \tag{1.68}$$

The equilibrium of $(x_{j_1}^*, x_{j_2}^*) = (a_{j_1j_11}, a_{j_1j_21})$ is a (UP:UP, nF)-negative hyperbolic-secant-to-hyperbolic flow.

(v_2) *For $a_{j_2j_1s_1} = a_{j_2j_1s_2} = a_{j_2j_11}$, the standard form is*

$$\begin{aligned}
\dot{x}_{j_2} &= a_{j_2j_20}(x_{j_1} - a_{j_2j_11})^3, \\
\dot{x}_{j_1} &= a_{j_1j_10}(x_{j_2} - a_{j_1j_21})(x_{j_1} - a_{j_1j_11})^2.
\end{aligned} \tag{1.69}$$

The first integral manifold for $a_{j_2j_11} \neq a_{j_1j_11}$ is

$$\frac{1}{2}\left[(x_{j_1} - a_{j_1j_11})^2 - (x_{j_10} - a_{j_1j_11})^2\right] + \frac{3}{2}(a_{j_1j_11} - a_{j_2j_11})(x_{j_1} - x_{j_10})$$

$$+ 3(a_{j_1j_11} - a_{j_2j_11})^2 \ln \frac{\mid x_{j_1} - a_{j_1j_11} \mid}{\mid x_{j_10} - a_{j_1j_11} \mid} - (a_{j_1j_11} - a_{j_2j_11})^3 \left(\frac{1}{x_{j_1} - a_{j_1j_11}} - \frac{1}{x_{j_10} - a_{j_1j_11}}\right)$$

$$= \frac{1}{2}\frac{a_{j_1j_10}}{a_{j_2j_20}}\left[(x_{j_2} - a_{j_1j_21})^2 - (x_{j_20} - a_{j_1j_21})^2\right]. \tag{1.70}$$

(v_{2a}) *The equilibrium of $(x_{j_2}^*, x_{j_1}^*) = (a_{j_1j_21}, a_{j_2j_11})$ has the following properties:*

- *For $a_{j_1j_10}(a_{j_2j_11} - a_{j_1j_11})^2 > 0$ and $a_{j_2j_20} > 0$,*

$$(a_{j_1j_21}, a_{j_2j_11}) = \underbrace{(\text{UP}_+, 3^{\text{rd}}\text{UP}_+)}_{\text{third-order positive saddle}} . \tag{1.71}$$

The equilibrium of $(x_{j_2}^*, x_{j_1}^*) = (a_{j_1j_21}, a_{j_2j_11})$ is a $(\text{UP}_+, 3^{\text{rd}}\text{UP}_+)$-third-order positive saddle.

- *For $a_{j_1j_10}(a_{j_2j_11} - a_{j_1j_11})^2 < 0$ and $a_{j_2j_20} > 0$,*

$$(a_{j_1j_21}, a_{j_2j_11}) = \underbrace{(\text{DP}_+, 3^{\text{rd}}\text{DP}_-)}_{\text{third-order CCW center}} . \tag{1.72}$$

The equilibrium of $(x_{j_2}^*, x_{j_1}^*) = (a_{j_1j_21}, a_{j_2j_11})$ is a $(\text{DP}_+, 3^{\text{rd}}\text{DP}_-)$-third-order counter-clockwise center.

- *For $a_{j_1j_10}(a_{j_2j_11} - a_{j_1j_11})^2 > 0$ and $a_{j_2j_20} < 0$,*

$$(a_{j_1j_21}, a_{j_2j_11}) = \underbrace{(\text{DP}_-, 3^{\text{rd}}\text{DP}_+)}_{\text{third-order CW center}}.\tag{1.73}$$

The equilibrium of $(x_{j_2}^*, x_{j_1}^*) = (a_{j_1j_21}, a_{j_2j_11})$ is a $(\text{DP}_-, 3^{\text{rd}}\text{DP}_+)$-third-order clockwise center.
- *For $a_{j_1j_10}(a_{j_2j_11} - a_{j_1j_11})^2 < 0$ and $a_{j_2j_20} < 0$,*

$$(a_{j_1j_21}, a_{j_2j_11}) = \underbrace{(\text{UP}_-, 3^{\text{rd}}\text{UP}_-)}_{\text{third-order negative saddle}}.\tag{1.74}$$

The equilibrium of $(x_{j_2}^*, x_{j_1}^*) = (a_{j_1j_21}, a_{j_2j_11})$ is a $(\text{UP}_-, 3^{\text{rd}}\text{UP}_-)$-third-order negative saddle.

- *The third-order saddles are the appearing and switching bifurcations of saddle, center, and saddle.*
- *The third-order centers are the appearing and switching bifurcations of center, saddle, and center.*

(v_{2b}) The equilibrium of $(x_{j_1}^*, x_{j_2}^*) = (a_{j_1j_11}, a_{j_1j_21})$ with $a_{j_1j_11} \neq a_{j_1j_2s_1}, a_{j_1j_2s_2}$ has the following properties:

- *For $a_{j_1j_10} > 0$ and $a_{j_2j_20}(a_{j_1j_11} - a_{j_2j_11})^3 > 0$,*

$$(a_{j_1j_11}, a_{j_1j_21}) = \underbrace{(\text{UP:UP, pF})}_{\text{hyperbolic-secant-to-hyperbolic flow } (+)}.\tag{1.75}$$

The equilibrium of $(x_{j_1}^*, x_{j_2}^*) = (a_{j_1j_11}, a_{j_1j_21})$ is a (UP:UP, pF)-positive hyperbolic-secant-to-hyperbolic flow.
- *For $a_{j_1j_10} < 0$ and $a_{j_2j_20}(a_{j_1j_11} - a_{j_2j_11})^3 > 0$,*

$$(a_{j_1j_11}, a_{j_1j_21}) = \underbrace{(\text{DP:DP, pF})}_{\text{hyperbolic-to-hyperbolic-secant flow } (+)}.\tag{1.76}$$

The equilibrium of $(x_{j_1}^*, x_{j_2}^*) = (a_{j_1j_11}, a_{j_1j_21})$ is a (DP:DP,pF)-positive hyperbolic-to-hyperbolic-secant flow.
- *For $a_{j_1j_10} > 0$ and $a_{j_2j_20}(a_{j_1j_11} - a_{j_2j_11})^3 < 0$,*

$$(a_{j_1j_11}, a_{j_1j_21}) = \underbrace{(\text{DP:DP, nF})}_{\text{hyperbolic-to-hyperbolic-secant flow } (-)}.\tag{1.77}$$

The equilibrium of $(x_{j_1}^*, x_{j_2}^*) = (a_{j_1j_11}, a_{j_1j_21})$ is a (DP:DP, nF)-negative hyperbolic-secant-to-hyperbolic flow.

- *For $a_{j_1j_10} < 0$ and $a_{j_2j_20}(a_{j_1j_11} - a_{j_2j_11})^3 < 0$,*

$$(a_{j_1j_11}, a_{j_1j_21}) = \underbrace{(\text{UP:UP}, \text{nF})}_{\text{hyperbolic-secant-to-hyperbolic flow } (-)}. \tag{1.78}$$

The equilibrium of $(x^*_{j_1}, x^*_{j_2}) = (a_{j_1j_11}, a_{j_1j_21})$ is a (UP:UP,nF)-negative hyperbolic-secant-to-hyperbolic flow.

(v_3) The first integral manifold for $a_{j_2j_1s_1} = a_{j_1j_11}$ is

$$\frac{1}{2}\left[(x_{j_1} - a_{j_1j_11})^2 - (x_{j_10} - a_{j_1j_11})^2\right] = \frac{1}{2}\frac{a_{j_1j_10}}{a_{j_2j_20}}\left[(x_{j_2} - a_{j_1j_21})^2 - (x_{j_20} - a_{j_1j_21})^2\right]. \tag{1.79}$$

(v_{3a}) The infinite-equilibrium of $x^*_{j_1} = a_{j_2j_1s_1} = a_{j_1j_11}$ with $\bar{x}_{j_2} \neq a_{j_1j_21}$ has the following properties:

- *For $a_{j_1j_10}(\bar{x}_{j_2} - a_{j_1j_21}) > 0$ and $a_{j_2j_20}(a_{j_2j_1s_1} - a_{j_2j_1s_2})^2 > 0$,*

$$(a_{j_1j_11}, \bar{x}_{j_2}) = \underbrace{(\text{US}, \text{DU})}_{\text{down-up upper-saddle}}. \tag{1.80}$$

The infinite-equilibrium of $x^*_{j_1} = a_{j_2j_1s_1} = a_{j_1j_11}$ is a (US,DU)-down-up asymptotic upper-saddle.

- *For $a_{j_1j_10}(\bar{x}_{j_2} - a_{j_1j_21}) < 0$ and $a_{j_2j_20}(a_{j_2j_1s_1} - a_{j_2j_1s_2})^2 > 0$,*

$$(a_{j_1j_11}, \bar{x}_{j_2}) = \underbrace{(\text{LS}, \text{UD})}_{\text{up-down lower-saddle}}. \tag{1.81}$$

The infinite-equilibrium of $x^*_{j_1} = a_{j_2j_1s_1} = a_{j_1j_11}$ is an (LS,UD)-up-down asymptotic lower-saddle.

- *For $a_{j_1j_10}(\bar{x}_{j_2} - a_{j_1j_21}) > 0$ and $a_{j_2j_20}(a_{j_2j_1s_1} - a_{j_2j_1s_2})^2 < 0$,*

$$(a_{j_1j_11}, \bar{x}_{j_2}) = \underbrace{(\text{US}, \text{UD})}_{\text{up-down upper-saddle}}. \tag{1.82}$$

The infinite-equilibrium of $x^*_{j_1} = a_{j_2j_1s_1} = a_{j_1j_11}$ is a (US,UD)-up-down asymptotic upper-saddle.

- *For $a_{j_1j_10}(\bar{x}_{j_2} - a_{j_1j_21}) < 0$ and $a_{j_2j_20}(a_{j_2j_1s_1} - a_{j_2j_1s_2})^2 < 0$,*

$$(a_{j_1j_11}, \bar{x}_{j_2}) = \underbrace{(\text{LS}, \text{DU})}_{\text{down-up lower-saddle}}. \tag{1.83}$$

The infinite-equilibrium of $x^*_{j_1} = a_{j_2j_1s_1} = a_{j_1j_11}$ is an (LS,DU)-down-up asymptotic lower-saddle.

(v_{3b}) The equilibrium of $(x_{j_2}^*, x_{j_1}^*) = (a_{j_1 j_2 1}, a_{j_2 j_1 s_1})$ with $a_{j_1 j_1 1} = a_{j_2 j_1 s_1}$ has the following properties:

- *For $a_{j_1 j_1 0} > 0$ and $a_{j_2 j_2 0}(a_{j_2 j_1 s_1} - a_{j_2 j_1 s_2})^2 > 0$,*

$$(a_{j_1 j_2 1}, a_{j_2 j_1 s_1}) = \underbrace{(_{\mathrm{UD}}\mathrm{LS}:_{\mathrm{DU}}\mathrm{US}, \mathrm{DP}_- : \mathrm{UP}_+)}_{\text{hyperbolic lower-to-upper saddle}}. \tag{1.84}$$

The equilibrium of $(x_{j_2}^*, x_{j_1}^*) = (a_{j_1 j_2 1}, a_{j_2 j_1 s_1})$ for $a_{j_1 j_1 1} = a_{j_2 j_1 s_1}$ is a $(_{\mathrm{UD}}\mathrm{LS}:_{\mathrm{DU}}\mathrm{US}, \mathrm{DP}_-:\mathrm{UP}_+)$-hyperbolic lower-to-upper saddle.
- *For $a_{j_1 j_1 0} < 0$ and $a_{j_2 j_2 0}(a_{j_2 j_1 s_1} - a_{j_2 j_1 s_2})^2 > 0$,*

$$(a_{j_1 j_2 1}, a_{j_2 j_1 s_1}) = \underbrace{(_{\mathrm{DU}}\mathrm{US}:_{\mathrm{UD}}\mathrm{LS}, \mathrm{UP}_- : \mathrm{DP}_+)}_{\text{hyperbolic-secant upper-to-lower saddle}}. \tag{1.85}$$

The equilibrium of $(x_{j_2}^*, x_{j_1}^*) = (a_{j_1 j_2 1}, a_{j_2 j_1 s_1})$ for $a_{j_1 j_1 1} = a_{j_2 j_1 s_1}$ is a $(_{\mathrm{DU}}\mathrm{US}:_{\mathrm{UD}}\mathrm{LS}, \mathrm{UP}_-:\mathrm{DP}_+)$-hyperbolic-secant upper-to-lower saddle.
- *For $a_{j_1 j_1 0} > 0$ and $a_{j_2 j_2 0}(a_{j_2 j_1 s_1} - a_{j_2 j_1 s_2})^2 < 0$,*

$$(a_{j_1 j_2 1}, a_{j_2 j_1 s_1}) = \underbrace{(_{\mathrm{DU}}\mathrm{LS}:_{\mathrm{UD}}\mathrm{US}, \mathrm{UP}_+ : \mathrm{DP}_-)}_{\text{hyperbolic-secant lower-to-upper saddle}}. \tag{1.86}$$

The equilibrium of $(x_{j_2}^*, x_{j_1}^*) = (a_{j_1 j_2 1}, a_{j_2 j_1 s_1})$ for $a_{j_1 j_1 1} = a_{j_2 j_1 s_1}$ is a $(_{\mathrm{DU}}\mathrm{LS}:_{\mathrm{UD}}\mathrm{US}, \mathrm{UP}_+:\mathrm{DP}_-)$-hyperbolic-secant lower-to-upper saddle.
- *For $a_{j_1 j_1 0} < 0$ and $a_{j_2 j_2 0}(a_{j_2 j_1 s_1} - a_{j_2 j_1 s_2})^2 < 0$,*

$$(a_{j_1 j_2 1}, a_{j_2 j_1 s_1}) = \underbrace{(_{\mathrm{UD}}\mathrm{US}:_{\mathrm{DU}}\mathrm{LS}, \mathrm{DP}_+ : \mathrm{UP}_-)}_{\text{hyperbolic upper-to-lower saddle}}. \tag{1.87}$$

The equilibrium of $(x_{j_2}^*, x_{j_1}^*) = (a_{j_1 j_2 1}, a_{j_2 j_1 s_1})$ for $a_{j_1 j_1 1} = a_{j_2 j_1 s_1}$ is a $(_{\mathrm{UD}}\mathrm{US}:_{\mathrm{DU}}\mathrm{LS}, \mathrm{DP}_+:\mathrm{UP}_-)$-hyperbolic upper-to-lower saddle.

(v_4) The first integral manifold for $a_{j_2 j_1 s_2} = a_{j_1 j_1 1}$ is

$$\frac{1}{2}\left[(x_{j_1} - a_{j_2 j_1 s_1})^2 - (x_{j_1 0} - a_{j_2 j_1 s_1})^2\right] = \frac{1}{2}\frac{a_{j_1 j_1 0}}{a_{j_2 j_2 0}}\left[(x_{j_2} - a_{j_1 j_2 1})^2 - (x_{j_2 0} - a_{j_1 j_2 1})^2\right]. \tag{1.88}$$

(v_{4a}) The infinite-equilibrium of $x_{j_1}^* = a_{j_1 j_1 1} = a_{j_2 j_1 s_2}$ with $\bar{x}_{j_2} \neq a_{j_1 j_2 1}$ has the following properties:

- *For $a_{j_1 j_1 0}(\bar{x}_{j_2} - a_{j_1 j_2 1}) > 0$ and $a_{j_2 j_2 0}(a_{j_2 j_1 s_2} - a_{j_2 j_1 s_1}) > 0$,*

$$(a_{j_1j_11}, \bar{x}_{j_2}) = \underbrace{(\text{US}, \text{II})}_{\text{increasing-inflection upper-saddle}} . \tag{1.89}$$

The infinite-equilibrium of $x_{j_1}^* = a_{j_1j_11} = a_{j_2j_1s_2}$ is a (US,II)-increasing-inflection upper-saddle.

- *For $a_{j_1j_10}(\bar{x}_{j_2} - a_{j_1j_21}) < 0$ and $a_{j_2j_20}(a_{j_2j_1s_2} - a_{j_2j_1s_1}) > 0$,*

$$(a_{j_1j_11}, \bar{x}_{j_2}) = \underbrace{(\text{LS}, \text{DI})}_{\text{decreasing-inflection lower-saddle}} . \tag{1.90}$$

The infinite-equilibrium of $x_{j_1}^* = a_{j_1j_11} = a_{j_2j_1s_2}$ is an (LS,DI)-decreasing-inflection lower-saddle.

- *For $a_{j_1j_10}(\bar{x}_{j_2} - a_{j_1j_21}) > 0$ and $a_{j_2j_20}(a_{j_2j_1s_2} - a_{j_2j_1s_1}) < 0$,*

$$(a_{j_1j_11}, \bar{x}_{j_2}) = \underbrace{(\text{US}, \text{DI})}_{\text{decreasing-inflection upper-saddle}} . \tag{1.91}$$

The infinite-equilibrium of $x_{j_1}^* = a_{j_1j_11} = a_{j_2j_1s_2}$ is a (US,DI)-decreasing-inflection upper-saddle.

- *For $a_{j_1j_10}(\bar{x}_{j_2} - a_{j_1j_21}) < 0$ and $a_{j_2j_20}(a_{j_2j_1s_2} - a_{j_2j_1s_1}) < 0$,*

$$(a_{j_1j_11}, \bar{x}_{j_2}) = \underbrace{(\text{LS}, \text{II})}_{\text{increasing-inflection lower-saddle}} . \tag{1.92}$$

The infinite-equilibrium of $x_{j_1}^* = a_{j_1j_11} = a_{j_2j_1s_2}$ is an (LS,II)-increasing-inflection lower-saddle.

(v_{4b}) The equilibrium of $(x_{j_2}^*, x_{j_1}^*) = (a_{j_1j_21}, a_{j_2j_1s_2})$ with $a_{j_1j_11} = a_{j_2j_1s_2}$ has the following properties:

- *For $a_{j_1j_10} > 0$ and $a_{j_2j_20}(a_{j_2j_1s_2} - a_{j_2j_1s_1}) > 0$,*

$$(a_{j_1j_21}, a_{j_2j_1s_2}) = \underbrace{(_{\text{DI:II}}\text{UP}, _{\text{LS:US}}\text{US})}_{\text{up-parabola upper-saddle}} . \tag{1.93}$$

The equilibrium of $(x_{j_2}^*, x_{j_1}^*) = (a_{j_1j_21}, a_{j_2j_1s_2})$ with $a_{j_1j_11} = a_{j_2j_1s_2}$ is a $(_{\text{DI:II}}\text{UP},_{\text{LS:}}$ $_{\text{US}}\text{US})$-up-parabola upper-saddle.

- *For $a_{j_1j_10} < 0$ and $a_{j_2j_20}(a_{j_2j_1s_2} - a_{j_2j_1s_1}) > 0$,*

$$(a_{j_1j_21}, a_{j_2j_1s_2}) = \underbrace{(_{\text{II:DI}}\text{DP}, _{\text{US:LS}}\text{US})}_{\text{down-parabola upper-saddle}} . \tag{1.94}$$

The equilibrium of $(x_{j_2}^*, x_{j_1}^*) = (a_{j_1j_21}, a_{j_2j_1s_2})$ with $a_{j_1j_11} = a_{j_2j_1s_2}$ is an $(_{\text{II:DI}}\text{DP},_{\text{US:}}$ $_{\text{LS}}\text{US})$-down-parabola upper-saddle.

- *For $a_{j_1j_10} > 0$ and $a_{j_2j_20}(a_{j_2j_1s_2} - a_{j_2j_1s_1}) < 0$,*

$$(a_{j_1j_21}, a_{j_2j_1s_2}) = \underbrace{(_{\mathrm{II:DI}}\mathrm{DP}, {}_{\mathrm{LS:US}}\mathrm{LS})}_{\text{down-parabola lower-saddle}} . \tag{1.95}$$

The equilibrium of $(x_{j_2}^*, x_{j_1}^*) = (a_{j_1j_21}, a_{j_2j_1s_2})$ with $a_{j_1j_11} = a_{j_2j_2s_2}$ is an $(_{\mathrm{II:DI}}\mathrm{DP}, {}_{\mathrm{LS:}}{}_{\mathrm{US}}\mathrm{LS})$-down-parabola lower-saddle.

- *For $a_{j_1j_10} < 0$ and $a_{j_2j_20}(a_{j_2j_1s_2} - a_{j_2j_1s_1}) < 0$,*

$$(a_{j_1j_21}, a_{j_2j_1s_2}) = \underbrace{(_{\mathrm{DI:II}}\mathrm{UP}, {}_{\mathrm{US:LS}}\mathrm{LS})}_{\text{up-parabola lower-saddle}}. \tag{1.96}$$

The equilibrium of $(x_{j_2}^*, x_{j_1}^*) = (a_{j_1j_21}, a_{j_2j_1s_2})$ with $a_{j_1j_11} = a_{j_2j_2s_2}$ is a $(_{\mathrm{DI:II}}\mathrm{UP}, {}_{\mathrm{US:}}{}_{\mathrm{LS}}\mathrm{LS})$-up-parabola lower-saddle.

(v_5) The first integral manifold for $a_{j_2j_11} = a_{j_1j_11}$ is

$$\frac{1}{2}\left[(x_{j_1} - a_{j_2j_11})^2 - (x_{j_10} - a_{j_2j_11})^2\right] = \frac{1}{2}\frac{a_{j_1j_10}}{a_{j_2j_20}}\left[(x_{j_2} - a_{j_1j_21})^2 - (x_{j_20} - a_{j_1j_21})^2\right]. \tag{1.97}$$

(v_{5a}) The infinite-equilibrium of $x_{j_1}^* = a_{j_2j_1s_1} = a_{j_1j_11}$ with $\bar{x}_{j_2} \neq a_{j_1j_21}$ has the following properties:

- *For $a_{j_1j_10}(\bar{x}_{j_2} - a_{j_1j_21}) > 0$ and $a_{j_2j_20} > 0$,*

$$(a_{j_1j_11}, \bar{x}_{j_2}) = \underbrace{(\mathrm{US}, \mathrm{UP})}_{\text{up-parabola upper-saddle}} . \tag{1.98}$$

The infinite-equilibrium of $x_{j_1}^* = a_{j_2j_11} = a_{j_1j_11}$ is a $(\mathrm{US},\mathrm{US})$-up-parabola upper-saddle.

- *For $a_{j_1j_10}(\bar{x}_{j_2} - a_{j_1j_21}) < 0$ and $a_{j_2j_20} > 0$,*

$$(a_{j_1j_11}, \bar{x}_{j_2}) = \underbrace{(\mathrm{LS}, \mathrm{DP})}_{\text{down-parabola lower-saddle}} . \tag{1.99}$$

The infinite-equilibrium of $x_{j_1}^* = a_{j_2j_11} = a_{j_1j_11}$ is an $(\mathrm{LS},\mathrm{DP})$-down-parabola lower-saddle.

- *For $a_{j_1j_10}(\bar{x}_{j_2} - a_{j_1j_21}) > 0$ and $a_{j_2j_20} < 0$,*

$$(a_{j_1j_11}, \bar{x}_{j_2}) = \underbrace{(\mathrm{US}, \mathrm{DP})}_{\text{down-parabola upper-saddle}} . \tag{1.100}$$

The infinite-equilibrium of $x_{j_1}^* = a_{j_2j_11} = a_{j_1j_11}$ is a $(\mathrm{US},\mathrm{DP})$-down-parabola upper-saddle.

- *For $a_{j_1j_10}(\overline{x}_{j_2} - a_{j_1j_21}) < 0$ and $a_{j_2j_20} < 0$,*

$$(a_{j_1j_11}, \overline{x}_{j_2}) = \underbrace{(\text{LS}, \text{UP})}_{\text{up-parabola lower-saddle}} . \tag{1.101}$$

The infinite-equilibrium of $x_{j_1}^* = a_{j_2j_11} = a_{j_1j_11}$ is an (LS,UP)-up-parabola lower-saddle.

(v_{5b}) The equilibrium of $(x_{j_2}^*, x_{j_1}^*) = (a_{j_1j_21}, a_{j_2j_11})$ with $a_{j_2j_11} = a_{j_1j_11}$ has the following properties:

- *For $a_{j_1j_10} > 0$ and $a_{j_2j_20} > 0$,*

$$(a_{j_1j_21}, a_{j_2j_11}) = \underbrace{(_{\text{DP}}\text{LS} : {}_{\text{UP}}\text{US}, \text{DP}_- : \text{UP}_+)}_{\text{positive hyperbolic lower-to-upper saddle}} . \tag{1.102}$$

The equilibrium of $(x_{j_2}^*, x_{j_1}^*) = (a_{j_1j_21}, a_{j_2j_11})$ with $a_{j_2j_11} = a_{j_1j_11}$ is a ($_{\text{DP}}$LS:$_{\text{UP}}$US, DP$_-$:UP$_+$)-positive hyperbolic lower-to-upper saddle.
- *For $a_{j_1j_10} < 0$ and $a_{j_2j_20} > 0$,*

$$(a_{j_1j_21}, a_{j_2j_11}) = \underbrace{(_{\text{UP}}\text{US} : {}_{\text{DP}}\text{LS}, \text{UP}_- : \text{DP}_+)}_{\text{CCW circular upper-to-lower saddle}}. \tag{1.103}$$

The equilibrium of $(x_{j_2}^*, x_{j_1}^*) = (a_{j_1j_21}, a_{j_2j_11})$ with $a_{j_2j_11} = a_{j_1j_11}$ is a ($_{\text{UP}}$US:$_{\text{DP}}$LS, UP$_-$:DP$_+$)-counter-clockwise circular upper-to-lower saddle.
- *For $a_{j_1j_10} > 0$ and $a_{j_2j_20} < 0$,*

$$(a_{j_1j_21}, a_{j_2j_11}) = \underbrace{(_{\text{UP}}\text{LS} : {}_{\text{DP}}\text{US}, \text{UP}_+ : \text{DP}_-)}_{\text{CW circular lower-to-upper saddle}}. \tag{1.104}$$

The equilibrium of $(x_{j_2}^*, x_{j_1}^*) = (a_{j_1j_21}, a_{j_2j_11})$ with $a_{j_2j_11} = a_{j_1j_11}$ is a ($_{\text{UP}}$LS:$_{\text{DP}}$US, UP$_+$:DP$_-$)-clockwise circular lower-to-upper saddle.
- *For $a_{j_1j_10} < 0$ and $a_{j_2j_20} < 0$,*

$$(a_{j_1j_21}, a_{j_2j_11}) = \underbrace{(_{\text{DP}}\text{US} : {}_{\text{UP}}\text{LS}, \text{DP}_+ : \text{UP}_-)}_{\text{negative hyperbolic upper-to-lower saddle}} . \tag{1.105}$$

The equilibrium of $(x_{j_2}^*, x_{j_1}^*) = (a_{j_1j_21}, a_{j_2j_11})$ with $a_{j_2j_11} = a_{j_1j_11}$ is a ($_{\text{DP}}$US:$_{\text{UP}}$LS, DP$_+$:UP$_-$)-negative hyperbolic upper-to-lower saddle.

(vi) For $\Delta_{j_1j_2} = 0$ and $\Delta_{j_2j_2} > 0$, the standard form is

$$\begin{aligned}
\dot{x}_{j_2} &= a_{j_2j_20}(x_{j_1} - a_{j_2j_11})(x_{j_1} - a_{j_2j_12})(x_{j_1} - a_{j_2j_13}), \\
\dot{x}_{j_1} &= a_{j_1j_10}(x_{j_2} - a_{j_1j_21})(x_{j_1} - a_{j_1j_11})^2
\end{aligned} \tag{1.106}$$

where

$$a_{j_1 j_1 1} = -\frac{1}{2} B_{j_1 j_1},$$

$$b_{j_2 j_1 2}, b_{j_2 j_1 3} = -\frac{1}{2}(B_{j_2 j_1} \pm \sqrt{\Delta_{j_2 j_1}}),$$

$$\{a_{j_2 j_1 1}, a_{j_2 j_1 2}, a_{j_2 j_1 3}\} = \text{sort}\{b_{j_2 j_1 1}, b_{j_2 j_1 2}, b_{j_2 j_1 3}\}, \quad (1.107)$$

$$a_{j_2 j_1 s_1} < a_{j_2 j_1 s_2}, s_1, s_2 \in \{1, 2, 3\}, s_1 < s_2.$$

(vi₁) The first integral manifold for $a_{j_2 j_1 s_1} \neq a_{j_1 j_1 1}$ ($s_1 = 1, 2, 3$) is

$$\frac{1}{2}\left[(x_{j_1} - a_{j_2 j_1 1})^2 - (x_{j_1 0} - a_{j_2 j_1 1})^2\right] + \sum\nolimits_{s_1 = 1}^{3}(a_{j_1 j_1 1} - a_{j_2 j_1 s_1})(x_{j_1} - x_{j_1 0})$$

$$+ \sum\nolimits_{s_1 = 1}^{3} \prod\nolimits_{s_2 = 1, s_2 \neq s_1}^{3}(a_{j_1 j_1 1} - a_{j_2 j_1 s_2}) \ln \frac{|x_{j_1} - a_{j_1 j_1 1}|}{|x_{j_1 0} - a_{j_1 j_1 1}|}$$

$$- \prod\nolimits_{s_1 = 1}^{3}(a_{j_1 j_1 1} - a_{j_2 j_1 s_1})\left(\frac{1}{x_{j_1} - a_{j_1 j_1 1}} - \frac{1}{x_{j_1 0} - a_{j_1 j_1 1}}\right) \quad (1.108)$$

$$= \frac{1}{2}\frac{a_{j_1 j_1 0}}{a_{j_2 j_2 0}}\left[(x_{j_2} - a_{j_1 j_2 1})^2 - (x_{j_2 0} - a_{j_1 j_2 1})^2\right].$$

(vi₁ₐ) The equilibrium of $(x_{j_1}^*, x_{j_1}^*) = (a_{j_1 j_2 1}, a_{j_2 j_1 s_1})$ ($s_1, s_2 \in \{1, 2, 3\}$, $s_1 \neq s_2$) has the following properties:

- *For $a_{j_1 j_1 0}(a_{j_2 j_1 s_1} - a_{j_1 j_1 1})^2 > 0$ and $a_{j_2 j_2 0}\prod_{s_2 = 1, s_2 \neq s_1}^{3}(a_{j_2 j_1 s_1} - a_{j_2 j_1 s_2}) > 0$,*

$$(a_{j_1 j_2 1}, a_{j_2 j_1 s_1}) = \underbrace{(\text{UP}_+, \text{UP}_+)}_{\text{positive saddle}}. \quad (1.109)$$

  The equilibrium of $(x_{j_2}^*, x_{j_1}^*) = (a_{j_1 j_2 1}, a_{j_2 j_1 s_1})$ is a (UP₊,UP₊)-positive saddle.
- *For $a_{j_1 j_1 0}(a_{j_2 j_1 s_1} - a_{j_1 j_1 1})^2 < 0$ and $a_{j_2 j_2 0}\prod_{s_2 = 1, s_2 \neq s_1}^{3}(a_{j_2 j_1 s_1} - a_{j_2 j_1 s_2}) > 0$,*

$$(a_{j_1 j_2 1}, a_{j_2 j_1 s_1}) = \underbrace{(\text{DP}_+, \text{DP}_-)}_{\text{CCW center}}. \quad (1.110)$$

  The equilibrium of $(x_{j_2}^*, x_{j_1}^*) = (a_{j_1 j_2 1}, a_{j_2 j_1 s_1})$ is a (DP₊,DP₋)-counter-clockwise center.
- *For $a_{j_1 j_1 0}(a_{j_2 j_1 s_1} - a_{j_1 j_1 1})^2 > 0$ and $a_{j_2 j_2 0}\prod_{s_2 = 1, s_2 \neq s_1}^{3}(a_{j_2 j_1 s_1} - a_{j_2 j_1 s_2}) < 0$,*

$$(a_{j_1 j_2 1}, a_{j_2 j_1 s_1}) = \underbrace{(\text{DP}_-, \text{DP}_+)}_{\text{CW center}}. \quad (1.111)$$

  The equilibrium of $(x_{j_2}^*, x_{j_1}^*) = (a_{j_1 j_2 1}, a_{j_2 j_1 s_1})$ is a (DP₋,DP₊)-clockwise center.

- *For $a_{j_1 j_1 0}(a_{j_2 j_1 s_1} - a_{j_1 j_1 1})^2 < 0$ and $a_{j_2 j_2 0}\prod_{s_2 = 1, s_2 \neq s_1}^{3}(a_{j_2 j_1 s_1} - a_{j_2 j_1 s_2}) < 0$,*

$$(a_{j_1 j_2 1}, a_{j_2 j_1 s_1}) = \underbrace{(\mathrm{UP}_-, \mathrm{UP}_-)}_{\text{negative saddle}}. \tag{1.112}$$

The equilibrium of $(x_{j_2}^*, x_{j_1}^*) = (a_{j_1 j_2 1}, a_{j_2 j_1 s_1})$ is a (UP$_-$,UP$_-$)-negative saddle.

(vi$_{Ib}$) The equilibrium of $(x_{j_1}^*, x_{j_2}^*) = (a_{j_1 j_1 1}, a_{j_1 j_2 1})$ with $a_{j_1 j_1 1} \neq a_{j_2 j_1 1}, a_{j_2 j_1 2}, a_{j_2 j_1 3}$ has the following properties:

- *For $a_{j_1 j_1 0} > 0$ and $a_{j_2 j_2 0}\prod_{s_1 = 1}^{3}(a_{j_1 j_1 1} - a_{j_2 j_1 s_1}) > 0$,*

$$(a_{j_1 j_1 1}, a_{j_1 j_2 1}) = \underbrace{(\mathrm{UP}{:}\mathrm{UP}, \mathrm{pF})}_{\text{hyperbolic-secant-to-hyperbolic flow } (+)}. \tag{1.113}$$

The equilibrium of $(x_{j_1}^*, x_{j_2}^*) = (a_{j_1 j_1 1}, a_{j_1 j_2 1})$ is a (UP:UP,pF)-positive hyperbolic-secant-to-hyperbolic flow.
- *For $a_{j_1 j_1 0} < 0$ and $a_{j_2 j_2 0}\prod_{s_1 = 1}^{3}(a_{j_1 j_1 1} - a_{j_2 j_1 s_1}) > 0$,*

$$(a_{j_1 j_1 1}, a_{j_1 j_2 1}) = \underbrace{(\mathrm{DP}{:}\mathrm{DP}, \mathrm{pF})}_{\text{hyperbolic-to-hyperbolic-secant flow } (+)}. \tag{1.114}$$

The equilibrium of $(x_{j_1}^*, x_{j_2}^*) = (a_{j_1 j_1 1}, a_{j_1 j_2 1})$ is a (DP:DP,pF)-positive hyperbolic-to-hyperbolic-secant flow.
- *For $a_{j_1 j_1 0} > 0$ and $a_{j_2 j_2 0}\prod_{s_1 = 1}^{3}(a_{j_1 j_1 1} - a_{j_2 j_1 s_1}) < 0$,*

$$(a_{j_1 j_1 1}, a_{j_1 j_2 1}) = \underbrace{(\mathrm{DP}{:}\mathrm{DP}, \mathrm{nF})}_{\text{hyperbolic-to-hyperbolic-secant flow } (-)}. \tag{1.115}$$

The equilibrium of $(x_{j_1}^*, x_{j_2}^*) = (a_{j_1 j_1 1}, a_{j_1 j_2 1})$ is a (DP:DP,nF)-negative hyperbolic-secant-to-hyperbolic flow.
- *For $a_{j_1 j_1 0} < 0$ and $a_{j_2 j_2 0}\prod_{s_1 = 1}^{3}(a_{j_1 j_1 1} - a_{j_2 j_1 s_1}) < 0$,*

$$(a_{j_1 j_1 1}, a_{j_1 j_2 1}) = \underbrace{(\mathrm{UP}{:}\mathrm{UP}, \mathrm{nF})}_{\text{hyperbolic-secant-to-hyperbolic flow } (-)}. \tag{1.116}$$

The equilibrium of $(x_{j_1}^*, x_{j_2}^*) = (a_{j_1 j_1 1}, a_{j_1 j_2 1})$ is a (UP:UP,nF)-negative hyperbolic-secant-to-hyperbolic flow.

(vi$_2$) The first integral manifold ($a_{j_2 j_1 s_1} = a_{j_1 j_1 1}$) is

$$\frac{1}{2}\left[(x_{j_1} - a_{j_1 j_1 1})^2 - (x_{j_1 0} - a_{j_1 j_1 1})^2\right] + \sum_{s_2 = 1, s_2 \neq s_1}^{3} (a_{j_1 j_1 1} - a_{j_2 j_1 s_2})(x_{j_1} - x_{j_1 0})$$

$$- \prod_{s_2 = 1, s_2 \neq s_1}^{3} (a_{j_1 j_1 1} - a_{j_2 j_1 s_2})(\frac{1}{x_{j_1} - a_{j_1 j_1 1}} - \frac{1}{x_{j_1 0} - a_{j_1 j_1 1}})$$

$$= \frac{1}{2}\frac{a_{j_1 j_1 0}}{a_{j_2 j_2 0}}\left[(x_{j_2} - a_{j_1 j_2 1})^2 - (x_{j_2 0} - a_{j_1 j_2 1})^2\right].$$

$$(1.117)$$

(vi$_{2a}$) The infinite-equilibrium of $x_{j_1}^* = a_{j_1 j_1 1} = a_{j_2 j_1 s_1}$ with $\overline{x}_{j_2} \neq a_{j_1 j_2 1}$ has the following properties:

- *For $a_{j_1 j_1 0}(\overline{x}_{j_2} - a_{j_1 j_2 1}) > 0$ and $a_{j_2 j_2 0}\prod_{s_2 = 1, s_2 \neq s_1}^{3} (a_{j_2 j_1 s_1} - a_{j_2 j_1 s_2}) > 0$,*

$$(a_{j_1 j_1 1}, \overline{x}_{j_2}) = \underbrace{(\text{US}, \text{DU})}_{\text{down-up upper-saddle}}.$$

$$(1.118)$$

  The infinite-equilibrium of $x_{j_1}^* = a_{j_1 j_1 1} = a_{j_2 j_1 s_1}$ is a (US,DU)-down-up asymptotic upper-saddle.
- *For $a_{j_1 j_1 0}(\overline{x}_{j_2} - a_{j_1 j_2 1}) < 0$ and $a_{j_2 j_2 0}\prod_{s_2 = 1, s_2 \neq s_1}^{3} (a_{j_2 j_1 s_1} - a_{j_2 j_1 s_2}) > 0$,*

$$(a_{j_1 j_1 1}, \overline{x}_{j_2}) = \underbrace{(\text{LS}, \text{UD})}_{\text{up-down lower-saddle}}.$$

$$(1.119)$$

  The infinite-equilibrium of $x_{j_1}^* = a_{j_1 j_1 1} = a_{j_2 j_1 s_1}$ is an (LS,UD)-up-down asymptotic lower-saddle.
- *For $a_{j_1 j_1 0}(\overline{x}_{j_2} - a_{j_1 j_2 1}) > 0$ and $a_{j_2 j_2 0}\prod_{s_2 = 1, s_2 \neq s_1}^{3} (a_{j_2 j_1 s_1} - a_{j_2 j_1 s_2}) < 0$,*

$$(a_{j_1 j_1 1}, \overline{x}_{j_2}) = \underbrace{(\text{US}, \text{UD})}_{\text{down-parabola upper-saddle}}.$$

$$(1.120)$$

  The infinite-equilibrium of $x_{j_1}^* = a_{j_1 j_1 1} = a_{j_2 j_1 s_1}$ is a (US,UD)-up-down asymptotic upper-saddle.
- *For $a_{j_1 j_1 0}(\overline{x}_{j_2} - a_{j_1 j_2 1}) < 0$ and $a_{j_2 j_2 0}\prod_{s_2 = 1, s_2 \neq s_1}^{3} (a_{j_2 j_1 s_1} - a_{j_2 j_1 s_2}) < 0$,*

$$(a_{j_1 j_1 1}, \overline{x}_{j_2}) = \underbrace{(\text{LS}, \text{DU})}_{\text{down-up lower-saddle}}.$$

$$(1.121)$$

  The infinite-equilibrium of $x_{j_1}^* = a_{j_1 j_1 1} = a_{j_2 j_1 s_1}$ is an (LS,DU)-down-up asymptotic lower-saddle.

(vi$_{2b}$) The equilibrium of $(x_{j_2}^*, x_{j_1}^*) = (a_{j_1 j_2 1}, a_{j_2 j_1 s_1})$ *with* $a_{j_1 j_1 1} = a_{j_2 j_1 s_1}$ *has the following properties:*

- *For* $a_{j_1 j_1 0} > 0$ *and* $a_{j_2 j_2 0}\prod_{s_2 = 1, s_2 \neq s_1}^{3}(a_{j_2 j_1 s_1} - a_{j_2 j_1 s_2}) > 0,$

$$(a_{j_1 j_2 1}, a_{j_2 j_1 s_1}) = \underbrace{(_{\mathrm{UD}}\mathrm{LS} : {}_{\mathrm{DU}}\mathrm{US}, \mathrm{DP}_- : \mathrm{UP}_+)}_{\text{hyperbolic lower-to-upper saddle}}. \tag{1.122}$$

The equilibrium of $(x_{j_2}^*, x_{j_1}^*) = (a_{j_1 j_2 1}, a_{j_2 j_1 s_1})$ *for* $a_{j_1 j_1 1} = a_{j_2 j_1 s_1}$ *is a* $(_{\mathrm{UD}}\mathrm{LS}: {}_{\mathrm{DU}}\mathrm{US},$ $\mathrm{DP}_-:\mathrm{UP}_+)$-*hyperbolic lower-to-upper saddle.*

- *For* $a_{j_1 j_1 0} < 0$ *and* $a_{j_2 j_2 0}\prod_{s_2 = 1, s_2 \neq s_1}^{3}(a_{j_2 j_1 s_1} - a_{j_2 j_1 s_2}) > 0,$

$$(a_{j_1 j_2 1}, a_{j_2 j_1 s_1}) = \underbrace{(_{\mathrm{DU}}\mathrm{US} : {}_{\mathrm{UD}}\mathrm{LS}, \mathrm{UP}_- : \mathrm{DP}_+)}_{\text{hyperbolic-secant upper-to-lower saddle}}. \tag{1.123}$$

The equilibrium of $(x_{j_2}^*, x_{j_1}^*) = (a_{j_1 j_2 1}, a_{j_2 j_1 s_1})$ *for* $a_{j_1 j_1 1} = a_{j_2 j_1 s_1}$ *is a* $(_{\mathrm{DU}}\mathrm{US}: {}_{\mathrm{UD}}\mathrm{LS},$ $\mathrm{UP}_-:\mathrm{DP}_+)$-*hyperbolic-secant upper-to-lower saddle.*

- *For* $a_{j_1 j_1 0} > 0$ *and* $a_{j_2 j_2 0}\prod_{s_2 = 1, s_2 \neq s_1}^{3}(a_{j_2 j_1 s_1} - a_{j_2 j_1 s_2}) < 0,$

$$(a_{j_1 j_2 1}, a_{j_2 j_1 s_1}) = \underbrace{(_{\mathrm{DU}}\mathrm{LS} : {}_{\mathrm{UD}}\mathrm{US}, \mathrm{UP}_+ : \mathrm{DP}_-)}_{\text{hyperbolic-secant lower-to-upper saddle}}. \tag{1.124}$$

The equilibrium of $(x_{j_2}^*, x_{j_1}^*) = (a_{j_1 j_2 1}, a_{j_2 j_1 s_1})$ *for* $a_{j_1 j_1 1} = a_{j_2 j_1 s_1}$ *is a* $(_{\mathrm{DU}}\mathrm{LS}: {}_{\mathrm{UD}}\mathrm{US},$ $\mathrm{UP}_-:\mathrm{DP}_+)$-*hyperbolic-secant lower-to-upper saddle.*

- *For* $a_{j_1 j_1 0} < 0$ *and* $a_{j_2 j_2 0}\prod_{s_2 = 1, s_2 \neq s_1}^{3}(a_{j_2 j_1 s_1} - a_{j_2 j_1 s_2}) < 0,$

$$(a_{j_1 j_2 1}, a_{j_2 j_1 s_1}) = \underbrace{(_{\mathrm{UD}}\mathrm{US} : {}_{\mathrm{DU}}\mathrm{LS}, \mathrm{DP}_+ : \mathrm{UP}_-)}_{\text{hyperbolic upper-to-lower saddle}}. \tag{1.125}$$

The equilibrium of $(x_{j_2}^*, x_{j_1}^*) = (a_{j_1 j_2 1}, a_{j_2 j_1 s_1})$ *for* $a_{j_1 j_1 1} = a_{j_2 j_1 s_1}$ *is a* $(_{\mathrm{UD}}\mathrm{US}: {}_{\mathrm{DU}}\mathrm{LS},$ $\mathrm{DP}_+:\mathrm{UP}_-)$-*hyperbolic upper-to-lower saddle.*

(vii) For $\Delta_{j_1 j_2} > 0$ *and* $\Delta_{j_2 j_1} < 0$, *the standard form is*

$$\begin{aligned}
\dot{x}_{j_2} &= a_{j_2 j_2 0}(x_{j_1} - a_{j_2 j_1 1})\left[(x_{j_1} - a_{j_2 j_1})^2 + b_{j_2 j_1}\right], \\
\dot{x}_{j_1} &= a_{j_1 j_1 0}(x_{j_2} - a_{j_1 j_2 1})(x_{j_1} - a_{j_1 j_1 1})(x_{j_1} - a_{j_1 j_1 2})
\end{aligned} \tag{1.126}$$

where

$$\begin{aligned}
b_{j_1 j_1 1}, b_{j_1 j_1 2} &= -\frac{1}{2}(B_{j_1 j_1} \pm \sqrt{\Delta_{j_1 j_1}}), \\
\{a_{j_1 j_1 1}, a_{j_1 j_1 2}\} &= \mathrm{sort}\{b_{j_1 j_1 1}, b_{j_1 j_1 2}\}, a_{j_1 j_2 1} < a_{j_1 j_2 2}; \\
a_{j_2 j_1 1} &= b_{j_2 j_1 1}, a_{j_2 j_1} = -\frac{1}{2}B_{j_2 j_1}, b_{j_2 j_1} = \frac{1}{4}(-\Delta_{j_2 j_1}).
\end{aligned} \tag{1.127}$$

The first integral manifold is

$$
\frac{1}{2}\left[(x_{j_1} - a_{j_2 j_1 1})^2 - (x_{j_1 0} - a_{j_2 j_1 1})^2\right] + (a_{j_1 j_1 2} + a_{j_1 j_1 1} - 2a_{j_2 j_1})(x_{j_1} - x_{j_1 0})
$$

$$
+ \frac{(a_{j_1 j_1 2} - a_{j_2 j_1 1})\left[(a_{j_1 j_1 2} - a_{j_2 j_1})^2 + b_{j_2 j_1}\right]}{a_{j_1 j_1 2} - a_{j_1 j_1 1}} \ln \frac{|x_{j_1} - a_{j_1 j_1 2}|}{|x_{j_1 0} - a_{j_1 j_1 2}|}
$$

$$
+ \frac{(a_{j_1 j_1 1} - a_{j_2 j_1 1})\left[(a_{j_1 j_1 1} - a_{j_2 j_1})^2 + b_{j_2 j_1}\right]}{a_{j_1 j_1 1} - a_{j_1 j_1 2}} \ln \frac{|x_{j_1} - a_{j_1 j_1 1}|}{|x_{j_1 0} - a_{j_1 j_1 1}|}
$$

$$
= \frac{1}{2}\frac{a_{j_1 j_1 0}}{a_{j_2 j_2 0}}\left[(x_{j_2} - a_{j_1 j_2 1})^2 - (x_{j_2 0} - a_{j_1 j_2 1})^2\right].
$$

$$(1.128)$$

(vii_{1a}) The equilibrium of $(x_{j_1}^*, x_{j_2}^*) = (a_{j_1 j_1 l_1}, a_{j_1 j_2 1})$ ($l_1, l_2 \in \{1,2\}$, $l_1 \neq l_2$) has the following properties:

- *For $a_{j_1 j_1 0}(a_{j_1 j_1 l_1} - a_{j_1 j_1 l_2}) > 0$ and $a_{j_2 j_2 0}(a_{j_1 j_1 l_1} - a_{j_2 j_1 1}) > 0$,*

$$
(a_{j_1 j_1 l_1}, a_{j_1 j_2 1}) = \underbrace{(\text{DP:UP}, \text{pF})}_{\text{hyperbolic flow } (+)}.
$$

$$(1.129)$$

The equilibrium of $(x_{j_1}^*, x_{j_2}^*) = (a_{j_1 j_1 l_1}, a_{j_1 j_2 1})$ is a (DP:UP,pF)-positive hyperbolic flow.

- *For $a_{j_1 j_1 0}(a_{j_1 j_1 l_1} - a_{j_1 j_1 l_2}) < 0$ and $a_{j_2 j_2 0}(a_{j_1 j_1 l_1} - a_{j_2 j_1 1}) > 0$,*

$$
(a_{j_1 j_1 l_1}, a_{j_1 j_2 1}) = \underbrace{(\text{UP:DP}, \text{pF})}_{\text{hyperbolic-secant flow } (+)}.
$$

$$(1.130)$$

The equilibrium of $(x_{j_1}^*, x_{j_2}^*) = (a_{j_1 j_1 l_1}, a_{j_1 j_2 1})$ is a (UP:DP,pF)-positive hyperbolic-secant flow.

- *For $a_{j_1 j_1 0}(a_{j_1 j_1 l_1} - a_{j_1 j_1 l_2}) > 0$ and $a_{j_2 j_2 0}(a_{j_1 j_1 l_1} - a_{j_2 j_1 1}) < 0$,*

$$
(a_{j_1 j_1 l_1}, a_{j_1 j_2 1}) = \underbrace{(\text{UP:DP}, \text{nF})}_{\text{hyperbolic-secant flow } (-)}.
$$

$$(1.131)$$

The equilibrium of $(x_{j_1}^*, x_{j_2}^*) = (a_{j_1 j_1 l_1}, a_{j_1 j_2 1})$ is a (UP:DP,nF)-negative hyperbolic-secant flow.

- *For $a_{j_1 j_1 0}(a_{j_1 j_1 l_1} - a_{j_1 j_1 l_2}) < 0$ and $a_{j_2 j_2 0}(a_{j_1 j_1 l_1} - a_{j_2 j_1 1}) < 0$,*

$$
(a_{j_1 j_1 l_1}, a_{j_1 j_2 1}) = \underbrace{(\text{DP:UP}, \text{nF})}_{\text{hyperbolic flow } (-)}.
$$

$$(1.132)$$

The equilibrium of $(x_{j_1}^*, x_{j_2}^*) = (a_{j_1 j_1 l_1}, a_{j_1 j_2 1})$ is a (DP:UP,nF)-negative hyperbolic flow.

(vii$_{1b}$) The equilibrium of $(x_{j_2}^*, x_{j_1}^*) = (a_{j_1j_21}, a_{j_2j_11})$ has the following properties:

- *For $a_{j_1j_10}\prod_{l_1=1}^{2}(a_{j_2j_11} - a_{j_1j_1l_1}) > 0$ and $a_{j_2j_20} > 0$,*

$$(a_{j_1j_21}, a_{j_2j_11}) = \underbrace{(\text{UP}_+, \text{UP}_+)}_{\text{positive saddle}}.$$ (1.133)

  The equilibrium of $(x_{j_2}^*, x_{j_1}^*) = (a_{j_1j_21}, a_{j_2j_11})$ is a (UP$_+$, UP$_+$)-positive saddle.
- *For $a_{j_1j_10}\prod_{l_1=1}^{2}(a_{j_2j_11} - a_{j_1j_1l_1}) < 0$ and $a_{j_2j_20} > 0$,*

$$(a_{j_1j_21}, a_{j_2j_11}) = \underbrace{(\text{DP}_+, \text{DP}_-)}_{\text{CCW center}}.$$ (1.134)

  The equilibrium of $(x_{j_2}^*, x_{j_1}^*) = (a_{j_1j_21}, a_{j_2j_11})$ is a (DP$_+$,DP$_-$)-counter-clockwise center.
- *For $a_{j_1j_10}\prod_{l_1=1}^{2}(a_{j_2j_11} - a_{j_1j_1l_1}) > 0$ and $a_{j_2j_20} < 0$,*

$$(a_{j_1j_21}, a_{j_2j_11}) = \underbrace{(\text{DP}_-, \text{DP}_+)}_{\text{CW center}}.$$ (1.135)

  The equilibrium of $(x_{j_2}^*, x_{j_1}^*) = (a_{j_1j_21}, a_{j_2j_11})$ is a (DP$_-$,DP$_+$)-clockwise center.
- *For $a_{j_1j_10}\prod_{l_1=1}^{2}(a_{j_2j_11} - a_{j_1j_1l_1}) < 0$ and $a_{j_2j_20} < 0$,*

$$(a_{j_1j_21}, a_{j_2j_11}) = \underbrace{(\text{UP}_-, \text{UP}_-)}_{\text{negative saddle}}.$$ (1.136)

  The equilibrium of $(x_{j_2}^*, x_{j_1}^*) = (a_{j_1j_21}, a_{j_2j_11})$ is a (UP$_-$,UP$_-$)-negative saddle.

(vii$_2$) The first integral manifold for $a_{j_2j_11} = a_{j_1j_1l_1}$ is

$$\frac{1}{2}\left[(x_{j_1} - a_{j_1j_1l_2})^2 - (x_{j_10} - a_{j_1j_1l_2})^2\right] + 2(a_{j_1j_1l_2} - a_{j_2j_1})(x_{j_1} - x_{j_10})$$
$$+ \left[(a_{j_1j_1l_2} - a_{j_2j_1})^2 + b_{j_2j_1}\right]\ln\frac{|x_{j_1} - a_{j_1j_1l_2}|}{|x_{j_10} - a_{j_1j_1l_2}|}$$ (1.137)
$$= \frac{1}{2}\frac{a_{j_1j_10}}{a_{j_2j_20}}\left[(x_{j_2} - a_{j_1j_21})^2 - (x_{j_20} - a_{j_1j_21})^2\right].$$

(vii$_{2a}$) The infinite-equilibrium of $x_{j_1}^* = a_{j_2j_11} = a_{j_1j_11}$ with $\bar{x}_{j_2} \neq a_{j_1j_21}$ has the following properties:

- *For $a_{j_1j_10}(a_{j_1j_1l_1} - a_{j_1j_1l_2})(\bar{x}_{j_2} - a_{j_1j_21}) > 0$ and $a_{j_2j_20} > 0$,*

$$(a_{j_1 j_1 l_1}, \bar{x}_{j_2}) = \underbrace{(SO, II)}_{\text{increasing-inflection source}} . \tag{1.138}$$

The infinite-equilibrium of $x_{j_1}^* = a_{j_2 j_1 1} = a_{j_1 j_1 l_1}$ is an (SO,II)-increasing-inflection source.

- *For $a_{j_1 j_1 0}(a_{j_1 j_1 l_1} - a_{j_1 j_1 l_2})(\bar{x}_{j_2} - a_{j_1 j_2 1}) < 0$ and $a_{j_2 j_2 0} > 0$,*

$$(a_{j_1 j_1 l_1}, \bar{x}_{j_2}) = \underbrace{(SI, DI)}_{\text{decreasing-inflection sink}} . \tag{1.139}$$

The infinite-equilibrium of $x_{j_1}^* = a_{j_2 j_1 1} = a_{j_1 j_1 l_1}$ is an (SI,DI)-decreasing-inflection sink.

- *For $a_{j_1 j_1 0}(a_{j_1 j_1 l_1} - a_{j_1 j_1 l_2})(\bar{x}_{j_2} - a_{j_1 j_2 1}) > 0$ and $a_{j_2 j_2 0} < 0$,*

$$(a_{j_1 j_1 l_1}, \bar{x}_{j_2}) = \underbrace{(SO, DI)}_{\text{decreasing-inflection source}} . \tag{1.140}$$

The infinite-equilibrium of $x_{j_1}^* = a_{j_2 j_1 1} = a_{j_1 j_1 l_1}$ is an (SO,DI)-decreasing-inflection source.

- *For $a_{j_1 j_1 0}(a_{j_1 j_1 l_1} - a_{j_1 j_1 l_2})(\bar{x}_{j_2} - a_{j_1 j_2 1}) < 0$ and $a_{j_2 j_2 0} < 0$,*

$$(a_{j_1 j_1 l_1}, \bar{x}_{j_2}) = \underbrace{(SI, II)}_{\text{increasing-inflection sink}} . \tag{1.141}$$

The infinite-equilibrium of $x_{j_1}^* = a_{j_2 j_1 1} = a_{j_1 j_1 l_1}$ is an (SI,II)-increasing-inflection sink.

(vii$_{2b}$) The equilibrium of $(x_{j_2}^*, x_{j_1}^*) = (a_{j_1 j_2 1}, a_{j_2 j_1 1})$ with $a_{j_2 j_1 1} = a_{j_1 j_1 l_1}$ has the following properties:

- *For $a_{j_1 j_1 0}(a_{j_1 j_1 l_1} - a_{j_1 j_1 l_2}) > 0$ and $a_{j_2 j_2 0} > 0$,*

$$(a_{j_1 j_2 1}, a_{j_2 j_1 1}) = \underbrace{(_{\text{DI:II}}UP, _{\text{SI:SO}}US)}_{\text{up-parabola upper-saddle}} . \tag{1.142}$$

The equilibrium of $(x_{j_2}^*, x_{j_1}^*) = (a_{j_1 j_2 1}, a_{j_2 j_1 1})$ is a $(_{\text{DI:II}}UP, _{\text{SI:SO}}US)$-up-parabola upper-saddle.

- *For $a_{j_1 j_1 0}(a_{j_1 j_1 l_1} - a_{j_1 j_1 l_2}) < 0$ and $a_{j_2 j_2 0} > 0$,*

$$(a_{j_1 j_2 1}, a_{j_2 j_1 1}) = \underbrace{(_{\text{II:DI}}DP, _{\text{SO:SI}}LS)}_{\text{down-parabola lower-saddle}} . \tag{1.143}$$

The equilibrium of $(x_{j_2}^*, x_{j_1}^*) = (a_{j_1 j_2 1}, a_{j_2 j_1 1})$ is an $(_{\text{II:DI}}DP, _{\text{SO:SI}}LS)$-down-parabola lower-saddle.

- *For $a_{j_1 j_1 0}(a_{j_1 j_1 l_1} - a_{j_1 j_1 l_2}) > 0$ and $a_{j_2 j_2 0} < 0$,*

$$(a_{j_1 j_2 1}, a_{j_2 j_1 1}) = \underbrace{({}_{\text{II:DI}}\text{DP}, {}_{\text{SI:SO}}\text{US})}_{\text{down-parabola upper-saddle}} .$$

(1.144)

The equilibrium of $(x^*_{j_2}, x^*_{j_1}) = (a_{j_1 j_2 1}, a_{j_2 j_1 1})$ is an $({}_{\text{II:DI}}\text{DP},{}_{\text{SI:SO}}\text{US})$-down-parabola upper-saddle.
- *For $a_{j_1 j_1 0}(a_{j_1 j_1 l_1} - a_{j_1 j_1 l_2}) < 0$ and $a_{j_2 j_2 0} < 0$,*

$$(a_{j_1 j_2 1}, a_{j_2 j_1 1}) = \underbrace{({}_{\text{DI:II}}\text{UP}, {}_{\text{SO:SI}}\text{LS})}_{\text{up-parabola lower-saddle}} .$$

(1.145)

The equilibrium of $(x^*_{j_2}, x^*_{j_1}) = (a_{j_1 j_2 1}, a_{j_2 j_1 1})$ is a $({}_{\text{DI:II}}\text{UP},{}_{\text{SO:SI}}\text{LS})$-up-parabola lower-saddle.

(viii) For $\Delta_{j_1 j_2} > 0$ and $\Delta_{j_2 j_1} = 0$, the standard form is

$$\begin{aligned}
\dot{x}_{j_2} &= a_{j_2 j_2 0}(x_{j_1} - a_{j_2 j_1 s_1})(x_{j_1} - a_{j_2 j_1 s_2})^2, \\
\dot{x}_{j_1} &= a_{j_1 j_1 0}(x_{j_2} - a_{j_1 j_2 1})(x_{j_1} - a_{j_1 j_1 1})(x_{j_2} - a_{j_1 j_1 2})
\end{aligned}$$

(1.146)

where

$$\begin{aligned}
b_{j_1 j_1 1}, b_{j_1 j_1 2} &= -\frac{1}{2}(B_{j_1 j_1} \pm \sqrt{\Delta_{j_1 j_1}}), \\
\{a_{j_1 j_1 1}, a_{j_1 j_1 2}\} &= \text{sort}\{b_{j_1 j_1 1}, b_{j_1 j_1 2}\}, a_{j_1 j_1 1} < a_{j_1 j_1 2}; \\
a_{j_2 j_1 s_2} &= b_{j_2 j_1 1}, a_{j_2 j_1 s_2} = -\frac{1}{2}B_{j_2 j_1}.
\end{aligned}$$

(1.147)

(viii$_1$) The first integral manifold for $a_{j_1 j_1 1}, a_{j_1 j_1 2} \neq a_{j_2 j_1 s_1}, a_{j_2 j_1 s_2}$ is

$$\begin{aligned}
&\frac{1}{2}\left[(x_{j_1} - a_{j_2 j_1 s_1})^2 - (x_{j_1 0} - a_{j_2 j_1 s_1})^2\right] \\
&+ (a_{j_1 j_1 2} + a_{j_1 j_1 1} - 2a_{j_2 j_1 s_2})(x_{j_1} - x_{j_1 0}) \\
&+ \frac{(a_{j_1 j_1 2} - a_{j_2 j_1 s_1})(a_{j_1 j_1 2} - a_{j_2 j_1 s_2})^2}{a_{j_1 j_1 2} - a_{j_1 j_1 1}} \ln \frac{|x_{j_1} - a_{j_1 j_1 2}|}{|x_{j_1 0} - a_{j_1 j_1 2}|} \\
&+ \frac{(a_{j_1 j_1 1} - a_{j_2 j_1 s_1})(a_{j_1 j_1 1} - a_{j_2 j_1 s_2})^2}{a_{j_1 j_1 1} - a_{j_1 j_1 2}} \ln \frac{|x_{j_1} - a_{j_1 j_1 1}|}{|x_{j_1 0} - a_{j_1 j_1 1}|} \\
&= \frac{1}{2}\frac{a_{j_1 j_1 0}}{a_{j_2 j_2 0}}\left[(x_{j_2} - a_{j_1 j_2 1})^2 - (x_{j_2 0} - a_{j_1 j_2 1})^2\right].
\end{aligned}$$

(1.148)

(viii$_{1a}$) The equilibrium of $(x^*_{j_2}, x^*_{j_1}) = (a_{j_1j_21}, a_{j_2j_1s_1})$ has the following properties:

- *For $a_{j_1j_10}\prod_{l_1=1}^2(a_{j_2j_1s_1} - a_{j_1j_1l_1}) > 0$ and $a_{j_2j_20}(a_{j_2j_1s_1} - a_{j_2j_1s_2})^2 > 0$,*

$$(a_{j_1j_21}, a_{j_2j_1s_1}) = \underbrace{(UP_+, UP_+)}_{\text{positive saddle}}. \tag{1.149}$$

The equilibrium of $(x^*_{j_2}, x^*_{j_1}) = (a_{j_1j_21}, a_{j_2j_1s_1})$ is a (UP_+, UP_+)-positive saddle.
- *For $a_{j_1j_10}\prod_{l_1=1}^2(a_{j_2j_1s_1} - a_{j_1j_1l_1}) < 0$ and $a_{j_2j_20}(a_{j_2j_1s_1} - a_{j_2j_1s_2})^2 > 0$,*

$$(a_{j_1j_21}, a_{j_2j_1s_1}) = \underbrace{(DP_+, DP_-)}_{\text{CCW center}}. \tag{1.150}$$

The equilibrium of $(x^*_{j_2}, x^*_{j_1}) = (a_{j_1j_21}, a_{j_2j_1s_1})$ is a (DP_+, DP_-)-counter-clockwise center.
- *For $a_{j_1j_10}\prod_{l_1=1}^2(a_{j_2j_1s_1} - a_{j_1j_1l_1}) > 0$ and $a_{j_2j_20}(a_{j_2j_1s_1} - a_{j_2j_1s_2})^2 < 0$,*

$$(a_{j_1j_21}, a_{j_2j_1s_1}) = \underbrace{(DP_-, DP_+)}_{\text{CW center}}. \tag{1.151}$$

The equilibrium of $(x^*_{j_2}, x^*_{j_1}) = (a_{j_1j_21}, a_{j_2j_1s_1})$ is a (DP_+, DP_-)-clockwise center.
- *For $a_{j_1j_10}\prod_{l_1=1}^2(a_{j_2j_1s_1} - a_{j_1j_1l_1}) < 0$ and $a_{j_2j_20}(a_{j_2j_1s_1} - a_{j_2j_1s_2})^2 < 0$,*

$$(a_{j_1j_21}, a_{j_2j_1s_1}) = \underbrace{(UP_-, UP_-)}_{\text{negative saddle}}. \tag{1.152}$$

The equilibrium of $(x^*_{j_2}, x^*_{j_1}) = (a_{j_1j_21}, a_{j_2j_1s_1})$ is a (UP_-, UP_-)-negative saddle.

(viii$_{1b}$) The equilibrium of $(x^*_{j_2}, x^*_{j_1}) = (a_{j_1j_21}, a_{j_2j_1s_2})$ has the following properties:

- *For $a_{j_1j_10}\prod_{l_1=1}^2(a_{j_2j_1s_2} - a_{j_1j_1l_1}) > 0$ and $a_{j_2j_20}(a_{j_2j_1s_2} - a_{j_2j_1s_1}) > 0$,*

$$(a_{j_1j_21}, a_{j_2j_1s_2}) = \underbrace{(UP, US)}_{\text{up-parabola upper-saddle}}. \tag{1.153}$$

The equilibrium of $(x^*_{j_2}, x^*_{j_1}) = (a_{j_1j_21}, a_{j_2j_1s_2})$ is a (UP,US)-up-parabola upper-saddle.
- *For $a_{j_1j_10}\prod_{l_1=1}^2(a_{j_2j_1s_2} - a_{j_1j_1l_1}) < 0$ and $a_{j_2j_20}(a_{j_2j_1s_2} - a_{j_2j_1s_1}) > 0$,*

$$(a_{j_1j_21}, a_{j_2j_1s_2}) = \underbrace{(DP, US)}_{\text{down-parabola upper-saddle}}. \tag{1.154}$$

The equilibrium of $(x^*_{j_2}, x^*_{j_1}) = (a_{j_1j_21}, a_{j_2j_1s_2})$ is a (DP,US)-down-parabola upper-saddle.

- For $a_{j_1j_10}\prod_{l_1=1}^{2}(a_{j_2j_1s_2} - a_{j_1j_1l_1}) > 0$ and $a_{j_2j_20}(a_{j_2j_1s_2} - a_{j_2j_1s_1}) < 0$,

$$(a_{j_1j_21}, a_{j_2j_1s_2}) = \underbrace{(\text{DP, LS})}_{\text{down-parabola upper-saddle}} . \qquad (1.155)$$

The equilibrium of $(x_{j_2}^*, x_{j_1}^*) = (a_{j_1j_21}, a_{j_2j_1s_2})$ is a (DP,LS)-down-parabola lower-saddle.

- For $a_{j_1j_10}\prod_{l_1=1}^{2}(a_{j_2j_1s_2} - a_{j_1j_1l_1}) < 0$ and $a_{j_2j_20}(a_{j_2j_1s_2} - a_{j_2j_1s_1}) < 0$,

$$(a_{j_1j_21}, a_{j_2j_1s_2}) = \underbrace{(\text{UP, LS})}_{\text{up-parabola lower-saddle}} . \qquad (1.156)$$

The equilibrium of $(x_{j_2}^*, x_{j_1}^*) = (a_{j_1j_21}, a_{j_2j_1s_2})$ is a (UP,LS)-up-parabola lower-saddle.

- The parabola upper-saddle and lower-saddle are the appearing bifurcations of a saddle and center.

$(viii_{1c})$ The equilibrium of $(x_{j_1}^*, x_{j_2}^*) = (a_{j_1j_1l_1}, a_{j_1j_21})$ $(l_1, l_2 \in \{1, 2\}, l_1 \neq l_2)$ with $a_{j_2j_1s_1} \neq a_{j_2j_1s_2}(s_1, s_2 \in \{1, 2\}, s_1 \neq s_2)$ has the following properties:

- For $a_{j_1j_10}(a_{j_1j_1l_1} - a_{j_1j_1l_2}) > 0$ and $a_{j_2j_20}(a_{j_1j_1l_1} - a_{j_2j_1s_1})(a_{j_1j_1l_1} - a_{j_2j_1s_2})^2 > 0$,

$$(a_{j_1j_1l_1}, a_{j_1j_21}) = \underbrace{(\text{DP:UP, pF})}_{\text{hyperbolic flow } (+)} . \qquad (1.157)$$

The equilibrium of $(x_{j_1}^*, x_{j_2}^*) = (a_{j_1j_1l_1}, a_{j_1j_21})$ is a (DP:UP,pF)-positive hyperbolic flow.

- For $a_{j_1j_10}(a_{j_1j_1l_1} - a_{j_1j_1l_2}) < 0$ and $a_{j_2j_20}(a_{j_1j_1l_1} - a_{j_2j_1s_1})(a_{j_1j_1l_1} - a_{j_2j_1s_2})^2 > 0$,

$$(a_{j_1j_1l_1}, a_{j_1j_21}) = \underbrace{(\text{UP:DP, pF})}_{\text{hyperbolic-secant flow } (+)} . \qquad (1.158)$$

The equilibrium of $(x_{j_1}^*, x_{j_2}^*) = (a_{j_1j_1l_1}, a_{j_1j_21})$ is a (UP:DP,pF)-positive hyperbolic-secant flow.

- For $a_{j_1j_10}(a_{j_1j_1l_1} - a_{j_1j_1l_2}) > 0$ and $a_{j_2j_20}(a_{j_1j_1l_1} - a_{j_2j_1s_1})(a_{j_1j_1l_1} - a_{j_2j_1s_2})^2 < 0$,

$$(a_{j_1j_1l_1}, a_{j_1j_21}) = \underbrace{(\text{UP:DP, nF})}_{\text{hyperbolic-secant flow } (-)} . \qquad (1.159)$$

The equilibrium of $(x_{j_1}^*, x_{j_2}^*) = (a_{j_1j_1l_1}, a_{j_1j_21})$ is a (UP:DP,nF)-negative hyperbolic-secant flow.

- *For $a_{j_1j_10}(a_{j_1j_1l_1} - a_{j_1j_1l_2}) < 0$ and $a_{j_2j_20}(a_{j_1j_1l_1} - a_{j_2j_1s_1})(a_{j_1j_1l_1} - a_{j_2j_1s_2})^2 < 0$,*

$$(a_{j_1j_1l_1}, a_{j_1j_21}) = \underbrace{(DP:UP, nF)}_{\text{hyperbolic flow } (-)} . \qquad (1.160)$$

The equilibrium of $(x_{j_1}^*, x_{j_2}^*) = (a_{j_1j_1l_1}, a_{j_1j_21})$ is a (DP:UP,nF)-negative hyperbolic flow.

(viii$_2$) For $a_{j_2j_1s_1} = a_{j_2j_1s_2} = a_{j_2j_11}$, the standard form is

$$\begin{aligned}
\dot{x}_{j_2} &= a_{j_2j_20}(x_{j_1} - a_{j_2j_11})^3, \\
\dot{x}_{j_1} &= a_{j_1j_10}(x_{j_2} - a_{j_1j_21})(x_{j_1} - a_{j_1j_11})(x_{j_1} - a_{j_1j_12}).
\end{aligned} \qquad (1.161)$$

The first integral manifold for $a_{j_2j_11} \neq a_{j_1j_11}$ is

$$\begin{aligned}
&\frac{1}{2}\left[(x_{j_1} - a_{j_2j_11})^2 - (x_{j_10} - a_{j_2j_11})^2\right] + (a_{j_1j_11} + a_{j_1j_12} - 2a_{j_2j_11})(x_{j_1} - x_{j_10}) \\
&+ \frac{(a_{j_1j_11} - a_{j_2j_11})^3}{(a_{j_1j_11} - a_{j_1j_12})} \ln \frac{|x_{j_1} - a_{j_1j_11}|}{|x_{j_10} - a_{j_1j_11}|} + \frac{(a_{j_1j_12} - a_{j_2j_11})^3}{(a_{j_1j_12} - a_{j_1j_11})} \ln \frac{|x_{j_1} - a_{j_1j_12}|}{|x_{j_10} - a_{j_1j_12}|} \qquad (1.162) \\
&= \frac{1}{2}\frac{a_{j_1j_10}}{a_{j_2j_20}}\left[(x_{j_2} - a_{j_1j_21})^2 - (x_{j_20} - a_{j_1j_21})^2\right].
\end{aligned}$$

(viii$_{2a}$) The equilibrium of $(x_{j_2}^*, x_{j_1}^*) = (a_{j_1j_2l_1}, a_{j_2j_11})$ has the following properties:

- *For $a_{j_1j_10}\prod_{l_1=1}^2(a_{j_2j_11} - a_{j_1j_1l_1}) > 0$ and $a_{j_2j_20} > 0$,*

$$(a_{j_1j_21}, a_{j_2j_11}) = \underbrace{(UP_+, 3^{\text{rd}}UP_+)}_{\text{third-order positive saddle}} . \qquad (1.163)$$

The equilibrium of $(x_{j_2}^*, x_{j_1}^*) = (a_{j_1j_21}, a_{j_2j_11})$ is a $(UP_+, 3^{\text{rd}}UP_+)$-third-order positive saddle.

- *For $a_{j_1j_10}\prod_{l_1=1}^2(a_{j_2j_11} - a_{j_1j_1l_1}) < 0$ and $a_{j_2j_20} > 0$,*

$$(a_{j_1j_21}, a_{j_2j_11}) = \underbrace{(DP_+, 3^{\text{rd}}DP_-)}_{\text{third-order CCW center}} . \qquad (1.164)$$

The equilibrium of $(x_{j_2}^*, x_{j_1}^*) = (a_{j_1j_21}, a_{j_2j_11})$ is a $(DP_+, 3^{\text{rd}}DP_-)$-third-order counter-clockwise center.

- *For $a_{j_1j_10}\prod_{l_1=1}^2(a_{j_2j_11} - a_{j_1j_1l_1}) > 0$ and $a_{j_2j_20} < 0$,*

$$(a_{j_1j_21}, a_{j_2j_11}) = \underbrace{(DP_-, 3^{rd}DP_+)}_{\text{third-order CW center}}.\qquad(1.165)$$

The equilibrium of $(x_{j_2}^*, x_{j_1}^*) = (a_{j_1j_21}, a_{j_2j_11})$ *is a* $(DP_-, 3^{rd}DP_+)$-*third-order clockwise center.*

- *For* $a_{j_1j_10}\prod_{l_1=1}^{2}(a_{j_2j_11} - a_{j_1j_1l_1}) < 0$ *and* $a_{j_2j_20} < 0,$

$$(a_{j_1j_21}, a_{j_2j_11}) = \underbrace{(UP_-, 3^{rd}UP_-)}_{\text{third-order negative saddle}}.\qquad(1.166)$$

The equilibrium of $(x_{j_2}^*, x_{j_1}^*) = (a_{j_1j_21}, a_{j_2j_11})$ *is a* $(UP_-, 3^{rd}UP_-)$-*third-order negative saddle.*

- *The third-order saddles are the appearing and switching bifurcations of saddle, center, and saddle.*
- *The third-order centers are the appearing and switching bifurcations of center, saddle, and center.*

$(viii_{2b})$ *The equilibrium of* $(x_{j_1}^*, x_{j_2}^*) = (a_{j_1j_1l_1}, a_{j_1j_21})$ *with* $a_{j_1j_1l_1} \neq a_{j_1j_1l_2}, a_{j_2j_11}$ *has the following properties:*

- *For* $a_{j_1j_10}(a_{j_1j_1l_1} - a_{j_1j_1l_2}) > 0$ *and* $a_{j_2j_20}(a_{j_1j_11} - a_{j_2j_11})^3 > 0,$

$$(a_{j_1j_1l_1}, a_{j_1j_21}) = \underbrace{(DP:UP, pF)}_{\text{hyperbolic flow }(+)}.\qquad(1.167)$$

The equilibrium of $(x_{j_1}^*, x_{j_2}^*) = (a_{j_1j_1l_1}, a_{j_1j_21})$ *is a* $(DP:UP, pF)$-*positive hyperbolic flow.*

- *For* $a_{j_1j_10}(a_{j_1j_1l_1} - a_{j_1j_1l_2}) < 0$ *and* $a_{j_2j_20}(a_{j_1j_11} - a_{j_2j_11})^3 > 0,$

$$(a_{j_1j_1l_1}, a_{j_1j_21}) = \underbrace{(UP:DP, pF)}_{\text{hyperbolic-secant flow }(+)}.\qquad(1.168)$$

The equilibrium of $(x_{j_1}^*, x_{j_2}^*) = (a_{j_1j_1l_1}, a_{j_1j_21})$ *is a* $(UP:DP,pF)$-*positive hyperbolic-secant flow.*

- *For* $a_{j_1j_10}(a_{j_1j_2l_1} - a_{j_1j_2l_2}) > 0$ *and* $a_{j_2j_20}(a_{j_1j_11} - a_{j_2j_11})^3 < 0,$

$$(a_{j_1j_1l_1}, a_{j_1j_21}) = \underbrace{(UP:DP, nF)}_{\text{hyperbolic-secant flow }(-)}.\qquad(1.169)$$

The equilibrium of $(x_{j_1}^*, x_{j_2}^*) = (a_{j_1j_1l_1}, a_{j_1j_21})$ *is a* $(UP:DP,nF)$-*negative hyperbolic-secant flow.*

- For $a_{j_1j_10}(a_{j_1j_2l_1} - a_{j_1j_2l_2}) < 0$ and $a_{j_2j_20}(a_{j_1j_11} - a_{j_2j_11})^3 < 0$,

$$(a_{j_1j_1l_1}, a_{j_1j_21}) = \underbrace{(DP{:}UP, nF)}_{\text{hyperbolic flow } (-)} . \tag{1.170}$$

The equilibrium of $(x_{j_1}^*, x_{j_2}^*) = (a_{j_1j_1l_1}, a_{j_1j_21})$ is a (DP:UP, nF)-negative hyperbolic flow.

(viii₃) The first integral manifold for $a_{j_2j_1s_1} = a_{j_1j_1l_1}$ is

$$\frac{1}{2}\left[(x_{j_1} - a_{j_2j_1s_2})^2 - (x_{j_10} - a_{j_2j_1s_2})^2\right] + 2(a_{j_1j_1s_2} - a_{j_2j_1s_2})(x_{j_1} - x_{j_10})$$

$$+ (a_{j_1j_1s_2} - a_{j_2j_1s_2})^2 \ln \frac{|x_{j_1} - a_{j_2j_1s_2}|}{|x_{j_10} - a_{j_2j_1s_2}|} \tag{1.171}$$

$$= \frac{1}{2}\frac{a_{j_1j_10}}{a_{j_2j_20}}\left[(x_{j_2} - a_{j_1j_21})^2 - (x_{j_20} - a_{j_1j_21})^2\right].$$

(viii₃ₐ) The infinite-equilibrium of $x_{j_1}^* = a_{j_2j_1s_1} = a_{j_1j_1l_1}$ with $\bar{x}_{j_2} \neq a_{j_1j_21}$ has the following properties:

- For $a_{j_1j_10}(\bar{x}_{j_2} - a_{j_1j_21})(a_{j_1j_1l_1} - a_{j_1j_1l_2}) > 0$ and $a_{j_2j_20}(a_{j_2j_1s_1} - a_{j_2j_1s_2})^2 > 0$,

$$(a_{j_1j_1l_1}, \bar{x}_{j_2}) = \underbrace{(SO, II)}_{\text{increasing-inflection source}} . \tag{1.172}$$

The infinite-equilibrium of $x_{j_1}^* = a_{j_2j_1s_1} = a_{j_1j_1l_1}$ is an (SO,II)-increasing-inflection source.

- For $a_{j_1j_10}(\bar{x}_{j_2} - a_{j_1j_21})(a_{j_1j_1l_1} - a_{j_1j_1l_2}) < 0$ and $a_{j_2j_20}(a_{j_2j_1s_1} - a_{j_2j_1s_2})^2 > 0$,

$$(a_{j_1j_1l_1}, \bar{x}_{j_2}) = \underbrace{(SI, DI)}_{\text{decreasing-inflection sink}} . \tag{1.173}$$

The infinite-equilibrium of $x_{j_1}^* = a_{j_2j_1s_1} = a_{j_1j_1l_1}$ is an (SI,DI)-decreasing-inflection sink.

- For $a_{j_1j_10}(\bar{x}_{j_2} - a_{j_1j_21})(a_{j_1j_1l_1} - a_{j_1j_1l_2}) > 0$ and $a_{j_2j_20}(a_{j_2j_1s_1} - a_{j_2j_1s_2})^2 < 0$,

$$(a_{j_1j_1l_1}, \bar{x}_{j_2}) = \underbrace{(SO, DI)}_{\text{decreasing-inflection source}} . \tag{1.174}$$

The infinite-equilibrium of $x_{j_1}^* = a_{j_2j_1s_1} = a_{j_1j_1l_1}$ is an (SO,DI)-decreasing-inflection source.

- For $a_{j_1j_10}(\bar{x}_{j_2} - a_{j_1j_21})(a_{j_1j_1l_1} - a_{j_1j_1l_2}) < 0$ and $a_{j_2j_20}(a_{j_2j_1s_1} - a_{j_2j_1s_2})^2 < 0$,

$$(a_{j_1 j_1 l_1}, \bar{x}_{j_2}) = \underbrace{(\text{SI, II})}_{\text{increasing-inflection sink}} .$$ (1.175)

The infinite-equilibrium of $x_{j_1}^* = a_{j_2 j_1 s_1} = a_{j_1 j_1 l_1}$ is an (SI,II)-increasing-inflection sink.

(viii$_{3b}$) The equilibrium of $(x_{j_2}^*, x_{j_1}^*) = (a_{j_1 j_2 1}, a_{j_2 j_1 s_1})$ with $a_{j_2 j_1 s_1} = a_{j_1 j_1 l_1}$ has the following properties:

- For $a_{j_1 j_1 0}(a_{j_1 j_1 l_1} - a_{j_1 j_1 l_2}) > 0$ and $a_{j_2 j_2 0}(a_{j_2 j_1 s_1} - a_{j_2 j_1 s_2})^2 > 0$,

$$(a_{j_1 j_2 1}, a_{j_2 j_1 s_1}) = \underbrace{(_{\text{DI:II}}\text{UP}, _{\text{SI:SO}}\text{US})}_{\text{up-parabola upper-saddle}} .$$ (1.176)

The equilibrium of $(x_{j_2}^*, x_{j_1}^*) = (a_{j_1 j_2 1}, a_{j_2 j_1 s_1})$ is a $(_{\text{DI:II}}\text{UP}, _{\text{SI:SO}}\text{US})$-up-parabola upper-saddle.
- For $a_{j_1 j_1 0}(a_{j_1 j_1 l_1} - a_{j_1 j_1 l_2}) < 0$ and $a_{j_2 j_2 0}(a_{j_2 j_1 s_1} - a_{j_2 j_1 s_2})^2 > 0$,

$$(a_{j_1 j_2 1}, a_{j_2 j_1 s_1}) = \underbrace{(_{\text{II:DI}}\text{DP}, _{\text{SO:SI}}\text{LS})}_{\text{down-parabola lower-saddle}} .$$ (1.177)

The equilibrium of $(x_{j_2}^*, x_{j_1}^*) = (a_{j_1 j_2 1}, a_{j_2 j_1 s_1})$ is an $(_{\text{II:DI}}\text{DP}, _{\text{SO:SI}}\text{LS})$-down-parabola lower-saddle.
- For $a_{j_1 j_1 0}(a_{j_1 j_1 l_1} - a_{j_1 j_1 l_2}) > 0$ and $a_{j_2 j_2 0}(a_{j_2 j_1 s_1} - a_{j_2 j_1 s_2})^2 < 0$,

$$(a_{j_1 j_2 1}, a_{j_2 j_1 s_1}) = \underbrace{(_{\text{II:DI}}\text{DP}, _{\text{SI:SO}}\text{US})}_{\text{down-parabola upper-saddle}} .$$ (1.178)

The equilibrium of $(x_{j_2}^*, x_{j_1}^*) = (a_{j_1 j_2 1}, a_{j_2 j_1 s_1})$ is an $(_{\text{II:DI}}\text{DP}, _{\text{SI:SO}}\text{US})$-down-parabola upper-saddle.
- For $a_{j_1 j_1 0}(a_{j_1 j_1 l_1} - a_{j_1 j_1 l_2}) < 0$ and $a_{j_2 j_2 0}(a_{j_2 j_1 s_1} - a_{j_2 j_1 s_2})^2 < 0$,

$$(a_{j_1 j_2 1}, a_{j_2 j_1 s_1}) = \underbrace{(_{\text{DI:II}}\text{UP}, _{\text{SO:SI}}\text{LS})}_{\text{up-parabola lower-saddle}} .$$ (1.179)

The equilibrium of $(x_{j_2}^*, x_{j_1}^*) = (a_{j_1 j_2 1}, a_{j_2 j_1 s_1})$ is a $(_{\text{DI:II}}\text{UP}, _{\text{SO:SI}}\text{LS})$-up-parabola lower-saddle.

(viii$_4$) The first integral manifold for $a_{j_1 j_1 1} = a_{j_2 j_1 s_2}$ ($s_1, s_2 \in \{1, 2\}$, $s_1 \neq s_2$) is

$$\frac{1}{2}\left[(x_{j_1} - a_{j_1j_1l_2})^2 - (x_{j_10} - a_{j_1j_1l_2})^2\right] + (2a_{j_1j_1l_2} - a_{j_2j_1s_1} - a_{j_2j_1s_2})(x_{j_1} - x_{j_10})$$

$$+(a_{j_1j_1l_2} - a_{j_2j_1s_1})(a_{j_1j_1l_2} - a_{j_2j_1s_2})\ln\frac{\mid x_{j_1} - a_{j_1j_1l_2}\mid}{\mid x_{j_10} - a_{j_1j_1l_2}\mid} \qquad (1.180)$$

$$=\frac{1}{2}\frac{a_{j_1j_10}}{a_{j_2j_20}}\left[(x_{j_2} - a_{j_1j_21})^2 - (x_{j_20} - a_{j_1j_21})^2\right].$$

(viii$_{4a}$) *The infinite-equilibrium of* $x_{j_1}^* = a_{j_1j_1l_1} = a_{j_2j_1s_2}$ *with* $\bar{x}_{j_2} \neq a_{j_1j_21}$ *has the following properties:*

- *For* $a_{j_1j_10}(\bar{x}_{j_2} - a_{j_1j_21})(a_{j_1j_1l_1} - a_{j_1j_1l_2}) > 0$ *and* $a_{j_2j_20}(a_{j_2j_1s_2} - a_{j_2j_1s_1}) > 0$,

$$(a_{j_1j_1l_1}, \bar{x}_{j_2}) = \underbrace{(\text{SO}, \text{UP})}_{\text{up-parabola source}}. \qquad (1.181)$$

   The infinite-equilibrium of $x_{j_1}^* = a_{j_1j_1l_1} = a_{j_2j_1s_2}$ *is an (SO,UP)-up-parabola source.*
- *For* $a_{j_1j_10}(\bar{x}_{j_2} - a_{j_1j_21})(a_{j_1j_1l_1} - a_{j_1j_1l_2}) < 0$ *and* $a_{j_2j_20}(a_{j_2j_1s_2} - a_{j_2j_1s_1}) > 0$,

$$(a_{j_1j_1l_1}, \bar{x}_{j_2}) = \underbrace{(\text{SI}, \text{DP})}_{\text{down-parabola sink}}. \qquad (1.182)$$

   The infinite-equilibrium of $x_{j_1}^* = a_{j_1j_1l_1} = a_{j_2j_1s_2}$ *is an (SI,DP)-down-parabola sink.*
- *For* $a_{j_1j_10}(\bar{x}_{j_2} - a_{j_1j_21})(a_{j_1j_1l_1} - a_{j_1j_1l_2}) > 0$ *and* $a_{j_2j_20}(a_{j_2j_1s_2} - a_{j_2j_1s_1}) < 0$,

$$(a_{j_1j_1l_1}, \bar{x}_{j_2}) = \underbrace{(\text{SO}, \text{DP})}_{\text{down-parabola sink}}. \qquad (1.183)$$

   The infinite-equilibrium of $x_{j_1}^* = a_{j_1j_1l_1} = a_{j_2j_1s_2}$ *is an (SO,DP)-down-parabola source.*
- *For* $a_{j_1j_10}(\bar{x}_{j_2} - a_{j_1j_21})(a_{j_1j_1l_1} - a_{j_1j_1l_2}) < 0$ *and* $a_{j_2j_20}(a_{j_2j_1s_2} - a_{j_2j_1s_1}) < 0$,

$$(a_{j_1j_1l_1}, \bar{x}_{j_2}) = \underbrace{(\text{SI}, \text{UP})}_{\text{up-parabola sink}}. \qquad (1.184)$$

   The infinite-equilibrium of $x_{j_1}^* = a_{j_1j_1l_1} = a_{j_2j_1s_2}$ *is an (SI,UP)-up-parabola sink.*

(viii$_{4b}$) *The equilibrium of* $(x_{j_2}^*, x_{j_1}^*) = (a_{j_1j_21}, a_{j_2j_1s_2})$ *with* $a_{j_1j_1l_1} = a_{j_2j_1s_2}$ *has the following properties:*

- *For* $a_{j_1j_10}(a_{j_1j_1l_1} - a_{j_1j_1l_2}) > 0$ *and* $a_{j_2j_20}(a_{j_2j_1s_2} - a_{j_2j_1s_1}) > 0$,

$$(a_{j_1 j_2 1}, a_{j_2 j_1 s_2}) = \underbrace{({}_{\text{DP}}\text{SI} : {}_{\text{UP}}\text{SO}, \text{DP}_+ : \text{UP}_+)}_{\text{hyperbolic sink-to-source}}. \tag{1.185}$$

The equilibrium of $(x_{j_2}^*, x_{j_1}^*) = (a_{j_1 j_2 1}, a_{j_2 j_1 s_2})$ *is a* $({}_{\text{DP}}\text{SI}:{}_{\text{UP}}\text{SO},\text{DP}_+:\text{UP}_+)$-*hyperbolic sink-to-source.*

- For $a_{j_1 j_1 0}(a_{j_1 j_1 l_1} - a_{j_1 j_1 l_2}) < 0$ and $a_{j_2 j_2 0}(a_{j_2 j_1 s_2} - a_{j_2 j_1 s_1}) > 0,$

$$(a_{j_1 j_2 1}, a_{j_2 j_1 s_2}) = \underbrace{({}_{\text{UP}}\text{SO} : {}_{\text{DP}}\text{SI}, \text{UP}_+ : \text{DP}_+)}_{\text{circular source-to-sink}}. \tag{1.186}$$

The equilibrium of $(x_{j_2}^*, x_{j_1}^*) = (a_{j_1 j_2 1}, a_{j_2 j_1 s_2})$ *is a* $({}_{\text{UP}}\text{SO}:{}_{\text{DP}}\text{SI},\text{UP}_+:\text{DP}_+)$-*circular source-to-sink.*

- For $a_{j_1 j_1 0}(a_{j_1 j_1 l_1} - a_{j_1 j_1 l_2}) > 0$ and $a_{j_2 j_2 0}(a_{j_2 j_1 s_2} - a_{j_2 j_1 s_1}) < 0,$

$$(a_{j_1 j_2 1}, a_{j_2 j_1 s_2}) = \underbrace{({}_{\text{UP}}\text{SI} : {}_{\text{DP}}\text{SO}, \text{UP}_- : \text{DP}_-)}_{\text{circular sink-to-source}}. \tag{1.187}$$

The equilibrium of $(x_{j_2}^*, x_{j_1}^*) = (a_{j_1 j_2 1}, a_{j_2 j_1 s_2})$ *is a* $({}_{\text{UP}}\text{SI}:{}_{\text{DP}}\text{SO},\text{UP}_-:\text{DP}_-)$-*circular sink-to-source.*

- For $a_{j_1 j_1 0}(a_{j_1 j_1 l_1} - a_{j_1 j_1 l_2}) < 0$ and $a_{j_2 j_2 0}(a_{j_2 j_1 s_2} - a_{j_2 j_1 s_1}) < 0,$

$$(a_{j_1 j_2 1}, a_{j_2 j_1 s_2}) = \underbrace{({}_{\text{DP}}\text{SO}:{}_{\text{UP}}\text{SI}, \text{DP}_- : \text{UP}_-)}_{\text{hyperbolic source-to-sink}}. \tag{1.188}$$

The equilibrium of $(x_{j_2}^*, x_{j_1}^*) = (a_{j_1 j_2 1}, a_{j_2 j_1 s_2})$ *is a* $({}_{\text{DP}}\text{SO}:{}_{\text{UP}}\text{SI},\text{DP}_-:\text{UP}_-)$-*hyperbolic source-to-sink.*

(viii$_5$) *The first integral manifold for* $a_{j_2 j_1 s_1} = a_{j_2 j_1 s_2} = a_{j_2 j_1 1} = a_{j_1 j_1 1}$ *is*

$$\frac{1}{2}\left[(x_{j_1} - a_{j_1 j_1 l_2})^2 - (x_{j_1 0} - a_{j_1 j_1 l_2})^2\right] + 2(a_{j_1 j_1 l_2} - a_{j_2 j_1 1})(x_{j_1} - x_{j_1 0})$$

$$+(a_{j_1 j_1 l_2} - a_{j_2 j_1 1})^2 \ln \frac{|x_{j_1} - a_{j_1 j_1 l_2}|}{|x_{j_1 0} - a_{j_1 j_1 l_2}|} \tag{1.189}$$

$$= \frac{1}{2}\frac{a_{j_1 j_1 0}}{a_{j_2 j_2 0}}\left[(x_{j_2} - a_{j_1 j_2 1})^2 - (x_{j_2 0} - a_{j_1 j_2 1})^2\right].$$

(viii$_{5a}$) *The infinite-equilibrium of* $x_{j_1}^* = a_{j_2 j_1 1} = a_{j_1 j_1 l_1}$ *with* $\bar{x}_{j_2} \neq a_{j_1 j_2 1}$ *has the following properties:*

- *For* $a_{j_1 j_1 0}(\bar{x}_{j_2} - a_{j_1 j_2 1})(a_{j_1 j_1 l_1} - a_{j_1 j_1 l_2}) > 0$ *and* $a_{j_2 j_2 0} > 0,$

$$(a_{j_1j_11}, \bar{x}_{j_2}) = \underbrace{(\text{SO}, 2^{\text{nd}}\text{II})}_{\text{second-order increasing-inflection source}} . \tag{1.190}$$

The infinite-equilibrium of $x_{j_1}^* = a_{j_2j_11} = a_{j_1j_1l_1}$ is an $(\text{SO},2^{\text{nd}}\text{II})$-second-order increasing-inflection source.

- For $a_{j_1j_10}(\bar{x}_{j_2} - a_{j_1j_21})(a_{j_1j_1l_1} - a_{j_1j_1l_2}) < 0$ and $a_{j_2j_20} > 0$,

$$(a_{j_1j_1l_1}, \bar{x}_{j_2}) = \underbrace{(\text{SI}, 2^{\text{nd}}\text{DI})}_{\text{second-order decreasing-inflection sink}} . \tag{1.191}$$

The infinite-equilibrium of $x_{j_1}^* = a_{j_2j_11} = a_{j_1j_1l_1}$ is an $(\text{SI},2^{\text{nd}}\text{DI})$-second-order decreasing-inflection sink.

- For $a_{j_1j_10}(\bar{x}_{j_2} - a_{j_1j_21})(a_{j_1j_1l_1} - a_{j_1j_1l_2}) > 0$ and $a_{j_2j_20} < 0$,

$$(a_{j_1j_1l_1}, \bar{x}_{j_2}) = \underbrace{(\text{SO}, 2^{\text{nd}}\text{DI})}_{\text{second-order decreasing-inflection source}} . \tag{1.192}$$

The infinite-equilibrium of $x_{j_1}^* = a_{j_2j_11} = a_{j_1j_1l_1}$ is an $(\text{SO},2^{\text{nd}}\text{DI})$-second-order decreasing-inflection source.

- For $a_{j_1j_10}(\bar{x}_{j_2} - a_{j_1j_21})(a_{j_1j_1l_1} - a_{j_1j_1l_2}) < 0$ and $a_{j_2j_20} < 0$,

$$(a_{j_1j_1l_1}, \bar{x}_{j_2}) = \underbrace{(\text{SI}, 2^{\text{nd}}\text{II})}_{\text{second-order increasing-inflection sink}} . \tag{1.193}$$

The infinite-equilibrium of $x_{j_1}^* = a_{j_2j_11} = a_{j_1j_1l_1}$ is an $(\text{SI},2^{\text{nd}}\text{II})$-second-order increasing-inflection sink.

$(viii_{5b})$ The equilibrium of $(x_{j_2}^*, x_{j_1}^*) = (a_{j_1j_21}, a_{j_2j_11})$ with $a_{j_2j_11} = a_{j_1j_1l_1}$ has the following properties:

- For $a_{j_1j_10}(a_{j_1j_1l_1} - a_{j_1j_1l_2}) > 0$ and $a_{j_2j_20} > 0$,

$$(a_{j_1j_21}, a_{j_2j_11}) = \underbrace{(_{2^{\text{nd}}(\text{DI:II})}\text{UP}, {}_{\text{SI:SO}}\text{US})}_{\text{second-order up-parabola upper-saddle}} . \tag{1.194}$$

The equilibrium of $(x_{j_2}^*, x_{j_1}^*) = (a_{j_1j_21}, a_{j_2j_11})$ is a $(_{2^{\text{nd}}(\text{DI:II})}\text{UP}, {}_{\text{SI:SO}}\text{US})$-second-order up-parabola upper-saddle.

- For $a_{j_1j_10}(a_{j_1j_1l_1} - a_{j_1j_1l_2}) < 0$ and $a_{j_2j_20} > 0$,

$$(a_{j_1j_21}, a_{j_2j_11}) = \underbrace{(_{2^{\text{nd}}(\text{II:DI})}\text{DP}, {}_{\text{SO:SI}}\text{US})}_{\text{second-order down-parabola upper-saddle}} . \tag{1.195}$$

The equilibrium of $(x_{j_2}^*, x_{j_1}^*) = (a_{j_1j_21}, a_{j_2j_11})$ is a $(\text{2nd}_{(\text{II:DI})}\text{DP},_{\text{SI:SO}}\text{US})$-second-order down-parabola upper-saddle.

- For $a_{j_1j_10}(a_{j_1j_1l_1} - a_{j_1j_1l_2}) > 0$ and $a_{j_2j_20} < 0$,

$$(a_{j_1j_21}, a_{j_2j_11}) = \underbrace{(_{\text{2nd}(\text{II:DI})}\text{DP, SO:SI}\text{LS})}_{\text{second-order down-parabola lower-saddle}} \qquad (1.196)$$

The equilibrium of $(x_{j_2}^*, x_{j_1}^*) = (a_{j_1j_21}, a_{j_2j_11})$ is a $(\text{2nd}_{(\text{II:DI})}\text{DP},_{\text{SO:SI}}\text{LS})$-second-order down-parabola lower-saddler.

- For $a_{j_1j_10}(a_{j_1j_1l_1} - a_{j_1j_1l_2}) < 0$ and $a_{j_2j_20} < 0$,

$$(a_{j_1j_21}, a_{j_2j_11}) = \underbrace{(_{\text{2nd}(\text{DI:II})}\text{UP, SO:SI}\text{LS})}_{\text{second-order up-parabola lower-saddle}} \qquad (1.197)$$

The equilibrium of $(x_{j_2}^*, x_{j_1}^*) = (a_{j_1j_21}, a_{j_2j_11})$ is a $(\text{2nd}_{(\text{DI:II})}\text{UP},_{\text{SO:SI}}\text{LS})$-second-order up-parabola lower-saddle.

(ix) For $\Delta_{j_1j_2} > 0$ and $\Delta_{j_2j_1} > 0$, the standard form is

$$\begin{aligned}
\dot{x}_{j_2} &= a_{j_2j_20}(x_{j_1} - a_{j_2j_11})(x_{j_1} - a_{j_2j_12})(x_{j_1} - a_{j_2j_13}), \\
\dot{x}_{j_1} &= a_{j_1j_10}(x_{j_1} - a_{j_1j_11})(x_{j_1} - a_{j_1j_12})(x_{j_2} - a_{j_1j_21})
\end{aligned} \qquad (1.198)$$

where

$$\begin{aligned}
& b_{j_1j_11}, b_{j_1j_12} = -\frac{1}{2}(B_{j_1j_1} \pm \sqrt{\Delta_{j_1j_1}}), \\
& \{a_{j_1j_11}, a_{j_1j_12}\} = \text{sort}\{b_{j_1j_11}, b_{j_1j_12}\}, a_{j_1j_11} < a_{j_1j_12}; \\
& b_{j_2j_12}, b_{j_2j_13} = -\frac{1}{2}(B_{j_2j_1} \pm \sqrt{\Delta_{j_2j_1}}), \\
& \{a_{j_2j_11}, a_{j_2j_12}, a_{j_2j_13}\} = \text{sort}\{b_{j_2j_11}, b_{j_2j_12}, b_{j_2j_13}\}, \\
& a_{j_2j_1s_1} < a_{j_2j_1s_2}, s_1, s_2 \in \{1, 2, 3\}, s_1 < s_2.
\end{aligned} \qquad (1.199)$$

(ix$_1$) The first integral manifold $(a_{j_2j_1s_1} \neq a_{j_1j_11}, s_1 = 1, 2, 3$ and $l_1, l_2 \in \{1, 2\}, l_1 \neq l_2)$ is

$$\begin{aligned}
& \frac{1}{2}\left[(x_{j_1} - a_{j_2j_1s_1})^2 - (x_{j_10} - a_{j_2j_1s_1})^2\right] \\
& + \sum_{l_1=1, l_1 \neq l_2}^{2} \frac{\prod_{s_2=1, s_2 \neq s_1}^{3}(a_{j_1j_1l_1} - a_{j_2j_1s_2})}{a_{j_1j_1l_2} - a_{j_1j_1l_1}}(x_{j_1} - x_{j_10}) \\
& + \sum_{l_1=1, l_1 \neq l_2}^{2} \frac{\prod_{s_1=1}^{3}(a_{j_1j_1l_1} - a_{j_2j_1s_1})}{a_{j_1j_1l_2} - a_{j_1j_1l_1}} \ln\frac{|x_{j_1} - a_{j_1j_1l_1}|}{|x_{j_10} - a_{j_1j_1l_1}|} \\
& = \frac{1}{2}\frac{a_{j_1j_10}}{a_{j_2j_20}}\left[(x_{j_2} - a_{j_1j_21})^3 - (x_{j_20} - a_{j_1j_21})^3\right].
\end{aligned} \qquad (1.200)$$

(ix$_{1a}$) The equilibrium of $(x_{j_2}^*, x_{j_1}^*) = (a_{j_1 j_2 1}, a_{j_2 j_1 s_1})(s_1, s_2 \in \{1,2,3\}, s_1 \neq s_2; l_1, l_2 \in \{1,2\}, l_1 \neq l_2)$ has the following properties:

- *For $a_{j_1 j_1 0} \prod_{l_1 = 1}^{2}(a_{j_2 j_1 s_1} - a_{j_1 j_1 l_1}) > 0$ and $a_{j_2 j_2 0} \prod_{s_2 = 1, s_2 \neq s_1}^{3}(a_{j_2 j_1 s_1} - a_{j_2 j_1 s_2}) > 0$,*

$$(a_{j_1 j_2 1}, a_{j_2 j_1 s_1}) = \underbrace{(\mathrm{UP}_+, \mathrm{UP}_+)}_{\text{positive saddle}}. \tag{1.201}$$

  The equilibrium of $(x_{j_2}^*, x_{j_1}^*) = (a_{j_1 j_2 1}, a_{j_2 j_1 s_1})$ is a (UP$_+$,UP$_+$)-positive saddle.
- *For $a_{j_1 j_1 0} \prod_{l_1 = 1}^{2}(a_{j_2 j_1 s_1} - a_{j_1 j_1 l_1}) < 0$ and $a_{j_2 j_2 0} \prod_{s_2 = 1, s_2 \neq s_1}^{3}(a_{j_2 j_1 s_1} - a_{j_2 j_1 s_2}) > 0$,*

$$(a_{j_1 j_2 1}, a_{j_2 j_1 s_1}) = \underbrace{(\mathrm{DP}_+, \mathrm{DP}_-)}_{\text{CCW center}}. \tag{1.202}$$

  The equilibrium of $(x_{j_2}^*, x_{j_1}^*) = (a_{j_1 j_2 1}, a_{j_2 j_1 s_1})$ is a (DP$_+$,DP$_-$)-counter-clockwise center.
- *For $a_{j_1 j_1 0} \prod_{l_1 = 1}^{2}(a_{j_2 j_1 s_1} - a_{j_1 j_1 l_1}) > 0$ and $a_{j_2 j_2 0} \prod_{s_2 = 1, s_2 \neq s_1}^{3}(a_{j_2 j_1 s_1} - a_{j_2 j_1 s_2}) < 0$,*

$$(a_{j_1 j_2 1}, a_{j_2 j_1 s_1}) = \underbrace{(\mathrm{DP}_-, \mathrm{DP}_+)}_{\text{CW center}}. \tag{1.203}$$

  The equilibrium of $(x_{j_2}^*, x_{j_1}^*) = (a_{j_1 j_2 1}, a_{j_2 j_1 s_1})$ is a (DP$_-$,DP$_+$)-clockwise center.
- *For $a_{j_1 j_1 0} \prod_{l_1 = 1}^{2}(a_{j_2 j_1 s_1} - a_{j_1 j_1 l_1}) < 0$ and $a_{j_2 j_2 0} \prod_{s_2 = 1, s_2 \neq s_1}^{3}(a_{j_2 j_1 s_1} - a_{j_2 j_1 s_2}) < 0$,*

$$(a_{j_1 j_2 1}, a_{j_2 j_1 s_1}) = \underbrace{(\mathrm{UP}_-, \mathrm{UP}_-)}_{\text{negative saddle}}. \tag{1.204}$$

  The equilibrium of $(x_{j_2}^*, x_{j_1}^*) = (a_{j_1 j_2 1}, a_{j_2 j_1 s_1})$ is a (UP$_-$,UP$_-$)-negative saddle.

(ix$_{1b}$) The equilibrium of $(x_{j_1}^*, x_{j_2}^*) = (a_{j_1 j_1 l_1}, a_{j_1 j_2 1})$ $(l_1, l_2 \in \{1,2\}, l_1 \neq l_2)$ with $a_{j_1 j_2 l_1} \neq a_{j_2 j_1 1}, a_{j_2 j_2 2}$ has the following properties:

- *For $a_{j_1 j_1 0}(a_{j_1 j_1 l_1} - a_{j_1 j_1 l_2}) < 0$ and $a_{j_2 j_2 0} \prod_{s_1 = 1}^{3}(a_{j_1 j_1 l_1} - a_{j_2 j_1 s_1}) > 0$,*

$$(a_{j_1 j_1 l_1}, a_{j_1 j_2 1}) = \underbrace{(\mathrm{DP}:\mathrm{UP}, \mathrm{pF})}_{\text{hyperbolic flow } (+)}. \tag{1.205}$$

  The equilibrium of $(x_{j_1}^*, x_{j_2}^*) = (a_{j_1 j_1 l_1}, a_{j_1 j_2 1})$ is a (DP:UP,pF)-positive hyperbolic flow.
- *For $a_{j_1 j_1 0}(a_{j_1 j_1 l_1} - a_{j_1 j_1 l_2}) < 0$ and $a_{j_2 j_2 0} \prod_{s_1 = 1}^{3}(a_{j_1 j_1 l_1} - a_{j_2 j_1 s_1}) > 0$,*

$$(a_{j_1 j_1 l_1}, a_{j_1 j_2 1}) = \underbrace{(\text{UP}:\text{DP},\text{pF})}_{\text{hyperbolic-secant flow } (+)} . \tag{1.206}$$

The equilibrium of $(x_{j_1}^*, x_{j_2}^*) = (a_{j_1 j_1 l_1}, a_{j_1 j_2 1})$ is a (UP:DP,pF)-positive hyperbolic-secant flow.

- For $a_{j_1 j_1 0}(a_{j_1 j_1 l_1} - a_{j_1 j_1 l_2}) > 0$ and $a_{j_2 j_2 0}\prod_{s_1=1}^{3}(a_{j_1 j_1 l_1} - a_{j_2 j_1 s_1}) < 0$,

$$(a_{j_1 j_1 l_1}, a_{j_1 j_2 1}) = \underbrace{(\text{UP}:\text{DP},\text{nF})}_{\text{hyperbolic-secant flow } (-)} . \tag{1.207}$$

The equilibrium of $(x_{j_1}^*, x_{j_2}^*) = (a_{j_1 j_1 l_1}, a_{j_1 j_2 1})$ is a (UP:DP,nF)-negative hyperbolic-secant flow.

- For $a_{j_1 j_1 0}(a_{j_1 j_1 l_1} - a_{j_1 j_1 l_2}) < 0$ and $a_{j_2 j_2 0}\prod_{s_1=1}^{3}(a_{j_1 j_1 l_1} - a_{j_2 j_1 s_1}) < 0$,

$$(a_{j_1 j_1 l_1}, a_{j_1 j_2 1}) = \underbrace{(\text{DP}:\text{UP},\text{nF})}_{\text{hyperbolic flow } (-)} . \tag{1.208}$$

The equilibrium of $(x_{j_1}^*, x_{j_2}^*) = (a_{j_1 j_1 l_1}, a_{j_1 j_2 1})$ is a (DP:UP,nF)-negative hyperbolic flow.

(ix₂) The first integral manifold $(a_{j_2 j_1 s_1} = a_{j_1 j_1 1}; s_1, s_2 \in \{1,2,3\}, s_1 \neq s_2; l_1, l_2 \in \{1,2\}, l_1 \neq l_2)$ is

$$\begin{aligned}
&\frac{1}{2}\left[(x_{j_1} - a_{j_1 j_1 l_2})^2 - (x_{j_1 0} - a_{j_1 j_1 l_2})^2\right] \\
&+ (2a_{j_1 j_1 l_2} - a_{j_2 j_1 s_2} - a_{j_2 j_1 s_3})(x_{j_1} - x_{j_1 0}) \\
&+ \prod_{s_2=1, s_2 \neq s_1}^{3}(a_{j_1 j_1 l_2} - a_{j_2 j_1 s_2})\ln\frac{|x_{j_1} - a_{j_1 j_1 l_2}|}{|x_{j_1 0} - a_{j_1 j_1 l_2}|} \\
&= \frac{1}{2}\frac{a_{j_1 j_1 0}}{a_{j_2 j_2 0}}\left[(x_{j_2} - a_{j_1 j_2 1})^2 - (x_{j_2 0} - a_{j_1 j_2 1})^2\right].
\end{aligned} \tag{1.209}$$

(ix₂ₐ) The infinite-equilibrium of $x_{j_1}^* = a_{j_2 j_1 s_1} = a_{j_1 j_1 l_1}$ $(l_1, l_2 \in \{1,2\}; l_1 \neq l_2)$ with $\bar{x}_{j_2} \neq a_{j_1 j_2 1}$ has the following properties:

- For $a_{j_1 j_1 0}(\bar{x}_{j_2} - a_{j_1 j_2 1})(a_{j_1 j_1 l_1} - a_{j_1 j_1 l_2}) > 0$ and $a_{j_2 j_2 0}\prod_{s_2=1, s_2 \neq s_1}^{3}(a_{j_2 j_1 s_1} - a_{j_2 j_1 s_2}) > 0$,

$$(a_{j_1 j_1 l_1}, \bar{x}_{j_2}) = \underbrace{(\text{SO},\text{II})}_{\text{increasing-inflection source}} . \tag{1.210}$$

The infinite-equilibrium of $x_{j_1}^* = a_{j_2 j_1 s_1} = a_{j_1 j_1 l_1}$ is an (SO,II)-increasing-inflection source.

- *For $a_{j_1j_10}(\bar{x}_{j_2} - a_{j_1j_21})(a_{j_1j_1l_1} - a_{j_1j_1l_2}) < 0$ and $a_{j_2j_20}\prod_{s_2=1,s_2\neq s_1}^{3}(a_{j_2j_1s_1} - a_{j_2j_1s_2}) > 0$,*

$$(a_{j_1j_1l_1}, \bar{x}_{j_2}) = \underbrace{\text{(SI, DI)}}_{\text{decreasing-inflection sink}}.\qquad(1.211)$$

The infinite-equilibrium of $x_{j_1}^* = a_{j_2j_1s_1} = a_{j_1j_1l_1}$ is an (SI,DI)-decreasing-inflection sink.

- *For $a_{j_1j_10}(\bar{x}_{j_2} - a_{j_1j_21})(a_{j_1j_1l_1} - a_{j_1j_1l_2}) > 0$ and $a_{j_2j_20}\prod_{s_2=1,s_2\neq s_1}^{3}(a_{j_2j_1s_1} - a_{j_2j_1s_2}) < 0$,*

$$(a_{j_1j_1l_1}, \bar{x}_{j_2}) = \underbrace{\text{(SO, DI)}}_{\text{decreasing-inflection source}}.\qquad(1.212)$$

The infinite-equilibrium of $x_{j_1}^* = a_{j_2j_1s_1} = a_{j_1j_1l_1}$ is an (SO,DI)-decreasing-inflection source.

- *For $a_{j_1j_10}(\bar{x}_{j_2} - a_{j_1j_21})(a_{j_1j_1l_1} - a_{j_1j_1l_2}) > 0$ and $a_{j_2j_20}\prod_{s_2=1,s_2\neq s_1}^{3}(a_{j_2j_1s_1} - a_{j_2j_1s_2}) < 0$,*

$$(a_{j_1j_1l_1}, \bar{x}_{j_2}) = \underbrace{\text{(SI, II)}}_{\text{increasing-inflection source}}.\qquad(1.213)$$

The infinite-equilibrium of $x_{j_1}^* = a_{j_2j_1s_1} = a_{j_1j_1l_1}$ is an (SI,II)-increasing-inflection sink.

(ix$_{2b}$) The equilibrium of $(x_{j_2}^*, x_{j_1}^*) = (a_{j_1j_21}, a_{j_2j_1s_1})$ with $a_{j_2j_1s_1} = a_{j_1j_1l_1}$ has the following properties:

- *For $a_{j_1j_10}(a_{j_1j_1l_1} - a_{j_1j_1l_2}) > 0$ and $a_{j_2j_20}\prod_{s_2=1,s_2\neq s_1}^{3}(a_{j_2j_1s_1} - a_{j_2j_1s_2}) > 0$,*

$$(a_{j_1j_21}, a_{j_2j_1s_1}) = \underbrace{(_{\text{DI:II}}\text{UP}, _{\text{SI:SO}}\text{US})}_{\text{up-parabola upper-saddle}}.\qquad(1.214)$$

The equilibrium of $(x_{j_2}^*, x_{j_1}^*) = (a_{j_1j_21}, a_{j_2j_1s_1})$ is a $(_{\text{DI:II}}\text{UP},_{\text{SI:SO}}\text{US})$-up-parabola upper-saddle.

- *For $a_{j_1j_10}(a_{j_1j_1l_1} - a_{j_1j_1l_2}) < 0$ and $a_{j_2j_20}\prod_{s_2=1,s_2\neq s_1}^{3}(a_{j_2j_1s_1} - a_{j_2j_1s_2}) > 0$,*

$$(a_{j_1j_21}, a_{j_2j_1s_1}) = \underbrace{(_{\text{II:DI}}\text{DP}, _{\text{SO:SI}}\text{LS})}_{\text{down-parabola lower-saddle}}.\qquad(1.215)$$

The equilibrium of $(x_{j_2}^*, x_{j_1}^*) = (a_{j_1j_21}, a_{j_2j_1s_1})$ is an $(_{\text{II:DI}}\text{DP},_{\text{SO:SI}}\text{LS})$-down-parabola lower-saddle.

- *For $a_{j_1j_10}(a_{j_1j_1l_1} - a_{j_1j_1l_2}) > 0$ and $a_{j_2j_20}\prod_{s_2=1,s_2\neq s_1}^{3}(a_{j_2j_1s_1} - a_{j_2j_1s_2}) < 0$,*

$$(a_{j_1 j_2 1}, a_{j_2 j_1 s_1}) = \underbrace{(_{\text{II:DI}}DP, {}_{\text{SI:SO}}US)}_{\text{down-parabola upper-saddle}} . \tag{1.216}$$

The equilibrium of $(x_{j_2}^*, x_{j_1}^*) = (a_{j_1 j_2 1}, a_{j_2 j_1 s_1})$ *is an* $(_{\text{II:DI}}DP, {}_{\text{SI:SO}}US)$-*down-parabola upper-saddle.*

- *For* $a_{j_1 j_1 0}(a_{j_1 j_1 l_1} - a_{j_1 j_1 l_2}) < 0$ *and* $a_{j_2 j_2 0} \prod_{s_2 = 1, s_2 \neq s_1}^{3} (a_{j_2 j_1 s_1} - a_{j_2 j_1 s_2}) < 0,$

$$(a_{j_1 j_2 1}, a_{j_2 j_1 s_1}) = \underbrace{(_{\text{DI:II}}UP, {}_{\text{SO:SI}}LS)}_{\text{up-parabola lower-saddle}} . \tag{1.217}$$

The equilibrium of $(x_{j_2}^*, x_{j_1}^*) = (a_{j_1 j_2 1}, a_{j_2 j_1 s_1})$ *is a* $(_{\text{DI:II}}UP, {}_{\text{SO:SI}}LS)$-*up-parabola upper-saddle.*

1.2 Proof of Theorem 1.1

Consider a two-dimensional, nonlinear dynamical system as

$$\dot{x}_{j_2} = a_{j_2 j_1 0}(x_{j_1} - b_{j_2 j_1 1})(x_{j_1}^2 + B_{j_2 j_1} x_{j_1} + C_{j_2 j_1}),$$
$$\dot{x}_{j_1} = a_{j_1 j_1 0}(x_{j_2} - a_{j_1 j_2 1})(x_{j_1}^2 + B_{j_1 j_1} x_{j_1} + C_{j_1 j_1}),$$

with

$$\Delta_{j_1 j_1} = B_{j_1 j_1}^2 - 4C_{j_1 j_1}, \Delta_{j_2 j_1} = B_{j_2 j_1}^2 - 4C_{j_2 j_1}.$$

(i) For $\Delta_{j_1 j_1} < 0$ and $\Delta_{j_2 j_1} < 0$, the standard form is

$$\dot{x}_{j_2} = a_{j_2 j_2 0}(x_{j_1} - a_{j_2 j_1 1})\left[(x_{j_1} - a_{j_2 j_1})^2 + b_{j_2 j_1}\right],$$
$$\dot{x}_{j_1} = a_{j_1 j_1 0}(x_{j_2} - a_{j_1 j_2 1})\left[(x_{j_1} - a_{j_1 j_1})^2 + b_{j_1 j_1}\right]$$

where

$$a_{j_1 j_1} = -\frac{1}{2}B_{j_1 j_1}, b_{j_1 j_1} = \frac{1}{4}(-\Delta_{j_1 j_1});$$
$$a_{j_2 j_1 1} = b_{j_2 j_1 1}, a_{j_2 j_1} = -\frac{1}{2}B_{j_2 j_1}, b_{j_2 j_1} = \frac{1}{4}(-\Delta_{j_2 j_1}).$$

In phase space,

$$\frac{dx_{j_1}}{dx_{j_2}} = \frac{a_{j_1j_10}}{a_{j_2j_20}} \frac{(x_{j_2} - a_{j_1j_21})\left[(x_{j_1} - a_{j_1j_1})^2 + b_{j_1j_1}\right]}{(x_{j_1} - a_{j_2j_11})\left[(x_{j_1} - a_{j_2j_1})^2 + b_{j_2j_1}\right]},$$

and the deformation of the foregoing equation is

$$\left\{(x_{j_1} - a_{j_2j_11}) + 2(a_{j_1j_1} - a_{j_2j_1})\right.$$

$$+ \frac{\left[(a_{j_1j_1} - a_{j_2j_1})^2 + b_{j_2j_1} - b_{j_1j_1}\right] + (a_{j_1j_1} - a_{j_2j_1})(a_{j_1j_1} - a_{j_2j_11})}{(x_{j_1} - a_{j_1j_1})^2 + b_{j_1j_1}}(x_{j_1} - a_{j_1j_1})$$

$$\left. + \frac{\left[(a_{j_1j_1} - a_{j_2j_1})^2 + b_{j_2j_1} - b_{j_1j_1}\right](a_{j_1j_1} - a_{j_2j_1}) - 2(a_{j_1j_1} - a_{j_2j_11})b_{j_1j_1}}{(x_{j_1} - a_{j_1j_1})^2 + b_{j_1j_1}}\right\}dx_{j_1}$$

$$= \frac{a_{j_1j_10}}{a_{j_2j_20}}(x_{j_2} - a_{j_1j_21})dx_{j_2}.$$

With an initial condition of (x_{j_10}, x_{j_20}) at $t = t_0$, the integration of the above equation gives

$$\frac{1}{2}\left[(x_{j_1} - a_{j_2j_11})^2 - (x_{j_10} - a_{j_2j_11})^2\right]$$

$$+ 2(a_{j_1j_1} - a_{j_2j_1})(x_{j_1} - x_{j_10}) + \left\{(a_{j_1j_1} - a_{j_2j_1})(a_{j_1j_1} - a_{j_2j_11})\right.$$

$$+ \frac{1}{2}\left[(a_{j_1j_1} - a_{j_2j_1})^2 + b_{j_2j_1} - b_{j_1j_1}\right]\right\}\ln\frac{\left|(x_{j_1} - a_{j_1j_1})^2 + b_{j_1j_1}\right|}{\left|(x_{j_10} - a_{j_1j_1})^2 + b_{j_1j_1}\right|}$$

$$+ \left\{(a_{j_1j_1} - a_{j_2j_11})\left[(a_{j_1j_1} - a_{j_2j_1})^2 + b_{j_2j_1} - b_{j_1j_1}\right]\right.$$

$$\left. - 2b_{j_1j_1}(a_{j_1j_1} - a_{j_2j_1})\right\}\frac{1}{\sqrt{b_{j_1j_1}}}(\arctan\frac{x_{j_1} - a_{j_1j_1}}{\sqrt{b_{j_1j_1}}} - \arctan\frac{x_{j_10} - a_{j_1j_1}}{\sqrt{b_{j_1j_1}}})$$

$$= \frac{1}{2}\frac{a_{j_1j_10}}{a_{j_2j_20}}\left[(x_{j_2} - a_{j_1j_21})^2 - (x_{j_20} - a_{j_1j_21})^2\right].$$

Consider two cases: (I)$x_{j_1}^* = a_{j_1j_21}$ and (II)$x_{j_1}^* = a_{j_2j_11}$.

(I) At $x_{j_2}^* = a_{j_1j_21}$ with $\bar{x}_{j_1} \neq a_{j_2j_11}$, in phase space,

$$\frac{dx_{j_1}}{dx_{j_2}}\Big|_{x_{j_2}^* = a_{j_1j_21}} = \frac{a_{j_1j_10}}{a_{j_2j_20}}\frac{(x_{j_2} - a_{j_1j_21})\left[(\bar{x}_{j_1} - a_{j_1j_1})^2 + b_{j_1j_1}\right]}{(\bar{x}_{j_1} - a_{j_2j_11})\left[(\bar{x}_{j_1} - a_{j_2j_1})^2 + b_{j_2j_1}\right]}\Big|_{x_{j_2}^* = a_{j_1j_21}} = 0.$$

If

$$\frac{d^2x_{j_1}}{dx_{j_2}^2}\Big|_{x_{j_2}^* = a_{j_1j_21}} = \frac{a_{j_1j_10}}{a_{j_2j_20}}\frac{(\bar{x}_{j_1} - a_{j_1j_1})^2 + b_{j_1j_1}}{(\bar{x}_{j_1} - a_{j_2j_11})\left[(\bar{x}_{j_1} - a_{j_2j_1})^2 + b_{j_2j_1}\right]} > 0,$$

there is an up-parabola flow. If

$$\frac{d^2 x_{j_1}}{dx_{j_2}^2}\bigg|_{x_{j_2}^* = a_{j_1 j_2 1}} = \frac{a_{j_1 j_1 0}}{a_{j_2 j_2 0}} \frac{(\bar{x}_{j_1} - a_{j_1 j_1})^2 + b_{j_1 j_1}}{(\bar{x}_{j_1} - a_{j_2 j_1 1})\left[(\bar{x}_{j_1} - a_{j_2 j_1})^2 + b_{j_2 j_1}\right]} < 0,$$

there is a down-parabola flow. Let

$$\dot{x}_{j_2} = a_{j_2 j_2 0}(\bar{x}_{j_1} - a_{j_2 j_1 1})\left[(\bar{x}_{j_1} - a_{j_2 j_1})^2 + b_{j_2 j_1}\right].$$

Because

$$(\bar{x}_{j_1} - a_{j_1 j_1})^2 + b_{j_1 j_1} > 0 \text{ and } (\bar{x}_{j_1} - a_{j_2 j_1})^2 + b_{j_2 j_1} > 0,$$

the parabola flows at $x_{j_2}^* = a_{j_1 j_2 1}$ are positive and negative for $a_{j_2 j_2 0}(\bar{x}_{j_1} - a_{j_2 j_1 1}) > 0$ and $a_{j_2 j_2 0}(\bar{x}_{j_1} - a_{j_2 j_1 1}) < 0$, respectively. Thus, the equilibrium of $x_{j_2}^* = a_{j_1 j_2 1}$ at $\bar{x}_{j_1} \neq a_{j_2 j_1 1}$ has the following properties.

- For $a_{j_1 j_1 0} > 0$ and $a_{j_2 j_2 0}(\bar{x}_{j_1} - a_{j_2 j_1 1}) > 0$,

$$(a_{j_1 j_2 1}, \dot{x}_{j_2}) = \underbrace{\text{(UP, pF)}}_{\text{up-parabola flow }(+)} .$$

- For $a_{j_1 j_1 0} < 0$ and $a_{j_2 j_2 0}(\bar{x}_{j_1} - a_{j_2 j_1 1}) > 0$,

$$(a_{j_1 j_2 1}, \dot{x}_{j_2}) = \underbrace{\text{(DP, pF)}}_{\text{down-parabola flow }(+)} .$$

- For $a_{j_1 j_1 0} > 0$ and $a_{j_2 j_2 0}(\bar{x}_{j_1} - a_{j_2 j_1 1}) < 0$,

$$(a_{j_1 j_2 1}, \dot{x}_{j_2}) = \underbrace{\text{(DP, nF)}}_{\text{down-parabola flow }(-)} .$$

- For $a_{j_1 j_1 0} < 0$ and $a_{j_2 j_2 0}(\bar{x}_{j_1} - a_{j_2 j_1 1}) < 0$,

$$(a_{j_1 j_2 1}, \dot{x}_{j_2}) = \underbrace{\text{(UP, nF)}}_{\text{up-parabola flow }(-)} .$$

(II) At $x_{j_1}^* = a_{j_2 j_1 1}$ and $\bar{x}_{j_2} \neq a_{j_1 j_2 1}$,

$$\frac{dx_{j_2}}{dx_{j_1}}\bigg|_{x_{j_1}^* = a_{j_2 j_1 1}} = \frac{a_{j_2 j_2 0}}{a_{j_1 j_1 0}} \frac{(x_{j_1} - a_{j_2 j_1 1})\left[(x_{j_1} - a_{j_2 j_1})^2 + b_{j_2 j_1}\right]}{(\bar{x}_{j_2} - a_{j_1 j_2 1})\left[(x_{j_1} - a_{j_1 j_1})^2 + b_{j_1 j_1}\right]}\bigg|_{x_{j_1}^* = a_{j_2 j_1 1}} = 0.$$

If

$$\frac{d^2 x_{j_2}}{dx_{j_1}^2}\bigg|_{x_{j_1}^* = a_{j_2 j_1} 1} = \frac{a_{j_2 j_2 0}}{a_{j_1 j_1 0}}\, \frac{(a_{j_2 j_1 1} - a_{j_2 j_1})^2 + b_{j_2 j_1}}{(\bar{x}_{j_2} - a_{j_1 j_2 1})\big[(a_{j_2 j_1 1} - a_{j_1 j_1})^2 + b_{j_1 j_1}\big]} > 0,$$

there is an up-parabola flow at $x_{j_1}^* = a_{j_2 j_1} 1$. If

$$\frac{d^2 x_{j_2}}{dx_{j_1}^2}\bigg|_{x_{j_1}^* = a_{j_2 j_1} 1} = \frac{a_{j_2 j_2 0}}{a_{j_1 j_1 0}}\, \frac{(a_{j_2 j_1 1} - a_{j_2 j_1})^2 + b_{j_2 j_1}}{(\bar{x}_{j_2} - a_{j_1 j_2 1})\big[(a_{j_2 j_1 1} - a_{j_1 j_1})^2 + b_{j_1 j_1}\big]} < 0,$$

there is a down-parabola flow at $x_{j_1}^* = a_{j_2 j_1} 1$. Let

$$\dot{x}_{j_1} = a_{j_1 j_1 0}(\bar{x}_{j_2} - a_{j_1 j_2 1})\big[(a_{j_2 j_1 1} - a_{j_1 j_1})^2 + b_{j_1 j_1}\big].$$

Because of

$$(a_{j_2 j_1 1} - a_{j_2 j_1})^2 + b_{j_2 j_1} > 0 \text{ and } (a_{j_2 j_1 1} - a_{j_1 j_1})^2 + b_{j_1 j_1} > 0,$$

the parabola flows at $x_{j_1}^* = a_{j_2 j_1} 1$ are positive and negative for $a_{j_1 j_1 0}(\bar{x}_{j_2} - a_{j_1 j_2 1}) > 0$ and $a_{j_1 j_1 0}(\bar{x}_{j_2} - a_{j_1 j_2 1}) < 0$.

Thus, the equilibrium of $x_{j_1}^* = a_{j_2 j_1} 1$ at $\bar{x}_{j_2} \neq a_{j_1 j_2 1}$ has the following properties:

- For $a_{j_1 j_1 0}(\bar{x}_{j_2} - a_{j_1 j_2 1}) > 0$ and $a_{j_2 j_2 0} > 0$,

$$(\dot{x}_{j_1}, a_{j_2 j_1} 1) = \underbrace{(\text{pF}, \text{UP})}_{\text{up-parabola flow } (+)}.$$

- For $a_{j_1 j_1 0}(\bar{x}_{j_2} - a_{j_1 j_2 1}) < 0$ and $a_{j_2 j_2 0} > 0$,

$$(\dot{x}_{j_1}, a_{j_2 j_1} 1) = \underbrace{(\text{nF}, \text{DP})}_{\text{down-parabola flow } (-)}.$$

- For $a_{j_1 j_1 0}(\bar{x}_{j_2} - a_{j_1 j_2 1}) > 0$ and $a_{j_2 j_2 0} < 0$,

$$(\dot{x}_{j_1}, a_{j_2 j_1} 1) = \underbrace{(\text{pF}, \text{DP})}_{\text{down-parabola flow } (+)}.$$

- For $a_{j_1 j_1 0}(\bar{x}_{j_2} - a_{j_1 j_2 1}) < 0$ and $a_{j_2 j_2 0} < 0$,

$$(\dot{x}_{j_1}, a_{j_2 j_1} 1) = \underbrace{(\text{nF}, \text{UP})}_{\text{up-parabola flow } (-)}.$$

Therefore, from cases I and II, the equilibrium of $(x_{j_2}^*, x_{j_1}^*) = (a_{j_1 j_2 1}, a_{j_2 j_1 1})$ has the following properties as in Eqs. (1.6)–(1.9):

- For $a_{j_1 j_1 0} > 0$ and $a_{j_2 j_2 0} > 0$,

$$(a_{j_1 j_2 1}, a_{j_2 j_1 1}) = \underbrace{(UP_+, UP_+)}_{\text{positive saddle}}.$$

The equilibrium of $(x_{j_2}^*, x_{j_1}^*) = (a_{j_1 j_2 1}, a_{j_2 j_1 1})$ is a (UP_+, UP_+)-positive saddle.
- For $a_{j_1 j_1 0} < 0$ and $a_{j_2 j_2 0} > 0$,

$$(a_{j_1 j_2 1}, a_{j_2 j_1 1}) = \underbrace{(DP_+, DP_-)}_{\text{CCW center}}.$$

The equilibrium of $(x_{j_2}^*, x_{j_1}^*) = (a_{j_1 j_2 1}, a_{j_2 j_1 1})$ is a (DP_+, DP_-)-counter-clockwise center.
- For $a_{j_1 j_1 0} > 0$ and $a_{j_2 j_2 0} < 0$,

$$(a_{j_1 j_2 1}, a_{j_2 j_1 1}) = \underbrace{(DP_-, DP_+)}_{\text{CW center}}.$$

The equilibrium of $(x_{j_2}^*, x_{j_1}^*) = (a_{j_1 j_2 1}, a_{j_2 j_1 1})$ is a (DP_-, DP_+)-clockwise center.
- For $a_{j_1 j_1 0} < 0$ and $a_{j_2 j_2 0} < 0$,

$$(a_{j_1 j_2 1}, a_{j_2 j_1 1}) = \underbrace{(UP_-, UP_-)}_{\text{negative saddle}}.$$

The equilibrium of $(x_{j_2}^*, x_{j_1}^*) = (a_{j_1 j_2 1}, a_{j_2 j_1 1})$ is a (UP_-, UP_-)-negative saddle.

(ii) For $\Delta_{j_1 j_2} < 0$ and $\Delta_{j_2 j_1} = 0$, the standard form is

$$
\begin{aligned}
\dot{x}_{j_2} &= a_{j_2 j_2 0}(x_{j_1} - a_{j_2 j_1 s_1})(x_{j_1} - a_{j_2 j_1 s_2})^2, \\
\dot{x}_{j_1} &= a_{j_1 j_1 0}(x_{j_2} - a_{j_1 j_2 1})\left[(x_{j_1} - a_{j_1 j_1})^2 + b_{j_1 j_1}\right]
\end{aligned}
$$

where

$$
\begin{aligned}
a_{j_1 j_1} &= -\frac{1}{2} B_{j_1 j_1}, \quad b_{j_1 j_1} = \frac{1}{4}(-\Delta_{j_1 j_1}), \\
a_{j_2 j_1 s_1} &= b_{j_2 j_1 1}, \quad a_{j_2 j_2 s_2} = -\frac{1}{2} B_{j_2 j_1}.
\end{aligned}
$$

(ii$_1$) In phase space,

$$\frac{dx_{j_1}}{dx_{j_2}} = \frac{a_{j_1j_10}}{a_{j_2j_20}} \frac{(x_{j_1} - a_{j_1j_21})\left[(x_{j_1} - a_{j_1j_1})^2 + b_{j_1j_1}\right]}{(x_{j_1} - a_{j_2j_1s_1})(x_{j_1} - a_{j_2j_1s_2})^2},$$

and the deformation of the foregoing equation is

$$\{(x_{j_1} - a_{j_2j_1s_1}) + 2(a_{j_1j_1} - a_{j_2j_1s_2})$$

$$+ \frac{\left[(a_{j_1j_1} - a_{j_2j_1s_2})^2 - b_{j_1j_1}\right] + 2(a_{j_1j_1} - a_{j_2j_1s_2})(a_{j_1j_1} - a_{j_2j_1s_1})}{(x_{j_1} - a_{j_1j_1})^2 + b_{j_1j_1}}(x_{j_1} - a_{j_1j_1})$$

$$+ \frac{\left[(a_{j_1j_1} - a_{j_2j_1s_2})^2 - b_{j_1j_1}\right](a_{j_1j_1} - a_{j_2j_1s_2}) - 2(a_{j_1j_1} - a_{j_2j_1s_1})b_{j_1j_1}}{(x_{j_1} - a_{j_1j_1})^2 + b_{j_1j_1}}\}dx_{j_1}$$

$$= \frac{a_{j_1j_10}}{a_{j_2j_20}}(x_{j_2} - a_{j_1j_21})dx_{j_2}.$$

With an initial condition of (x_{j_10}, x_{j_20}) at $t = t_0$, the integration of the above equation gives

$$\frac{1}{2}\left[(x_{j_1} - a_{j_2j_1s_1})^2 - (x_{j_10} - a_{j_2j_1s_1})^2\right]$$

$$+ 2(a_{j_1j_1} - a_{j_2j_1s_2})(x_{j_1} - x_{j_10}) + \{(a_{j_1j_1} - a_{j_2j_1s_2})(a_{j_1j_1} - a_{j_2j_1s_1})$$

$$+ \frac{1}{2}\left[(a_{j_1j_1} - a_{j_2j_1s_2})^2 - b_{j_1j_1}\right]\}\ln\frac{|(x_{j_1} - a_{j_1j_1})^2 + b_{j_1j_1}|}{|(x_{j_10} - a_{j_1j_1})^2 + b_{j_1j_1}|}$$

$$+ \{(a_{j_1j_1} - a_{j_2j_1s_1})\left[(a_{j_1j_1} - a_{j_2j_1s_2})^2 - b_{j_1j_1}\right]$$

$$- 2b_{j_1j_1}(a_{j_1j_1} - a_{j_2j_1s_2})\}\frac{1}{\sqrt{b_{j_1j_1}}}(\arctan\frac{x_{j_1} - a_{j_1j_1}}{\sqrt{b_{j_1j_1}}} - \arctan\frac{x_{j_10} - a_{j_1j_1}}{\sqrt{b_{j_1j_1}}})$$

$$= \frac{1}{2}\frac{a_{j_1j_10}}{a_{j_2j_20}}\left[(x_{j_2} - a_{j_1j_21})^2 - (x_{j_20} - a_{j_1j_21})^2\right].$$

(ii$_{1a}$) Based on the proof of case (i) with $b_{j_2j_1} = 0$, $a_{j_2j_11} = a_{j_2j_1s_1}$ and $a_{j_2j_1} = a_{j_2j_1s_2}$, the equilibrium of $(x_{j_2}^*, x_{j_1}^*) = (a_{j_1j_21}, a_{j_2j_1s_1})$ has the following properties as in Eqs. (1.13)–(1.16):

• For $a_{j_1j_10} > 0$ and $a_{j_2j_20}(a_{j_2j_1s_1} - a_{j_2j_1s_2})^2 > 0$,

$$(a_{j_1j_21}, a_{j_2j_1s_1}) = \underbrace{(\text{UP}_+, \text{UP}_+)}_{\text{positive saddle}}.$$

The equilibrium of $(x_{j_2}^*, x_{j_1}^*) = (a_{j_1j_21}, a_{j_2j_1s_1})$ is a $(\text{UP}_+, \text{UP}_+)$-positive saddle.

- For $a_{j_1 j_1 0} < 0$ and $a_{j_2 j_2 0}(a_{j_2 j_1 s_1} - a_{j_2 j_1 s_2})^2 > 0$,

$$(a_{j_1 j_2 1}, a_{j_2 j_1 s_1}) = \underbrace{(\text{DP}_+, \text{DP}_-)}_{\text{CCW center}}.$$

The equilibrium of $(x_{j_2}^*, x_{j_1}^*) = (a_{j_1 j_2 1}, a_{j_2 j_1 s_1})$ is a $(\text{DP}_+, \text{DP}_-)$-counter-clockwise center.

- For $a_{j_1 j_1 0} > 0$ and $a_{j_2 j_2 0}(a_{j_2 j_1 s_1} - a_{j_2 j_1 s_2})^2 < 0$,

$$(a_{j_1 j_2 1}, a_{j_2 j_1 s_1}) = \underbrace{(\text{DP}_-, \text{DP}_+)}_{\text{CW center}}.$$

The equilibrium of $(x_{j_2}^*, x_{j_1}^*) = (a_{j_1 j_2 1}, a_{j_2 j_1 s_1})$ is a $(\text{DP}_-, \text{DP}_+)$-clockwise center.

- For $a_{j_1 j_1 0} < 0$ and $a_{j_2 j_2 0}(a_{j_2 j_1 s_1} - a_{j_2 j_1 s_2})^2 < 0$,

$$(a_{j_1 j_2 1}, a_{j_2 j_1 s_1}) = \underbrace{(\text{UP}_-, \text{UP}_-)}_{\text{negative saddle}}.$$

The equilibrium of $(x_{j_2}^*, x_{j_1}^*) = (a_{j_1 j_2 1}, a_{j_2 j_1 s_1})$ is a $(\text{UP}_-, \text{UP}_-)$-negative saddle.

(ii$_{1b}$) In phase space, at $x_{j_1}^* = a_{j_2 j_1 s_2}$ and $\bar{x}_{j_2} \neq a_{j_1 j_2 1}$,

$$\frac{dx_{j_2}}{dx_{j_1}}\Big|_{x_{j_1}^* = a_{j_2 j_1 s_2}} = \frac{a_{j_2 j_2 0}}{a_{j_1 j_1 0}} \frac{(x_{j_1} - a_{j_2 j_1 s_1})(x_{j_1} - a_{j_2 j_1 s_2})^2}{(\bar{x}_{j_2} - a_{j_1 j_2 1})[(x_{j_1} - a_{j_1 j_1})^2 + b_{j_1 j_1}]}\Big|_{x_{j_1}^* = a_{j_2 j_1 s_2}} = 0$$

$$\frac{d^2 x_{j_2}}{dx_{j_1}^2}\Big|_{x_{j_1}^* = a_{j_2 j_1 s_2}} = 0.$$

If

$$\frac{d^3 x_{j_2}}{dx_{j_1}^3}\Big|_{x_{j_1}^* = a_{j_2 j_1 s_2}} = \frac{a_{j_2 j_2 0}}{a_{j_1 j_1 0}} \frac{2(a_{j_2 j_1 s_2} - a_{j_2 j_1 s_1})}{(\bar{x}_{j_2} - a_{j_1 j_2 1})[(a_{j_2 j_1 s_2} - a_{j_1 j_1})^2 + b_{j_1 j_1}]} > 0,$$

there is an increasing-inflection flow. If

$$\frac{d^3 x_{j_2}}{dx_{j_1}^3}\Big|_{x_{j_1}^* = a_{j_2 j_1 s_2}} = \frac{a_{j_2 j_2 0}}{a_{j_1 j_1 0}} \frac{2(a_{j_2 j_1 s_2} - a_{j_2 j_1 s_1})}{(\bar{x}_{j_2} - a_{j_1 j_2 1})[(a_{j_2 j_1 s_2} - a_{j_1 j_1})^2 + b_{j_1 j_1}]} < 0,$$

there is a decreasing-inflection flow. Let

$$\dot{x}_{j_1} = a_{j_1 j_1 0}(\bar{x}_{j_2} - a_{j_1 j_2 1})\left[(a_{j_2 j_1 s_2} - a_{j_1 j_1})^2 + b_{j_1 j_1}\right].$$

Because of

$$(a_{j_2 j_1 s_2} - a_{j_1 j_1})^2 + b_{j_1 j_1} > 0,$$

the inflection flows for $a_{j_1 j_1 0}(\bar{x}_{j_2} - a_{j_1 j_2 1}) > 0$ and $a_{j_1 j_1 0}(\bar{x}_{j_2} - a_{j_1 j_2 1}) < 0$ are positive and negative at $x_{j_1}^* = a_{j_2 j_1 s_2}$, respectively.

Thus, the equilibrium of $x_{j_1}^* = a_{j_2 j_1 s_2}$ at $\bar{x}_{j_2} \neq a_{j_1 j_2 1}$ has the following properties:

- For $a_{j_1 j_1 0}(\bar{x}_{j_2} - a_{j_1 j_2 1}) > 0$ and $a_{j_2 j_2 0}(a_{j_2 j_1 s_2} - a_{j_2 j_1 s_1}) > 0$,

$$(\dot{x}_{j_1}, a_{j_2 j_1 s_2}) = \underbrace{(\text{pF, II})}_{\text{increasing-inflection flow } (+)}.$$

- For $a_{j_1 j_1 0}(\bar{x}_{j_2} - a_{j_1 j_2 1}) < 0$ and $a_{j_2 j_2 0}(a_{j_2 j_1 s_2} - a_{j_2 j_1 s_1}) > 0$,

$$(\dot{x}_{j_1}, a_{j_2 j_1 s_2}) = \underbrace{(\text{nF, DI})}_{\text{decreasing-inflection flow } (-)}.$$

- For $a_{j_1 j_1 0}(\bar{x}_{j_2} - a_{j_1 j_2 1}) > 0$ and $a_{j_2 j_2 0}(a_{j_2 j_1 s_2} - a_{j_2 j_1 s_1}) < 0$,

$$(\dot{x}_{j_1}, a_{j_2 j_1 s_2}) = \underbrace{(\text{pF, DI})}_{\text{decreasing-inflection flow } (+)}.$$

- For $a_{j_1 j_1 0}(\bar{x}_{j_2} - a_{j_1 j_2 1}) < 0$ and $a_{j_2 j_2 0}(a_{j_2 j_1 s_2} - a_{j_2 j_1 s_1}) < 0$,

$$(\dot{x}_{j_1}, a_{j_2 j_1 s_2}) = \underbrace{(\text{nF, II})}_{\text{increasing-inflection flow } (-)}.$$

From the proof (I) of case (i) with $b_{j_2 j_1} = 0$ and $a_{j_2 j_1} = a_{j_2 j_1 s_2}$ plus the above analysis, therefore, the equilibrium of $(x_{j_2}^*, x_{j_1}^*) = (a_{j_1 j_2 1}, a_{j_2 j_1 s_2})$ has the following properties as in Eqs. (1.17)–(1.20):

- For $a_{j_1 j_1 0} > 0$ and $a_{j_2 j_2 0}(a_{j_2 j_1 s_2} - a_{j_2 j_1 s_1}) > 0$,

$$(a_{j_1 j_2 1}, a_{j_2 j_1 s_2}) = \underbrace{(\text{UP, US})}_{\text{up-parabola upper-saddle}}.$$

The equilibrium of $(x_{j_2}^*, x_{j_1}^*) = (a_{j_1 j_2 1}, a_{j_2 j_1 s_2})$ is a (UP,US)-up-parabola upper-saddle.

- For $a_{j_1 j_1 0} < 0$ and $a_{j_2 j_2 0}(a_{j_2 j_1 s_2} - a_{j_2 j_1 s_1}) > 0$,

$$(a_{j_1 j_2 1}, a_{j_2 j_1 s_2}) = \underbrace{(\text{DP, US})}_{\text{down-parabola upper-saddle}} .$$

The equilibrium of $(x_{j_2}^*, x_{j_1}^*) = (a_{j_1 j_2 1}, a_{j_2 j_1 s_2})$ is a (DP,US)-down-parabola upper-saddle.

- For $a_{j_1 j_1 0} > 0$ and $a_{j_2 j_2 0}(a_{j_2 j_1 s_2} - a_{j_2 j_1 s_1}) < 0$,

$$(a_{j_1 j_2 1}, a_{j_2 j_1 s_2}) = \underbrace{(\text{DP, LS})}_{\text{down-parabola lower-saddle}} .$$

The equilibrium of $(x_{j_2}^*, x_{j_1}^*) = (a_{j_1 j_2 1}, a_{j_2 j_1 s_2})$ is a (DP,LS)-down-parabola lower-saddle.

- For $a_{j_1 j_1 0} < 0$ and $a_{j_2 j_2 0}(a_{j_2 j_1 s_2} - a_{j_2 j_1 s_1}) < 0$,

$$(a_{j_1 j_2 1}, a_{j_2 j_1 s_2}) = \underbrace{(\text{UP, LS})}_{\text{up-parabola lower-saddle}} .$$

The equilibrium of $(x_{j_2}^*, x_{j_1}^*) = (a_{j_1 j_2 1}, a_{j_2 j_1 s_2})$ is a (UP,LS)-up-parabola lower-saddle.

(ii_2) For $a_{j_2 j_1 s_1} = a_{j_2 j_1 s_2} = a_{j_2 j_1 1}$, the standard form is

$$\dot{x}_{j_2} = a_{j_2 j_2 0}(x_{j_1} - a_{j_2 j_1 1})^3,$$
$$\dot{x}_{j_1} = a_{j_1 j_1 0}(x_{j_2} - a_{j_1 j_2 1})\big[(x_{j_1} - a_{j_1 j_1})^2 + b_{j_1 j_1}\big].$$

For $a_{j_1 j_1 1} \neq a_{j_2 j_1 1}$,

$$\frac{dx_{j_1}}{dx_{j_2}} = \frac{a_{j_1 j_1 0}}{a_{j_2 j_2 0}} \frac{(x_{j_2} - a_{j_1 j_2 1})\big[(x_{j_1} - a_{j_1 j_1})^2 + b_{j_1 j_1}\big]}{(x_{j_1} - a_{j_2 j_1 1})^3},$$

and the deformation of the foregoing equation is

$$\left\{ (x_{j_1} - a_{j_2 j_1 1}) + 2(a_{j_1 j_1} - a_{j_2 j_1 1}) + \frac{3(a_{j_1 j_1} - a_{j_2 j_1 1})^2 - b_{j_1 j_1}}{(x_{j_1} - a_{j_1 j_1})^2 + b_{j_1 j_1}}(x_{j_1} - a_{j_1 j_1}) \right.$$
$$\left. + \frac{(a_{j_1 j_1} - a_{j_2 j_1 1})\big[(a_{j_1 j_1} - a_{j_2 j_1 1})^2 - 3b_{j_1 j_1}\big]}{(x_{j_1} - a_{j_1 j_1})^2 + b_{j_1 j_1}} \right\} dx_{j_1} = \frac{a_{j_1 j_1 0}}{a_{j_2 j_2 0}}(x_{j_2} - a_{j_1 j_2 1}) dx_{j_2}.$$

With an initial condition of $(x_{j_1 0}, x_{j_2 0})$ at $t = t_0$, the integration of the above equation gives the first integral manifold for $a_{j_1 j_1 1} \neq a_{j_2 j_1 1}$ as

$$\frac{1}{2}\left[(x_{j_1} - a_{j_2 j_1 1})^2 - (x_{j_1 0} - a_{j_2 j_1 1})^2\right] + 2(a_{j_1 j_1} - a_{j_2 j_1 1})(x_{j_1} - x_{j_1 0})$$

$$+\frac{1}{2}\left[3(a_{j_1 j_1} - a_{j_2 j_1 1})^2 - b_{j_1 j_1}\right] \ln \frac{|(x_{j_1} - a_{j_1 j_1})^2 + b_{j_1 j_1}|}{|(x_{j_1 0} - a_{j_1 j_1})^2 + b_{j_1 j_1}|}$$

$$+(a_{j_1 j_1} - a_{j_2 j_1 1})\left[(a_{j_1 j_1} - a_{j_2 j_1 1})^2 - 3b_{j_1 j_1}\right]\frac{1}{\sqrt{b_{j_1 j_1}}}(\arctan\frac{x_{j_1} - a_{j_1 j_1}}{\sqrt{b_{j_1 j_1}}} - \arctan\frac{x_{j_1 0} - a_{j_1 j_1}}{\sqrt{b_{j_1 j_1}}})$$

$$= \frac{1}{2}\frac{a_{j_1 j_1 0}}{a_{j_2 j_2 0}}\left[(x_{j_2} - a_{j_1 j_2 1})^2 - (x_{j_2 0} - a_{j_1 j_2 1})^2\right].$$

At $x_{j_1}^* = a_{j_2 j_1 1}$ and $\bar{x}_{j_2} \neq a_{j_1 j_2 1}$,

$$\frac{dx_{j_2}}{dx_{j_1}}\bigg|_{x_{j_1}^* = a_{j_2 j_1 1}} = \frac{a_{j_2 j_2 0}}{a_{j_1 j_1 0}}\frac{(x_{j_1} - a_{j_2 j_1 1})^3}{(\bar{x}_{j_2} - a_{j_1 j_2 1})[(x_{j_1} - a_{j_1 j_1})^2 + b_{j_1 j_1}]}\bigg|_{x_{j_1}^* = a_{j_2 j_1 1}} = 0,$$

$$\frac{d^2 x_{j_2}}{dx_{j_1}^2}\bigg|_{x_{j_1}^* = a_{j_2 j_1 1}} = \frac{d^3 x_{j_2}}{dx_{j_1}^3}\bigg|_{x_{j_1}^* = a_{j_2 j_1 1}} = 0.$$

If

$$\frac{d^4 x_{j_2}}{dx_{j_1}^4}\bigg|_{x_{j_1}^* = a_{j_2 j_1 1}} = \frac{a_{j_2 j_2 0}}{a_{j_1 j_1 0}}\frac{6}{(\bar{x}_{j_2} - a_{j_1 j_2 1})[(a_{j_2 j_1 1} - a_{j_1 j_1})^2 + b_{j_1 j_1}]} > 0,$$

there is a third-order up-parabola flow at $x_{j_1}^* = a_{j_2 j_1 1}$. If

$$\frac{d^4 x_{j_2}}{dx_{j_1}^4}\bigg|_{x_{j_1}^* = a_{j_2 j_1 1}} = \frac{a_{j_2 j_2 0}}{a_{j_1 j_1 0}}\frac{6}{(\bar{x}_{j_2} - a_{j_1 j_2 1})[(a_{j_2 j_1 1} - a_{j_1 j_1})^2 + b_{j_1 j_1}]} < 0,$$

there is a third-order down-parabola flow at $x_{j_1}^* = a_{j_2 j_1 1}$. Let

$$\dot{x}_{j_1} = a_{j_1 j_1 0}(\bar{x}_{j_2} - a_{j_1 j_2 1})\left[(a_{j_2 j_1 1} - a_{j_1 j_1})^2 + b_{j_1 j_1}\right].$$

Because of

$$(a_{j_2 j_1 1} - a_{j_1 j_1})^2 + b_{j_1 j_1} > 0,$$

the third-order parabola flows at $x_{j_1}^* = a_{j_2j_11}$ in the x_{j_2}-direction are positive and negative for $a_{j_1j_10}(\bar{x}_{j_2} - a_{j_1j_21}) > 0$ and $a_{j_1j_10}(\bar{x}_{j_2} - a_{j_1j_21}) < 0$.

Thus, the equilibrium of $x_{j_1}^* = a_{j_2j_11}$ has the following properties:

- For $a_{j_1j_10}(\bar{x}_{j_2} - a_{j_1j_21}) > 0$ and $a_{j_2j_20} > 0$,

$$(\dot{x}_{j_1}, a_{j_2j_11}) = \underbrace{(\text{pF}, 3^{\text{rd}}\text{UP})}_{\text{third-order up-parabola flow } (+)} .$$

- For $a_{j_1j_10}(\bar{x}_{j_2} - a_{j_1j_21}) < 0$ and $a_{j_2j_20} > 0$,

$$(\dot{x}_{j_1}, a_{j_2j_11}) = \underbrace{(\text{nF}, 3^{\text{rd}}\text{DP})}_{\text{third-order down-parabola flow } (-)} .$$

- For $a_{j_1j_10}(\bar{x}_{j_2} - a_{j_1j_21}) > 0$ and $a_{j_2j_20} < 0$,

$$(\dot{x}_{j_1}, a_{j_2j_11}) = \underbrace{(\text{pF}, 3^{\text{rd}}\text{DP})}_{\text{third-order down-parabola flow } (+)} .$$

- For $a_{j_1j_10}(\bar{x}_{j_2} - a_{j_1j_21}) < 0$ and $a_{j_2j_20} < 0$,

$$(\dot{x}_{j_1}, a_{j_2j_11}) = \underbrace{(\text{nF}, 3^{\text{rd}}\text{UP})}_{\text{third-order up-parabola flow } (-)} .$$

Therefore, from case (I) of the proof of (ii$_{1a}$) and the above analysis, the equilibrium of $(x_{j_2}^*, x_{j_1}^*) = (a_{j_1j_21}, a_{j_2j_1s_2})$ has the following properties as in Eqs. (1.23)–(1.26):

- For $a_{j_1j_10} > 0$ and $a_{j_2j_20} > 0$,

$$(a_{j_1j_21}, a_{j_2j_11}) = \underbrace{(\text{UP}_+, 3^{\text{rd}}\text{UP}_+)}_{\text{third-order positive saddle}} .$$

The equilibrium of $(x_{j_2}^*, x_{j_1}^*) = (a_{j_1j_21}, a_{j_2j_11})$ is a $(\text{UP}_+, 3^{\text{rd}}\text{UP}_+)$-third-order positive saddle.

- For $a_{j_1j_10} < 0$ and $a_{j_2j_20} > 0$,

$$(a_{j_1j_21}, a_{j_2j_11}) = \underbrace{(\text{DP}_+, 3^{\text{rd}}\text{DP}_-)}_{\text{third-order CCW center}} .$$

The equilibrium of $(x_{j_2}^*, x_{j_1}^*) = (a_{j_1 j_2 1}, a_{j_2 j_1 1})$ is a (DP$_+$,3rdDP$_-$)-third-order counter-clockwise center.

- For $a_{j_1 j_1 0} > 0$ and $a_{j_2 j_2 0} < 0$,

$$(a_{j_1 j_2 1}, a_{j_2 j_1 1}) = \underbrace{(\text{DP}_-, 3^{\text{rd}}\text{DP}_+)}_{\text{third-order CW center}} .$$

The equilibrium of $(x_{j_2}^*, x_{j_1}^*) = (a_{j_1 j_2 1}, a_{j_2 j_1 1})$ is a (DP$_-$,3rdDP$_+$)-third-order clockwise center.

- For $a_{j_1 j_1 0} < 0$ and $a_{j_2 j_2 0} < 0$,

$$(a_{j_1 j_2 1}, a_{j_2 j_1 1}) = \underbrace{(\text{UP}_-, 3^{\text{rd}}\text{UP}_-)}_{\text{third-order negative saddle}} .$$

The equilibrium of $(x_{j_2}^*, x_{j_1}^*) = (a_{j_1 j_2 1}, a_{j_2 j_1 1})$ is a (UP$_-$,3rdUP$_-$)-third-order negative saddle.

(iii) For $\Delta_{j_1 j_1} < 0$ and $\Delta_{j_2 j_1} > 0$, the standard form is

$$\dot{x}_{j_2} = a_{j_2 j_2 0}(x_{j_1} - a_{j_2 j_1 1})(x_{j_1} - a_{j_2 j_1 2})(x_{j_1} - a_{j_2 j_1 3}),$$
$$\dot{x}_{j_1} = a_{j_1 j_1 0}(x_{j_2} - a_{j_1 j_2 1})\left[(x_{j_1} - a_{j_1 j_1})^2 + b_{j_1 j_1}\right]$$

where

$$a_{j_1 j_1} = -\frac{1}{2}B_{j_1 j_1}, b_{j_1 j_1} = \frac{1}{4}(-\Delta_{j_1 j_1}),$$
$$b_{j_2 j_1 2}, b_{j_2 j_1 3} = -\frac{1}{2}(B_{j_2 j_1} \pm \sqrt{\Delta_{j_2 j_1}}),$$
$$\{a_{j_2 j_1 1}, a_{j_2 j_1 2}, a_{j_2 j_1 3}\} = \text{sort}\{b_{j_2 j_1 1}, b_{j_2 j_1 2}, b_{j_2 j_1 2}\},$$
$$a_{j_2 j_1 s_1} < a_{j_2 j_1 s_2}, s_1, s_2 \in \{1, 2, 3\}, s_1 < s_2.$$

In phase space, for $a_{j_1 j_1 1} \neq a_{j_2 j_1 s_1}$ $(s_1, s_2 \in \{1, 2, 3\}; s_1 \neq s_2)$,

$$\frac{dx_{j_1}}{dx_{j_2}} = \frac{a_{j_1 j_1 0}}{a_{j_2 j_2 0}} \frac{(x_{j_2} - a_{j_1 j_2 1})\left[(x_{j_1} - a_{j_1 j_1})^2 + b_{j_1 j_1}\right]}{\prod_{s_1 = 1}^{s}(x_{j_1} - a_{j_2 j_1 s_1})},$$

and the deformation of the foregoing equation is

$$((x_{j_1} - a_{j_2 j_1 s_1}) + (2a_{j_1 j_1} - a_{j_2 j_1 s_2} - a_{j_2 j_1 s_3})$$

$$+ \{(2a_{j_1 j_1} - a_{j_2 j_1 s_2} - a_{j_2 j_1 s_3})(a_{j_1 j_1} - a_{j_2 j_1 s_1})$$

$$+ [(a_{j_1 j_1} - a_{j_2 j_1 s_2})(a_{j_1 j_1} - a_{j_2 j_1 s_3}) - b_{j_1 j_1}]\} \frac{(x_{j_1} - a_{j_1 j_1})}{(x_{j_1} - a_{j_1 j_1})^2 + b_{j_1 j_1}}$$

$$+ \{(a_{j_1 j_1} - a_{j_2 j_1 s_1})[(a_{j_1 j_1} - a_{j_2 j_1 s_2})(a_{j_1 j_1} - a_{j_2 j_1 s_3}) - b_{j_1 j_1}]$$

$$- b_{j_1 j_1}(2a_{j_1 j_1} - a_{j_2 j_1 s_2} - a_{j_2 j_1 s_3})\} \frac{1}{(x_{j_1} - a_{j_1 j_1})^2 + b_{j_1 j_1}}) dx_{j_1}$$

$$= \frac{a_{j_1 j_1 0}}{a_{j_2 j_2 0}}(x_{j_2} - a_{j_1 j_2 1}) dx_{j_2}.$$

With an initial condition of $(x_{j_1 0}, x_{j_2 0})$ at $t = t_0$, the integration of the above equation gives

$$\frac{1}{2}[(x_{j_1} - a_{j_2 j_1 s_1})^2 - (x_{j_1 0} - a_{j_2 j_1 s_1})^2] + (2a_{j_1 j_1} - a_{j_2 j_1 s_2} - a_{j_2 j_1 s_3})(x_{j_1} - x_{j_1 0})$$

$$+ \frac{1}{2}\{(2a_{j_1 j_1} - a_{j_2 j_1 s_2} - a_{j_2 j_1 s_3})(a_{j_1 j_1} - a_{j_2 j_1 s_1})$$

$$+ [(a_{j_1 j_1} - a_{j_2 j_1 s_2})(a_{j_1 j_1} - a_{j_2 j_1 s_3}) - b_{j_1 j_1}]\} \ln \frac{|(x_{j_1} - a_{j_1 j_1})^2 + b_{j_1 j_1}|}{|(x_{j_1 0} - a_{j_1 j_1})^2 + b_{j_1 j_1}|}$$

$$+ \{(a_{j_1 j_1} - a_{j_2 j_1 s_1})[(a_{j_1 j_1} - a_{j_2 j_1 s_2})(a_{j_1 j_1} - a_{j_2 j_1 s_3}) - b_{j_1 j_1}]$$

$$- b_{j_1 j_1}(2a_{j_1 j_1} - a_{j_2 j_1 s_2} - a_{j_2 j_1 s_3})\} \frac{1}{\sqrt{b_{j_1 j_1}}}(\arctan \frac{x_{j_1} - a_{j_1 j_1}}{\sqrt{b_{j_1 j_1}}} - \arctan \frac{x_{j_1 0} - a_{j_1 j_1}}{\sqrt{b_{j_1 j_1}}})$$

$$= \frac{1}{2}\frac{a_{j_1 j_1 0}}{a_{j_2 j_2 0}}[(x_{j_2} - a_{j_1 j_2 1})^2 - (x_{j_2 0} - a_{j_1 j_2 1})^2].$$

Consider two cases: (I) $x_{j_1}^* = a_{j_1 j_2 1}$ and (II) $x_{j_1}^* = a_{j_2 j_1 s_1}$.

(I) At $x_{j_1}^* = a_{j_1 j_2 1}$ with $\bar{x}_{j_1} \neq a_{j_2 j_1 s_1}$ ($s_1 = 1, 2, 3$), in phase space,

$$\frac{dx_{j_1}}{dx_{j_2}}\Big|_{x_{j_1}^* = a_{j_1 j_2 1}} = \frac{a_{j_1 j_1 0}}{a_{j_2 j_2 0}}\frac{(x_{j_2} - a_{j_1 j_2 1})[(\bar{x}_{j_1} - a_{j_1 j_1})^2 + b_{j_1 j_1}]}{\prod_{s_1 = 1}^{3}(\bar{x}_{j_1} - a_{j_2 j_1 s_1})} = 0.$$

If

$$\frac{d^2 x_{j_1}}{dx_{j_2}^2}\Big|_{x_{j_2}^* = a_{j_1 j_2 1}} = \frac{a_{j_1 j_1 0}}{a_{j_2 j_2 0}}\frac{(\bar{x}_{j_1} - a_{j_1 j_1})^2 + b_{j_1 j_1}}{\prod_{s_1 = 1}^{3}(\bar{x}_{j_1} - a_{j_2 j_1 s_1})} > 0,$$

there is an up-parabola flow. If

$$\frac{d^2 x_{j_1}}{dx_{j_2}^2}\Big|_{x_{j_1}^* = a_{j_1 j_2 1}} = \frac{a_{j_1 j_1 0}}{a_{j_2 j_2 0}} \frac{(\bar{x}_{j_1} - a_{j_1 j_1})^2 + b_{j_1 j_1}}{\prod_{s_1 = 1}^{3}(\bar{x}_{j_1} - a_{j_2 j_1 s_1})} < 0,$$

there is a down-parabola flow. Let

$$\dot{x}_{j_2} = a_{j_2 j_2 0} \prod_{s_1 = 1}^{3}(\bar{x}_{j_1} - a_{j_2 j_1 s_1}).$$

Because of

$$(\bar{x}_{j_1} - a_{j_1 j_1})^2 + b_{j_1 j_1} > 0,$$

the parabola flows at $x_{j_1}^* = a_{j_1 j_2 1}$ are positive and negative for $\dot{x}_{j_2} > 0$ and $\dot{x}_{j_2} < 0$, respectively. Thus, the equilibrium of $x_{j_1}^* = a_{j_1 j_2 1}$ at $\bar{x}_{j_1} \neq a_{j_2 j_1 s_1}$ $(s_1 = 1, 2, 3)$ has the following properties:

- For $a_{j_1 j_1 0} > 0$ and $a_{j_2 j_2 0}\prod_{s_1 = 1}^{3}(\bar{x}_{j_1} - a_{j_2 j_1 s_1}) > 0$,

$$(a_{j_1 j_2 1}, \dot{x}_{j_2}) = \underbrace{(\text{UP}, \text{pF})}_{\text{up-parabola flow }(+)}.$$

- For $a_{j_1 j_1 0} < 0$ and $a_{j_2 j_2 0}\prod_{s_1 = 1}^{3}(\bar{x}_{j_1} - a_{j_2 j_1 s_1}) > 0$,

$$(a_{j_1 j_2 1}, \dot{x}_{j_2}) = \underbrace{(\text{DP}, \text{pF})}_{\text{down-parabola flow }(+)}.$$

- For $a_{j_1 j_1 0} > 0$ and $a_{j_2 j_2 0}\prod_{s_1 = 1}^{3}(\bar{x}_{j_1} - a_{j_2 j_1 s_1}) < 0$,

$$(a_{j_1 j_2 1}, \dot{x}_{j_2}) = \underbrace{(\text{DP}, \text{nF})}_{\text{down-parabola flow }(-)}.$$

- For $a_{j_1 j_1 0} < 0$ and $a_{j_2 j_2 0}\prod_{s_1 = 1}^{3}(\bar{x}_{j_1} - a_{j_2 j_1 s_1}) < 0$,

$$(a_{j_1 j_2 1}, \dot{x}_{j_2}) = \underbrace{(\text{UP}, \text{nF})}_{\text{up-parabola flow }(-)}.$$

(II) For $x_{j_1}^* = a_{j_2 j_1 s_1}$ $(s_1, s_2 \in \{1, 2, 3\}, s_1 \neq s_2)$ and $\bar{x}_{j_2} \neq a_{j_1 j_2 1}$, in phase space,

$$\frac{dx_{j_2}}{dx_{j_1}}\Big|_{x_{j_1}^* = a_{j_2 j_1 s_1}} = \frac{a_{j_2 j_2 0}}{a_{j_1 j_1 0}} \frac{(x_{j_1} - a_{j_2 j_1 s_1})\prod_{s_2 = 1, s_2 \neq s_1}^{3}(x_{j_1} - a_{j_2 j_1 s_2})}{(\bar{x}_{j_2} - a_{j_1 j_2 1})[(x_{j_1} - a_{j_1 j_1})^2 + b_{j_1 j_1}]}\Big|_{x_{j_1}^* = a_{j_2 j_1 s_1}} = 0.$$

If

$$\frac{d^2 x_{j_2}}{dx_{j_1}^2}\bigg|_{x_{j_1}^* = a_{j_2 j_1 s_1}} = \frac{a_{j_2 j_2 0}}{a_{j_1 j_1 0}} \frac{\prod_{s_2=1, s_2 \neq s_1}^{3}(a_{j_2 j_1 s_1} - a_{j_2 j_1 s_2})}{(\bar{x}_{j_2} - a_{j_1 j_2 1})[(a_{j_2 j_1 s_1} - a_{j_1 j_1})^2 + b_{j_1 j_1}]} > 0,$$

there is an up-parabola flow. If

$$\frac{d^2 x_{j_2}}{dx_{j_1}^2}\bigg|_{x_{j_1}^* = a_{j_2 j_1 s_1}} = \frac{a_{j_2 j_2 0}}{a_{j_1 j_1 0}} \frac{\prod_{s_2=1, s_2 \neq s_1}^{3}(a_{j_2 j_1 s_1} - a_{j_2 j_1 s_2})}{(\bar{x}_{j_2} - a_{j_1 j_2 1})[(a_{j_2 j_1 s_1} - a_{j_1 j_1})^2 + b_{j_1 j_1}]} < 0,$$

there is a down-parabola flow. Let

$$\dot{x}_{j_1} = a_{j_1 j_1 0}(\bar{x}_{j_2} - a_{j_1 j_2 1})[(a_{j_2 j_1 s_1} - a_{j_1 j_1})^2 + b_{j_1 j_1}].$$

Because of

$$(a_{j_2 j_1 s_1} - a_{j_1 j_1})^2 + b_{j_1 j_1} > 0,$$

the parabola flows for $a_{j_1 j_1 0}(\bar{x}_{j_2} - a_{j_1 j_2 1}) > 0$ and $a_{j_1 j_1 0}(\bar{x}_{j_2} - a_{j_1 j_2 1}) < 0$ at $x_{j_1}^* = a_{j_2 j_1 s_1}$ are positive and negative, respectively.

Therefore, the equilibrium of $x_{j_1}^* = a_{j_2 j_1 s_1}$ ($s_1, s_2 \in \{1, 2\}$, $s_1 \neq s_2$) at $\bar{x}_{j_2} \neq a_{j_1 j_2 1}$ has the following properties:

- For $a_{j_1 j_1 0}(\bar{x}_{j_2} - a_{j_1 j_2 1}) > 0$ and $a_{j_2 j_2 0}\prod_{s_2=1, s_2 \neq s_1}^{3}(a_{j_2 j_1 s_1} - a_{j_2 j_1 s_2}) > 0$,

$$(\dot{x}_{j_1}, a_{j_2 j_1 s_1}) = \underbrace{(\text{pF, UP})}_{\text{up-parabola flow } (+)}.$$

- For $a_{j_1 j_1 0}(\bar{x}_{j_2} - a_{j_1 j_2 1}) < 0$ and $a_{j_2 j_2 0}\prod_{s_2=1, s_2 \neq s_1}^{3}(a_{j_2 j_1 s_1} - a_{j_2 j_1 s_2}) > 0$,

$$(\dot{x}_{j_1}, a_{j_2 j_1 s_1}) = \underbrace{(\text{nF, DP})}_{\text{down-paabola flow } (-)}.$$

- For $a_{j_1 j_1 0}(\bar{x}_{j_2} - a_{j_1 j_2 1}) > 0$ and $a_{j_2 j_2 0}\prod_{s_2=1, s_2 \neq s_1}^{3}(a_{j_2 j_1 s_1} - a_{j_2 j_1 s_2}) < 0$,

$$(\dot{x}_{j_1}, a_{j_2 j_1 s_1}) = \underbrace{(\text{pF, DP})}_{\text{down-parabola flow } (+)}.$$

- For $a_{j_1 j_1 0}(\bar{x}_{j_2} - a_{j_1 j_2 1}) < 0$ and $a_{j_2 j_2 0}\prod_{s_2=1, s_2 \neq s_1}^{3}(a_{j_2 j_1 s_1} - a_{j_2 j_1 s_2}) < 0$,

$$(\dot{x}_{j_1}, a_{j_2 j_1 s_1}) = \underbrace{(\text{nF}, \text{UP})}_{\text{up-parabola flow } (-)}.$$

Therefore, from cases (I) and (II), the equilibrium of $(x^*_{j_2}, x^*_{j_1}) = (a_{j_1 j_2 1}, a_{j_2 j_1 s_1})$ ($s_1 = 1, 2, 3$) has the following properties as in Eqs. (1.30)–(1.33):

- For $a_{j_1 j_1 0} > 0$ and $a_{j_2 j_2 0} \prod_{s_2 = 1, s_2 \neq s_1}^{3} (a_{j_2 j_1 s_1} - a_{j_2 j_1 s_2}) > 0$,

$$(a_{j_1 j_2 1}, a_{j_2 j_1 1}) = \underbrace{(\text{UP}_+, \text{UP}_+)}_{\text{positive saddle}}.$$

The equilibrium of $(x^*_{j_2}, x^*_{j_1}) = (a_{j_1 j_2 1}, a_{j_2 j_1 s_1})$ is a $(\text{UP}_+, \text{UP}_+)$-positive saddle.

- For $a_{j_1 j_1 0} < 0$ and $a_{j_2 j_2 0} \prod_{s_2 = 1, s_2 \neq s_1}^{3} (a_{j_2 j_1 s_1} - a_{j_2 j_1 s_2}) > 0$,

$$(a_{j_1 j_2 1}, a_{j_2 j_1 1}) = \underbrace{(\text{DP}_+, \text{DP}_-)}_{\text{CCW center}}.$$

The equilibrium of $(x^*_{j_2}, x^*_{j_1}) = (a_{j_1 j_2 1}, a_{j_2 j_1 s_1})$ is a $(\text{DP}_+, \text{DP}_-)$-counter-clockwise center.

- For $a_{j_1 j_1 0} > 0$ and $a_{j_2 j_2 0} \prod_{s_2 = 1, s_2 \neq s_1}^{3} (a_{j_2 j_1 s_1} - a_{j_2 j_1 s_2}) < 0$,

$$(a_{j_1 j_2 1}, a_{j_2 j_1 1}) = \underbrace{(\text{DP}_-, \text{DP}_+)}_{\text{CW center}}.$$

The equilibrium of $(x^*_{j_2}, x^*_{j_1}) = (a_{j_1 j_2 1}, a_{j_2 j_1 s_1})$ is a $(\text{DP}_-, \text{DP}_+)$-clockwise center.

- For $a_{j_1 j_1 0} < 0$ and $a_{j_2 j_2 0} \prod_{s_2 = 1, s_2 \neq s_1}^{3} (a_{j_2 j_1 s_1} - a_{j_2 j_1 s_2}) < 0$,

$$(a_{j_1 j_2 1}, a_{j_2 j_1 1}) = \underbrace{(\text{UP}_-, \text{UP}_-)}_{\text{negative saddle}}.$$

The equilibrium of $(x^*_{j_2}, x^*_{j_1}) = (a_{j_1 j_2 1}, a_{j_2 j_1 s_1})$ is a $(\text{UP}_-, \text{UP}_-)$-negative saddle.

(iv) For $\Delta_{j_1 j_2} = 0$ and $\Delta_{j_2 j_1} < 0$, the standard form is

$$\dot{x}_{j_2} = a_{j_2 j_2 0}(x_{j_1} - a_{j_2 j_1 1})\left[(x_{j_1} - a_{j_2 j_1})^2 + b_{j_2 j_1}\right],$$
$$\dot{x}_{j_1} = a_{j_1 j_1 0}(x_{j_2} - a_{j_1 j_2 1})(x_{j_1} - a_{j_1 j_1 1})^2$$

where

$$a_{j_1j_1 1} = -\frac{1}{2}B_{j_1j_1}, \quad a_{j_2j_1 1} = b_{j_2j_1 1},$$

$$a_{j_2j_1} = -\frac{1}{2}B_{j_2j_1}, \quad b_{j_2j_1} = \frac{1}{4}(-\Delta_{j_2j_1}).$$

In phase space,

$$\frac{dx_{j_1}}{dx_{j_2}} = \frac{a_{j_1j_1 0}}{a_{j_2j_2 0}} \frac{(x_{j_1} - a_{j_1j_2 1})(x_{j_2} - a_{j_1j_1 1})^2}{(x_{j_1} - a_{j_2j_1 1})[(x_{j_1} - a_{j_2j_1})^2 + b_{j_2j_1}]},$$

and the deformation of the foregoing equation is

$$((x_{j_1} - a_{j_2j_1 1}) + 2(a_{j_1j_1 1} - a_{j_2j_1})$$
$$+2(a_{j_1j_1 1} - a_{j_2j_1}) + \{2(a_{j_1j_1 1} - a_{j_2j_1})(a_{j_1j_1 1} - a_{j_2j_1 1})$$
$$+ [(a_{j_1j_1 1} - a_{j_2j_1})^2 + b_{j_1j_1}]\}\frac{1}{x_{j_1} - a_{j_1j_1 1}}$$
$$+ (a_{j_1j_1 1} - a_{j_2j_1 1})[(a_{j_1j_1 1} - a_{j_2j_1})^2 + b_{j_2j_1}]\frac{1}{(x_{j_1} - a_{j_1j_1 1})^2})dx_{j_1}$$
$$= \frac{a_{j_1j_1 0}}{a_{j_2j_2 0}}(x_{j_2} - a_{j_1j_2 1})dx_{j_2}.$$

With an initial condition of $(x_{j_1 0}, x_{j_2 0})$ at $t = t_0$, the integration of the above equation gives the first integral manifold for $a_{j_1j_1 1} \neq a_{j_2j_1 1}$ as

$$\frac{1}{2}\left[(x_{j_1} - a_{j_2j_1 1})^2 - (x_{j_1 0} - a_{j_2j_1 1})^2\right]$$
$$+ 2(a_{j_1j_1 1} - a_{j_2j_1})(x_{j_1} - x_{j_1 0}) + \{2(a_{j_1j_1 1} - a_{j_2j_1})(a_{j_1j_1 1} - a_{j_2j_1 1})$$
$$+ [(a_{j_1j_1 1} - a_{j_2j_1})^2 + b_{j_2j_1}]\} \ln\frac{|x_{j_1} - a_{j_1j_1 1}|}{|x_{j_1 0} - a_{j_1j_1 1}|}$$
$$- (a_{j_1j_1 1} - a_{j_2j_1 1})[(a_{j_1j_1 1} - a_{j_2j_1})^2 + b_{j_2j_1}](\frac{1}{x_{j_1} - a_{j_1j_1 1}} - \frac{1}{x_{j_1 0} - a_{j_1j_1 1}})$$
$$= \frac{1}{2}\frac{a_{j_1j_1 0}}{a_{j_2j_2 0}}\left[(x_{j_2} - a_{j_1j_2 1})^2 - (x_{j_2 0} - a_{j_1j_2 1})^2\right].$$

(iv$_{1a}$) Consider two cases: (I) $x_{j_1}^* = a_{j_1j_2 1}$ and (II) $x_{j_1}^* = a_{j_2j_1 1}$.

(I) At $x_{j_1}^* = a_{j_1j_2 1}$ with $\bar{x}_{j_1} \neq a_{j_2j_1 1}, a_{j_1j_1 1}$, in phase space,

$$\frac{dx_{j_1}}{dx_{j_2}}\Big|_{x_{j_2}^* = a_{j_1j_2 1}} = \frac{a_{j_1j_1 0}}{a_{j_2j_2 0}}\frac{(x_{j_2} - a_{j_1j_2 1})(\bar{x}_{j_1} - a_{j_1j_1 1})^2}{(\bar{x}_{j_1} - a_{j_2j_1 1})[(\bar{x}_{j_1} - a_{j_2j_1})^2 + b_{j_2j_1}]}\Big|_{x_{j_2}^* = a_{j_1j_2 1}} = 0.$$

If

$$\frac{d^2 x_{j_1}}{dx_{j_2}^2}\bigg|_{x_{j_2}^* = a_{j_1 j_2 1}} = \frac{a_{j_1 j_1 0}}{a_{j_2 j_2 0}} \frac{(\bar{x}_{j_1} - a_{j_1 j_1 1})^2}{(\bar{x}_{j_1} - a_{j_2 j_1 1})\left[(\bar{x}_{j_1} - a_{j_2 j_1})^2 + b_{j_2 j_1}\right]} > 0,$$

there is an up-parabola flow. If

$$\frac{d^2 x_{j_1}}{dx_{j_2}^2}\bigg|_{x_{j_2}^* = a_{j_1 j_2 1}} = \frac{a_{j_1 j_1 0}}{a_{j_2 j_2 0}} \frac{(\bar{x}_{j_1} - a_{j_1 j_1 1})^2}{(\bar{x}_{j_1} - a_{j_2 j_1 1})\left[(\bar{x}_{j_1} - a_{j_2 j_1})^2 + b_{j_2 j_1}\right]} < 0,$$

there is a down-parabola flow.

Let

$$\dot{x}_{j_2} = a_{j_2 j_2 0}(\bar{x}_{j_1} - a_{j_2 j_1 1})\left[(\bar{x}_{j_1} - a_{j_2 j_1})^2 + b_{j_2 j_1}\right].$$

For $\bar{x}_{j_1} \neq a_{j_1 j_1 1}$, because of

$$(\bar{x}_{j_1} - a_{j_1 j_1 1})^2 > 0 \text{ and } (\bar{x}_{j_1} - a_{j_2 j_1})^2 + b_{j_2 j_1} > 0,$$

the parabola flows at $x_{j_1}^* = a_{j_1 j_2 1}$ are positive and negative for $a_{j_2 j_2 0}(\bar{x}_{j_1} - a_{j_2 j_1 1}) > 0$ and $a_{j_2 j_2 0}(\bar{x}_{j_1} - a_{j_2 j_1 1}) < 0$, respectively. Thus, the parabola flows at $x_{j_1}^* = a_{j_1 j_2 1}$ are summarized as follows.

- For $a_{j_1 j_1 0}(\bar{x}_{j_1} - a_{j_1 j_1 1})^2 > 0$ and $a_{j_2 j_2 0}(\bar{x}_{j_1} - a_{j_2 j_1 1}) > 0$,

$$(a_{j_1 j_2 1}, \dot{x}_{j_2}) = \underbrace{(\text{UP}, \text{pF})}_{\text{up-parabola flow } (+)}.$$

- For $a_{j_1 j_1 0}(\bar{x}_{j_1} - a_{j_1 j_1 1})^2 < 0$ and $a_{j_2 j_2 0}(\bar{x}_{j_1} - a_{j_2 j_1 1}) > 0$,

$$(a_{j_1 j_2 1}, \dot{x}_{j_2}) = \underbrace{(\text{DP}, \text{pF})}_{\text{down-parabola flow } (+)}.$$

- For $a_{j_1 j_1 0}(\bar{x}_{j_1} - a_{j_1 j_1 1})^2 > 0$ and $a_{j_2 j_2 0}(\bar{x}_{j_1} - a_{j_2 j_1 1}) < 0$,

$$(a_{j_1 j_2 1}, \dot{x}_{j_2}) = \underbrace{(\text{DP}, \text{nF})}_{\text{down-parabola flow } (-)}.$$

- For $a_{j_1 j_1 0}(\bar{x}_{j_1} - a_{j_1 j_1 1})^2 < 0$ and $a_{j_2 j_2 0}(\bar{x}_{j_1} - a_{j_2 j_1 1}) < 0$,

$$(a_{j_1 j_2 1}, \dot{x}_{j_2}) = \underbrace{(\text{UP}, \text{nF})}_{\text{up-parabola flow } (-)}.$$

(II) At $x_{j_1}^* = a_{j_2 j_1 1}$ and $\bar{x}_{j_2} \neq a_{j_1 j_2 1}$,

$$\frac{dx_{j_2}}{dx_{j_1}}\bigg|_{x_{j_1}^*=a_{j_2j_11}} = \frac{a_{j_2j_20}}{a_{j_1j_10}} \frac{(x_{j_1}-a_{j_2j_11})\left[(x_{j_1}-a_{j_2j_1})^2+b_{j_2j_1}\right]}{(\overline{x}_{j_2}-a_{j_1j_21})(x_{j_1}-a_{j_1j_11})^2}\bigg|_{x_{j_1}^*=a_{j_2j_11}}=0.$$

If

$$\frac{d^2x_{j_2}}{dx_{j_1}^2}\bigg|_{x_{j_1}^*=a_{j_2j_11}} = \frac{a_{j_2j_20}}{a_{j_1j_10}} \frac{(a_{j_2j_11}-a_{j_2j_1})^2+b_{j_2j_1}}{\left((\overline{x}_{j_2}-a_{j_1j_21})(a_{j_2j_11}-a_{j_1j_11})\right)^2}>0,$$

there is an up-parabola flow at $x_{j_1}^*=a_{j_2j_11}$. If

$$\frac{d^2x_{j_2}}{dx_{j_1}^2}\bigg|_{x_{j_1}^*=a_{j_2j_11}} = \frac{a_{j_2j_20}}{a_{j_1j_10}} \frac{(a_{j_2j_11}-a_{j_2j_1})^2+b_{j_2j_1}}{(\overline{x}_{j_2}-a_{j_1j_21})(a_{j_2j_11}-a_{j_1j_11})^2}<0,$$

there is a down-parabola flow at $x_{j_1}^*=a_{j_2j_11}$. Let

$$\dot{x}_{j_1} = a_{j_1j_10}(\overline{x}_{j_2}-a_{j_1j_21})(a_{j_2j_11}-a_{j_1j_11})^2.$$

Because of

$$(a_{j_2j_11}-a_{j_2j_1})^2+b_{j_2j_1}>0, (a_{j_2j_11}-a_{j_1j_11})^2>0,$$

the parabola flows at $x_{j_1}^*=a_{j_2j_11}$ are positive and negative for $\dot{x}_{j_1}>0$ and $\dot{x}_{j_1}<0$.

Thus, the equilibrium of $x_{j_1}^*=a_{j_2j_11}$ has the following properties:

- For $a_{j_1j_10}(\overline{x}_{j_2}-a_{j_1j_21})(a_{j_2j_11}-a_{j_1j_11})^2>0$ and $a_{j_2j_20}>0$,

$$(\dot{x}_{j_1},a_{j_2j_11})= \underbrace{(pF,UP)}_{\text{up-parabola flow }(+)}.$$

- For $a_{j_1j_10}(\overline{x}_{j_2}-a_{j_1j_21})(a_{j_2j_11}-a_{j_1j_11})^2<0$ and $a_{j_2j_20}>0$,

$$(\dot{x}_{j_1},a_{j_2j_11})= \underbrace{(nF,DP)}_{\text{down-parabola flow }(-)}.$$

- For $a_{j_1j_10}(\overline{x}_{j_2}-a_{j_1j_21})(a_{j_2j_11}-a_{j_1j_11})^2>0$ and $a_{j_2j_20}<0$,

$$(\dot{x}_{j_1},a_{j_2j_11})= \underbrace{(pF,DP)}_{\text{down-parabola flow }(+)}.$$

- For $a_{j_1j_10}(\overline{x}_{j_2}-a_{j_1j_21})(a_{j_2j_11}-a_{j_1j_11})^2<0$ and $a_{j_2j_20}<0$,

$$(\dot{x}_{j_1}, a_{j_2 j_1} 1) = \underbrace{(\text{nF}, \text{UP})}_{\text{up-parabola flow } (-)}.$$

Therefore, from cases (I) and (II), the equilibrium of $(x_{j_2}^*, x_{j_1}^*) = (a_{j_1 j_2} 1, a_{j_2 j_1} 1)$ has the following properties as in Eqs. (1.37)–(1.40):

- For $a_{j_1 j_1 0}(a_{j_2 j_1} 1 - a_{j_1 j_1} 1)^2 > 0$ and $a_{j_2 j_2 0} > 0$,

$$(a_{j_1 j_2} 1, a_{j_2 j_1} 1) = \underbrace{(\text{UP}_+, \text{UP}_+)}_{\text{positive saddle}}.$$

The equilibrium of $(x_{j_2}^*, x_{j_1}^*) = (a_{j_1 j_2} 1, a_{j_2 j_1} 1)$ is a $(\text{UP}_+, \text{UP}_+)$-positive saddle.
- For $a_{j_1 j_1 0}(a_{j_2 j_1} 1 - a_{j_1 j_1} 1)^2 < 0$ and $a_{j_2 j_2 0} > 0$,

$$(a_{j_1 j_2} 1, a_{j_2 j_1} 1) = \underbrace{(\text{DP}_+, \text{DP}_-)}_{\text{CCW center}}.$$

The equilibrium of $(x_{j_2}^*, x_{j_1}^*) = (a_{j_1 j_2} 1, a_{j_2 j_1} 1)$ is a $(\text{DP}_+, \text{DP}_-)$-counter-clockwise center.
- For $a_{j_1 j_1 0}(a_{j_2 j_1} 1 - a_{j_1 j_1} 1)^2 > 0$ and $a_{j_2 j_2 0} < 0$,

$$(a_{j_1 j_2} 1, a_{j_2 j_1} 1) = \underbrace{(\text{DP}_-, \text{DP}_+)}_{\text{CW center}}.$$

The equilibrium of $(x_{j_2}^*, x_{j_1}^*) = (a_{j_1 j_2} 1, a_{j_2 j_1} 1)$ is a $(\text{DP}_-, \text{DP}_+)$-clockwise center.
- For $a_{j_1 j_1 0}(a_{j_2 j_1} 1 - a_{j_1 j_1} 1)^2 < 0$ and $a_{j_2 j_2 0} < 0$,

$$(a_{j_1 j_2} 1, a_{j_2 j_1} 1) = \underbrace{(\text{UP}_-, \text{UP}_-)}_{\text{negative saddle}}.$$

The equilibrium of $(x_{j_2}^*, x_{j_1}^*) = (a_{j_1 j_2} 1, a_{j_2 j_1} 1)$ is a $(\text{UP}_-, \text{UP}_-)$-negative saddle.

(iv$_{1b}$) In phase space, at $x_{j_2}^* = a_{j_1 j_2} 1$ with $\bar{x}_{j_1} \neq a_{j_1 j_1} 1, a_{j_2 j_1} 1$,

$$\frac{dx_{j_1}}{dx_{j_2}}\bigg|_{x_{j_2}^* = a_{j_1 j_2} 1} = \frac{a_{j_1 j_1 0}}{a_{j_2 j_2 0}} \frac{(x_{j_2} - a_{j_1 j_2} 1)(\bar{x}_{j_1} - a_{j_1 j_1} 1)^2}{(\bar{x}_{j_1} - a_{j_2 j_1} 1)[(\bar{x}_{j_1} - a_{j_2 j_1})^2 + b_{j_2 j_1}]}\bigg|_{x_{j_2}^* = a_{j_1 j_2} 1} = 0.$$

If

$$\frac{d^2 x_{j_1}}{dx_{j_2}^2}\bigg|_{x_{j_2}^* = a_{j_1 j_2} 1} = \frac{a_{j_1 j_1 0}}{a_{j_2 j_2 0}} \frac{(\bar{x}_{j_1} - a_{j_1 j_1} 1)^2}{(x_{j_1} - a_{j_2 j_1} 1)[(\bar{x}_{j_1} - a_{j_2 j_1})^2 + b_{j_2 j_1}]} > 0,$$

there is an up-parabola flow at $x_{j_2}^* = a_{j_1 j_2} 1$ with $\bar{x}_{j_1} \neq a_{j_1 j_1} 1$. If

$$\frac{d^2 x_{j_1}}{dx_{j_2}^2}\bigg|_{x_{j_2}^* = a_{j_1 j_2 1}} = \frac{a_{j_1 j_1 0}}{a_{j_2 j_2 0}} \frac{(\bar{x}_{j_1} - a_{j_1 j_1 1})^2}{(x_{j_2} - a_{j_1 j_2 1})\left[(\bar{x}_{j_1} - a_{j_1 j_1})^2 + b_{j_1 j_1}\right]} < 0,$$

there is a down-parabola flow at $x_{j_2}^* = a_{j_1 j_2 1}$ with $\bar{x}_{j_1} \neq a_{j_1 j_1 1}$. Let

$$\dot{x}_{j_2} = a_{j_2 j_2 0}(\bar{x}_{j_1} - a_{j_2 j_1 1})\left[(\bar{x}_{j_1} - a_{j_2 j_1})^2 + b_{j_2 j_1}\right],$$

with

$$(\bar{x}_{j_1} - a_{j_2 j_1})^2 + b_{j_2 j_1} > 0.$$

The inflection flows at $x_{j_2}^* = a_{j_1 j_2 1}$ with $\bar{x}_{j_1} \neq a_{j_1 j_1 1}$ are positive and negative for $a_{j_2 j_2 0}(\bar{x}_{j_1} - a_{j_2 j_1 1}) > 0$ and $a_{j_2 j_2 0}(\bar{x}_{j_1} - a_{j_2 j_1 1}) < 0$, respectively.

Thus, the equilibrium of $x_{j_2}^* = a_{j_1 j_2 1}$ with $\bar{x}_{j_1} \neq a_{j_1 j_1 1}, a_{j_2 j_1 1}$ has the following properties:

- For $a_{j_1 j_1 0}(\bar{x}_{j_1} - a_{j_1 j_1 1})^2 > 0$ and $a_{j_2 j_2 0}(\bar{x}_{j_1} - a_{j_2 j_1 1}) > 0$,

$$(a_{j_1 j_2 1}, \dot{x}_{j_2}) = \underbrace{\text{(UP, pF)}}_{\text{up-parabola flow } (+)} .$$

- For $a_{j_1 j_1 0}(\bar{x}_{j_1} - a_{j_1 j_1 1})^2 < 0$ and $a_{j_2 j_2 0}(\bar{x}_{j_1} - a_{j_2 j_1 1}) > 0$,

$$(a_{j_1 j_2 1}, \dot{x}_{j_2}) = \underbrace{\text{(DP, pF)}}_{\text{down-parabola flow } (+)} .$$

- For $a_{j_1 j_1 0}(\bar{x}_{j_1} - a_{j_1 j_1 1})^2 > 0$ and $a_{j_2 j_2 0}(\bar{x}_{j_1} - a_{j_2 j_1 1}) < 0$,

$$(a_{j_1 j_2 1}, \dot{x}_{j_2}) = \underbrace{\text{(DP, nF)}}_{\text{down-parabola flow } (-)} .$$

- For $a_{j_1 j_1 0}(\bar{x}_{j_1} - a_{j_1 j_1 1})^2 < 0$ and $a_{j_2 j_2 0}(\bar{x}_{j_1} - a_{j_2 j_1 1}) < 0$,

$$(a_{j_1 j_2 1}, \dot{x}_{j_2}) = \underbrace{\text{(UP, nF)}}_{\text{up-parabola flow } (-)} .$$

The variational equation at $x_{j_1}^* = a_{j_1 j_1 1}$ with $\bar{x}_{j_2} \neq a_{j_1 j_2 1}$ is

$$\Delta \dot{x}_{j_1} = a_{j_1 j_1 0}(\bar{x}_{j_2} - a_{j_1 j_2 1})(\Delta x_{j_1})^2.$$

The flows at $x_{j_1}^* = a_{j_1 j_1 1}$ are upper-saddle and lower-saddle flows in the x_{j_1}-direction for $a_{j_1 j_1 0}(\bar{x}_{j_2} - a_{j_1 j_2 1}) > 0$ and $a_{j_1 j_1 0}(\bar{x}_{j_2} - a_{j_1 j_2 1}) < 0$, respectively. Let

$$\dot{x}_{j_2} = a_{j_2 j_2 0}(a_{j_1 j_1 1} - a_{j_2 j_1 1})\left[(a_{j_1 j_1 1} - a_{j_2 j_1 1})^2 + b_{j_2 j_1}\right].$$

The upper-saddle and lower-saddle flows are positive and negative in the x_{j_2}-direction for $a_{j_2 j_2 0}(a_{j_1 j_1 1} - a_{j_2 j_1 1}) > 0$ and $a_{j_2 j_2 0}(a_{j_1 j_1 1} - a_{j_2 j_1 1}) < 0$, respectively.

Thus, the equilibrium of $x_{j_1}^* = a_{j_1 j_1 1}$ with $\bar{x}_{j_2} \neq a_{j_1 j_2 1}$ has the following properties:

- For $a_{j_1 j_1 0}(\bar{x}_{j_2} - a_{j_1 j_2 1})^2 > 0$ and $a_{j_2 j_2 0}(a_{j_1 j_1 1} - a_{j_2 j_1 1}) > 0$,

$$(a_{j_1 j_1 1}, \dot{x}_{j_2}) = \underbrace{(\mathrm{US, pF})}_{\text{upper-saddle flow } (+)}.$$

- For $a_{j_1 j_1 0}(\bar{x}_{j_2} - a_{j_1 j_2 1})^2 < 0$ and $a_{j_2 j_2 0}(a_{j_1 j_1 1} - a_{j_2 j_1 1}) > 0$,

$$(a_{j_1 j_1 1}, \dot{x}_{j_2}) = \underbrace{(\mathrm{LS, pF})}_{\text{lower-saddle flow } (+)}.$$

- For $a_{j_1 j_1 0}(\bar{x}_{j_2} - a_{j_1 j_2 1})^2 > 0$ and $a_{j_2 j_2 0}(a_{j_1 j_1 1} - a_{j_2 j_1 1}) < 0$,

$$(a_{j_1 j_1 1}, \dot{x}_{j_2}) = \underbrace{(\mathrm{US, nF})}_{\text{upper-saddle flow } (-)}.$$

- For $a_{j_1 j_1 0}(\bar{x}_{j_2} - a_{j_1 j_2 1})^2 < 0$ and $a_{j_2 j_2 0}(a_{j_1 j_1 1} - a_{j_2 j_1 1}) < 0$,

$$(a_{j_1 j_1 1}, \dot{x}_{j_2}) = \underbrace{(\mathrm{LS, nF})}_{\text{lower-saddle flow } (-)}.$$

Therefore, the equilibrium of $(x_{j_1}^*, x_{j_2}^*) = (a_{j_1 j_1 1}, a_{j_1 j_2 1})$ has the following properties as in Eqs. (1.41)–(1.44):

- For $a_{j_1 j_1 0} > 0$ and $a_{j_2 j_2 0}(a_{j_1 j_1 1} - a_{j_2 j_1 1}) > 0$,

$$(a_{j_1 j_1 1}, a_{j_1 j_2 1}) = \underbrace{(\mathrm{UP:UP, pF})}_{\text{hyperbolic-secant to hyperbolic flow } (+)}.$$

The equilibrium of $(x_{j_1}^*, x_{j_2}^*) = (a_{j_1 j_1 1}, a_{j_1 j_2 1})$ is a (UP:UP, pF)-positive hyperbolic-secant-to-hyperbolic flow.

- For $a_{j_1 j_1 0} < 0$ and $a_{j_2 j_2 0}(a_{j_1 j_1 1} - a_{j_2 j_1 1}) > 0$,

$$(a_{j_1 j_1 1}, a_{j_1 j_2 1}) = \qquad \underbrace{(\text{DP:DP}, \text{pF})}_{\text{hyperbolic to hyperbolic-secant flow }(+)} \qquad .$$

The equilibrium of $(x_{j_1}^*, x_{j_2}^*) = (a_{j_1 j_1 1}, a_{j_1 j_2 1})$ is a (DP:DP, pF)-positive hyperbolic-to-hyperbolic-secant flow.

- For $a_{j_1 j_1 0} > 0$ and $a_{j_2 j_2 0}(a_{j_1 j_1 1} - a_{j_2 j_1 1}) < 0$,

$$(a_{j_1 j_1 1}, a_{j_1 j_2 1}) = \qquad \underbrace{(\text{DP:DP}, \text{nF})}_{\text{hyperbolic to hyperbolic-secant flow }(-)} \qquad .$$

The equilibrium of $(x_{j_1}^*, x_{j_2}^*) = (a_{j_1 j_1 1}, a_{j_1 j_2 1})$ is a (DP:DP, nF)-negative hyperbolic-secant-to-hyperbolic flow.

- For $a_{j_1 j_1 0} < 0$ and $a_{j_2 j_2 0}(a_{j_1 j_1 1} - a_{j_2 j_1 1}) < 0$,

$$(a_{j_1 j_1 1}, a_{j_1 j_2 1}) = \qquad \underbrace{(\text{UP:UP}, \text{nF})}_{\text{hyperbolic-secant to hyperbolic flow }(-)} \qquad .$$

The equilibrium of $(x_{j_1}^*, x_{j_2}^*) = (a_{j_1 j_1 1}, a_{j_1 j_2 1})$ is a (UP:UP, nF)-negative hyperbolic-secant-to-hyperbolic flow.

(iv$_2$) For $a_{j_2 j_1 1} = a_{j_1 j_1 1}$, in phase space,

$$\frac{dx_{j_1}}{dx_{j_2}} = \frac{a_{j_1 j_1 0}}{a_{j_2 j_2 0}} \frac{(x_{j_2} - a_{j_1 j_2 1})(x_{j_1} - a_{j_1 j_1 1})}{(x_{j_1} - a_{j_2 j_1 1})^2 + b_{j_2 j_1}},$$

and the deformation of the foregoing equation is

$$\frac{(x_{j_1} - a_{j_2 j_1 1})^2 + b_{j_2 j_1}}{x_{j_1} - a_{j_1 j_1 1}} dx_{j_1} = \frac{a_{j_1 j_1 0}}{a_{j_2 j_2 0}} (x_{j_2} - a_{j_1 j_2 1}) dx_{j_2}.$$

With an initial condition of $(x_{j_1 0}, x_{j_2 0})$ at $t = t_0$, the integration of the above equation gives the first integral manifold for $a_{j_1 j_1 1} \neq a_{j_2 j_1 1}$ as

$$\frac{1}{2} \left[(x_{j_1} - a_{j_1 j_1 1})^2 - (x_{j_1 0} - a_{j_1 j_1 1})^2 \right] + 2(a_{j_1 j_1 1} - a_{j_2 j_1 1})(x_{j_1} - x_{j_1 0})$$

$$+ \left[(a_{j_1 j_1 1} - a_{j_2 j_1 1})^2 + b_{j_2 j_1} \right] \ln \frac{|x_{j_1} - a_{j_1 j_1 1}|}{|x_{j_1 0} - a_{j_1 j_1 1}|}$$

$$= \frac{1}{2} \frac{a_{j_1 j_1 0}}{a_{j_2 j_2 0}} \left[(x_{j_2} - a_{j_1 j_2 1})^2 - (x_{j_2 0} - a_{j_1 j_2 1})^2 \right].$$

In phase space, at $x_{j_1}^* = a_{j_2 j_1 1} = a_{j_1 j_1 1}$ with $\bar{x}_{j_2} \neq a_{j_1 j_2 1}$,

$$\frac{dx_{j_2}}{dx_{j_1}}\Big|_{x_{j_1}=a_{j_2j_1}1\pm\varepsilon} = \pm\frac{a_{j_2j_20}}{a_{j_1j_10}}\frac{(a_{j_2j_1}1-a_{j_2j_1})^2+b_{j_2j_1}}{(\bar{x}_{j_2}-a_{j_1j_21})\varepsilon},$$

with

$$(a_{j_2j_1}1-a_{j_2j_1})^2+b_{j_2j_1}>0.$$

If

$$\frac{a_{j_2j_20}}{a_{j_1j_10}(\bar{x}_{j_2}-a_{j_1j_21})}>0,$$

there is a down-up asymptotic flow. If

$$\frac{a_{j_2j_20}}{a_{j_1j_10}(\bar{x}_{j_2}-a_{j_1j_21})}<0,$$

there is an up-down flow at $x_{j_1}^*=a_{j_2j_1}1=a_{j_1j_1}1$ with $\bar{x}_{j_2}\neq a_{j_1j_21}$. The variational equation is

$$\Delta\dot{x}_{j_1}=a_{j_1j_10}(\bar{x}_{j_2}-a_{j_1j_21})(\Delta x_{j_1})^2.$$

The up-down and down-up asymptotic flows at $x_{j_1}^*=a_{j_2j_1}1=a_{j_1j_1}1$ with $\bar{x}_{j_2}\neq a_{j_1j_21}$ for $a_{j_1j_10}(\bar{x}_{j_2}-a_{j_1j_21})>0$ and $a_{j_1j_10}(\bar{x}_{j_2}-a_{j_1j_21})<0$ are upper-saddle and lower-saddle, respectively.

Therefore, the infinite equilibrium of $x_{j_1}^*=a_{j_2j_1}1=a_{j_1j_1}1$ with $\bar{x}_{j_2}\neq a_{j_1j_21}$ has the following properties as in Eqs. (1.46)–(1.49):

- For $a_{j_1j_10}(\bar{x}_{j_2}-a_{j_1j_21})>0$ and $a_{j_2j_20}>0$,

$$(a_{j_1j_1}1,\bar{x}_{j_2})=\underbrace{(\text{US},\text{DU})}_{\text{down-up upper-saddle}}.$$

  The infinite-equilibrium of $x_{j_1}^*=a_{j_2j_1}1=a_{j_1j_1}1$ is a (US,DU)-down-up asymptotic upper-saddle.

- For $a_{j_1j_10}(\bar{x}_{j_2}-a_{j_1j_21})<0$ and $a_{j_2j_20}>0$,

$$(a_{j_1j_1}1,\bar{x}_{j_2})=\underbrace{(\text{LS},\text{UD})}_{\text{up-down lower-saddle}}.$$

  The infinite-equilibrium of $x_{j_1}^*=a_{j_2j_1}1=a_{j_1j_1}1$ is an (LS,UD)-up-down asymptotic lower-saddle.

- For $a_{j_1j_10}(\bar{x}_{j_2}-a_{j_1j_21})>0$ and $a_{j_2j_20}<0$,

$$(a_{j_1 j_1 1}, \bar{x}_{j_2}) = \underbrace{(\text{US}, \text{UD})}_{\text{up-down upper-saddle}} .$$

The infinite-equilibrium of $x_{j_1}^* = a_{j_2 j_1 1} = a_{j_1 j_1 1}$ is a (US,UD)-up-down asymptotic upper-saddle.

- For $a_{j_1 j_1 0}(\bar{x}_{j_2} - a_{j_1 j_2 1}) < 0$ and $a_{j_2 j_2 0} < 0$,

$$(a_{j_1 j_1 1}, \bar{x}_{j_2}) = \underbrace{(\text{LS}, \text{DU})}_{\text{down-up lower-saddle}} .$$

The infinite-equilibrium of $x_{j_1}^* = a_{j_2 j_1 1} = a_{j_1 j_1 1}$ is an (LS,DU)-down-up asymptotic lower-saddle.

(iv$_{2b}$) From case (iv$_{2a}$) with $\bar{x}_{j_2} = a_{j_1 j_2 1}$, the equilibrium of $(x_{j_2}^*, x_{j_1}^*) = (a_{j_1 j_2 1}, a_{j_2 j_1 1})$ with $a_{j_1 j_1 1} = a_{j_2 j_1 1}$ has the following properties as in Eqs. (1.50)–(1.53):

- For $a_{j_1 j_1 0} > 0$ and $a_{j_2 j_2 0} > 0$,

$$(a_{j_1 j_2 1}, a_{j_2 j_1 1}) = \underbrace{({}_{\text{UD}}\text{LS}: {}_{\text{DU}}\text{US}, \text{DP}_-:\text{UP}_+)}_{\text{hyperbolic lower-to-upper saddle}}.$$

The equilibrium of $(x_{j_2}^*, x_{j_1}^*) = (a_{j_1 j_2 1}, a_{j_2 j_1 1})$ for $a_{j_1 j_1 1} = a_{j_2 j_1 1}$ is a $({}_{\text{UD}}\text{LS}: {}_{\text{DU}}\text{US},$ $\text{DP}_-:\text{UP}_+)$-hyperbolic lower-to-upper saddle.
- For $a_{j_1 j_1 0} < 0$ and $a_{j_2 j_2 0} > 0$,

$$(a_{j_1 j_2 1}, a_{j_2 j_1 1}) = \underbrace{({}_{\text{DU}}\text{US}: {}_{\text{UD}}\text{LS}, \text{UP}_-:\text{DP}_+)}_{\text{hyperbolic-secant upper-to-lower saddle}} .$$

The equilibrium of $(x_{j_2}^*, x_{j_1}^*) = (a_{j_1 j_2 1}, a_{j_2 j_1 1})$ for $a_{j_1 j_1 1} = a_{j_2 j_1 1}$ is a $({}_{\text{DU}}\text{US}: {}_{\text{UD}}\text{LS},$ $\text{UP}_-:\text{DP}_+)$-hyperbolic-secant upper-to-lower saddle.
- For $a_{j_1 j_1 0} > 0$ and $a_{j_2 j_2 0} < 0$,

$$(a_{j_1 j_2 1}, a_{j_2 j_1 1}) = \underbrace{({}_{\text{DU}}\text{LS}: {}_{\text{UD}}\text{US}, \text{UP}_+:\text{DP}_-)}_{\text{hyperbolic-secant lower-to-upper saddle}} .$$

The equilibrium of $(x_{j_2}^*, x_{j_1}^*) = (a_{j_1 j_2 1}, a_{j_2 j_1 1})$ for $a_{j_1 j_1 1} = a_{j_2 j_1 1}$ is a $({}_{\text{DU}}\text{LS}: {}_{\text{UD}}\text{US},$ $\text{UP}_+:\text{DP}_-)$-hyperbolic-secant lower-to-upper saddle.
- For $a_{j_1 j_1 0} < 0$ and $a_{j_2 j_2 0} < 0$,

$$(a_{j_1 j_2 1}, a_{j_2 j_1 1}) = \underbrace{({}_{\text{UD}}\text{US}: {}_{\text{DU}}\text{LS}, \text{DP}_+:\text{UP}_-)}_{\text{hyperbolic upper-to-lower saddle}}.$$

The equilibrium of $(x_{j_2}^*, x_{j_1}^*) = (a_{j_1 j_2 1}, a_{j_2 j_1 1})$ for $a_{j_1 j_1 1} = a_{j_2 j_1 1}$ is a ($_{\mathrm{UD}}$US: $_{\mathrm{DU}}$LS, DP$_+$:UP$_-$)-hyperbolic upper-to-lower saddle.

(v) For $\Delta_{j_1 j_2} = 0$ and $\Delta_{j_2 j_1} = 0$, the standard form is

$$\dot{x}_{j_2} = a_{j_2 j_2 0}(x_{j_1} - a_{j_2 j_1 s_1})(x_{j_1} - a_{j_2 j_1 s_2})^2,$$
$$\dot{x}_{j_1} = a_{j_1 j_1 0}(x_{j_1} - a_{j_1 j_1 1})^2 (x_{j_2} - a_{j_1 j_2 1})$$

where

$$a_{j_1 j_1 1} = -\frac{1}{2} B_{j_1 j_1}, \, a_{j_2 j_1 s_1} = b_{j_2 j_1 1}, a_{j_2 j_1 s_2} = -\frac{1}{2} B_{j_2 j_1}.$$

(v$_1$) For $a_{j_2 j_1 1} \neq a_{j_1 j_1 1}$, in phase space,

$$\frac{dx_{j_1}}{dx_{j_2}} = \frac{a_{j_1 j_1 0}}{a_{j_2 j_2 0}} \frac{(x_{j_1} - a_{j_1 j_1 1})^2 (x_{j_2} - a_{j_1 j_2 1})}{(x_{j_1} - a_{j_2 j_1 s_1})(x_{j_1} - a_{j_2 j_1 s_2})^2},$$

and the deformation of the foregoing equation is

$$\{(x_{j_1} - a_{j_2 j_1 s_1}) + 2(a_{j_1 j_1 1} - a_{j_2 j_1 s_2})$$
$$+ [2(a_{j_1 j_1 1} - a_{j_2 j_1 s_2})(a_{j_1 j_1 1} - a_{j_2 j_1 s_1}) + (a_{j_1 j_1 1} - a_{j_2 j_1 s_2})^2] \frac{1}{x_{j_1} - a_{j_1 j_1 1}}$$
$$+ (a_{j_1 j_1 1} - a_{j_2 j_1 s_1})(a_{j_1 j_1 1} - a_{j_2 j_1 s_2})^2 \frac{1}{(x_{j_1} - a_{j_1 j_1 1})^2} \} dx_{j_1}$$
$$= \frac{a_{j_1 j_1 0}}{a_{j_2 j_2 0}}(x_{j_2} - a_{j_1 j_2 1}) dx_{j_2}.$$

With an initial condition of $(x_{j_1 0}, x_{j_2 0})$ at $t = t_0$, the integration of the above equation gives

$$\frac{1}{2}[(x_{j_1} - a_{j_2 j_1 s_1})^2 - (x_{j_1 0} - a_{j_2 j_1 s_1})^2]$$
$$+ 2(a_{j_1 j_1 1} - a_{j_2 j_1 s_2})(x_{j_1} - x_{j_1 0})$$
$$+ \{2(a_{j_1 j_1 1} - a_{j_2 j_1 s_1})(a_{j_1 j_1 1} - a_{j_2 j_1 s_2}) + (a_{j_1 j_1 1} - a_{j_2 j_1 s_2})^2\} \ln \frac{|x_{j_1} - a_{j_1 j_1 1}|}{|x_{j_1 0} - a_{j_1 j_1 1}|}$$
$$- (a_{j_1 j_1 1} - a_{j_2 j_1 s_1})(a_{j_1 j_1 1} - a_{j_2 j_1 s_2})^2 (\frac{1}{x_{j_1} - a_{j_1 j_1 1}} - \frac{1}{x_{j_1 0} - a_{j_1 j_1 1}})$$
$$= \frac{1}{2}\frac{a_{j_1 j_1 0}}{a_{j_2 j_2 0}}[(x_{j_2} - a_{j_1 j_2 1})^2 - (x_{j_2 0} - a_{j_1 j_2 1})^2].$$

(v_{1a}) Consider two cases: (I) $x_{j_1}^* = a_{j_2 j_1 s_1}$ and (II) $x_{j_2}^* = a_{j_1 j_2 1}$.

Case (I): At $x_{j_1}^* = a_{j_2 j_1 s_1}$ and $\bar{x}_{j_2} \neq a_{j_1 j_2 1}$,

$$\frac{dx_{j_2}}{dx_{j_1}}\bigg|_{x_{j_1}^* = a_{j_2 j_1 s_1}} = \frac{a_{j_2 j_2 0}}{a_{j_1 j_1 0}} \frac{(x_{j_1} - a_{j_2 j_1 s_1})(x_{j_1} - a_{j_2 j_1 s_2})^2}{(x_{j_1} - a_{j_1 j_1 1})^2 (\bar{x}_{j_2} - a_{j_1 j_2 1})}\bigg|_{x_{j_1}^* = a_{j_2 j_1 s_1}} = 0.$$

If

$$\frac{d^2 x_{j_2}}{dx_{j_1}^2}\bigg|_{x_{j_1}^* = a_{j_2 j_1 s_1}} = \frac{a_{j_2 j_2 0}}{a_{j_1 j_1 0}} \frac{(a_{j_2 j_1 s_1} - a_{j_2 j_1 s_2})^2}{(a_{j_2 j_1 s_1} - a_{j_1 j_1 1})^2 (\bar{x}_{j_2} - a_{j_1 j_2 1})} > 0,$$

there is an up-parabola flow at $x_{j_1}^* = a_{j_2 j_1 s_1}$. If

$$\frac{d^2 x_{j_2}}{dx_{j_1}^2}\bigg|_{x_{j_1}^* = a_{j_2 j_1 s_1}} = \frac{a_{j_2 j_2 0}}{a_{j_1 j_1 0}} \frac{(a_{j_2 j_1 s_1} - a_{j_2 j_1 s_2})^2}{(a_{j_2 j_1 s_1} - a_{j_1 j_1 1})^2 (\bar{x}_{j_2} - a_{j_1 j_2 1})} < 0,$$

there is a down-parabola flow at $x_{j_1}^* = a_{j_2 j_1 s_1}$. Let

$$\dot{x}_{j_1} = a_{j_1 j_1 0}(a_{j_2 j_1 s_1} - a_{j_1 j_1 1})^2 (\bar{x}_{j_2} - a_{j_1 j_2 1}).$$

Because of

$$(a_{j_2 j_1 s_1} - a_{j_2 j_1 s_2})^2 > 0, (a_{j_2 j_1 s_1} - a_{j_1 j_1 1})^2 > 0,$$

the parabola flows at $x_{j_1}^* = a_{j_2 j_1 s_1}$ are positive and negative for $\dot{x}_{j_1} > 0$ and $\dot{x}_{j_1} < 0$. Thus the equilibrium of $x_{j_1}^* = a_{j_2 j_1 s_1}$ has the following proportions:

- For $a_{j_1 j_1 0}(a_{j_2 j_1 s_1} - a_{j_1 j_1 1})^2 (\bar{x}_{j_2} - a_{j_1 j_2 1}) > 0$ and $a_{j_2 j_2 0}(a_{j_2 j_1 s_1} - a_{j_2 j_1 s_2})^2 > 0$,

$$(\dot{x}_{j_1}, a_{j_2 j_1 1}) = \underbrace{(\text{pF, UP})}_{\text{up-parabola flow } (+)}.$$

- For $a_{j_1 j_1 0}(\bar{x}_{j_2} - a_{j_1 j_2 1})(a_{j_2 j_1 s_1} - a_{j_1 j_1 1})^2 < 0$ and $a_{j_2 j_2 0}(a_{j_2 j_1 s_1} - a_{j_2 j_1 s_2})^2 > 0$,

$$(\dot{x}_{j_1}, a_{j_2 j_1 1}) = \underbrace{(\text{nF, DP})}_{\text{down-parabola flow } (-)}.$$

- For $a_{j_1 j_1 0}(\bar{x}_{j_2} - a_{j_1 j_2 1})(a_{j_2 j_1 s_1} - a_{j_1 j_1 1})^2 > 0$ and $a_{j_2 j_2 0}(a_{j_2 j_1 s_1} - a_{j_2 j_1 s_2})^2 < 0$,

$$(\dot{x}_{j_1}, a_{j_2 j_1}1) = \underbrace{(pF, DP)}_{\text{down-parabola flow }(+)}.$$

- For $a_{j_1 j_1 0}(\bar{x}_{j_2} - a_{j_1 j_2}1)(a_{j_2 j_1 s_1} - a_{j_1 j_1}1)^2 < 0$ and $a_{j_2 j_2 0}(a_{j_2 j_1 s_1} - a_{j_2 j_1 s_2})^2 < 0$,

$$(\dot{x}_{j_1}, a_{j_2 j_1}1) = \underbrace{(nF, UP)}_{\text{up-parabola flow }(-)}.$$

Case (II): At $x_{j_2}^* = a_{j_1 j_2}1$ and $\bar{x}_{j_1} \neq a_{j_1 j_1}1, a_{j_2 j_1 s_1}, a_{j_2 j_1 s_2}$,

$$\frac{dx_{j_1}}{dx_{j_2}}\bigg|_{x_{j_2}^* = a_{j_1 j_2}1} = \frac{a_{j_1 j_1 0}}{a_{j_2 j_2 0}} \frac{(\bar{x}_{j_1} - a_{j_1 j_1}1)^2 (x_{j_2} - a_{j_1 j_2}1)}{(\bar{x}_{j_1} - a_{j_2 j_1 s_1})(\bar{x}_{j_1} - a_{j_2 j_1 s_2})^2}\bigg|_{x_{j_2}^* = a_{j_1 j_2}1} = 0.$$

If

$$\frac{d^2 x_{j_1}}{dx_{j_2}^2}\bigg|_{x_{j_2}^* = a_{j_1 j_2}1} = \frac{a_{j_1 j_1 0}}{a_{j_2 j_2 0}} \frac{(\bar{x}_{j_1} - a_{j_1 j_1}1)^2}{(\bar{x}_{j_1} - a_{j_2 j_1 s_1})(\bar{x}_{j_1} - a_{j_2 j_1 s_2})^2} > 0,$$

there is an up-parabola flow at $x_{j_2}^* = a_{j_1 j_2}1$. If

$$\frac{d^2 x_{j_1}}{dx_{j_2}^2}\bigg|_{x_{j_2}^* = a_{j_1 j_2}1} = \frac{a_{j_1 j_1 0}}{a_{j_2 j_2 0}} \frac{(\bar{x}_{j_1} - a_{j_1 j_1}1)^2}{(\bar{x}_{j_1} - a_{j_2 j_1 s_1})(\bar{x}_{j_1} - a_{j_2 j_1 s_2})^2} < 0,$$

there is a down-parabola flow at $x_{j_2}^* = a_{j_1 j_2}1$. Let

$$\dot{x}_{j_2} = a_{j_2 j_2 0}(\bar{x}_{j_1} - a_{j_2 j_1 s_1})(\bar{x}_{j_1} - a_{j_2 j_1 s_2})^2.$$

The parabola flows at $x_{j_1}^* = a_{j_2 j_1}1$ are positive and negative for $\dot{x}_{j_2} > 0$ and $\dot{x}_{j_2} < 0$. Thus, the equilibrium of $x_{j_2}^* = a_{j_1 j_2}1$ has the following properties:

- For $a_{j_1 j_1 0}(\bar{x}_{j_1} - a_{j_1 j_1}1)^2 > 0$ and $a_{j_2 j_2 0}(\bar{x}_{j_1} - a_{j_1 j_1 s_1})(\bar{x}_{j_1} - a_{j_2 j_1 s_2})^2 > 0$,

$$(a_{j_1 j_2}1, \dot{x}_{j_2}) = \underbrace{(UP, pF)}_{\text{up-parobla flow }(+)}.$$

- For $a_{j_1 j_1 0}(\bar{x}_{j_1} - a_{j_1 j_1}1)^2 < 0$ and $a_{j_2 j_2 0}(\bar{x}_{j_1} - a_{j_1 j_1 s_1})(\bar{x}_{j_1} - a_{j_2 j_1 s_2})^2 > 0$,

$$(a_{j_1 j_2}1, \dot{x}_{j_2}) = \underbrace{(DP, pF)}_{\text{down-parobla flow }(+)}.$$

- For $a_{j_1 j_1 0}(\bar{x}_{j_1} - a_{j_1 j_1 1})^2 > 0$ and $a_{j_2 j_2 0}(\bar{x}_{j_1} - a_{j_1 j_1 s_1})(\bar{x}_{j_1} - a_{j_2 j_1 s_2})^2 < 0$,

$$(a_{j_1 j_2 1}, \dot{x}_{j_2}) = \underbrace{\text{(DP, nF)}}_{\text{down-parobla flow }(-)}.$$

- For $a_{j_1 j_1 0}(\bar{x}_{j_1} - a_{j_1 j_1 1})^2 < 0$ and $a_{j_2 j_2 0}(\bar{x}_{j_1} - a_{j_1 j_1 s_1})(\bar{x}_{j_1} - a_{j_2 j_1 s_2})^2 < 0$,

$$(a_{j_1 j_2 1}, \dot{x}_{j_2}) = \underbrace{\text{(UP, nF)}}_{\text{up-parobla flow }(-)}.$$

Therefore, from case (I) for $\bar{x}_{j_2} = a_{j_1 j_2 1}$ and case (II) for $\bar{x}_{j_1} = a_{j_2 j_1 s_1}$, the equilibrium of $(x_{j_2}^*, x_{j_1}^*) = (a_{j_1 j_2 s_1}, a_{j_2 j_1 1})$ has the following properties as in Eqs. (1.57)–(1.60):

- For $a_{j_1 j_1 0}(a_{j_2 j_1 s_1} - a_{j_1 j_1 1})^2 > 0$ and $a_{j_2 j_2 0}(a_{j_2 j_1 s_1} - a_{j_2 j_1 s_2})^2 > 0$,

$$(a_{j_1 j_2 1}, a_{j_2 j_1 s_1}) = \underbrace{(\text{UP}_+, \text{UP}_+)}_{\text{positive saddle}}.$$

The equilibrium of $(x_{j_2}^*, x_{j_1}^*) = (a_{j_1 j_2 1}, a_{j_2 j_1 s_1})$ is a (UP$_+$,UP$_+$)-positive saddle.
- For $a_{j_1 j_1 0}(a_{j_2 j_1 s_1} - a_{j_1 j_1 1})^2 < 0$ and $a_{j_2 j_2 0}(a_{j_2 j_1 s_1} - a_{j_2 j_1 s_2})^2 > 0$,

$$(a_{j_1 j_2 1}, a_{j_2 j_1 s_1}) = \underbrace{(\text{DP}_+, \text{DP}_-)}_{\text{CCW center}}.$$

The equilibrium of $(x_{j_2}^*, x_{j_1}^*) = (a_{j_1 j_2 1}, a_{j_2 j_1 s_1})$ is a (DP$_+$,DP$_-$)-counter-clockwise center.
- For $a_{j_1 j_1 0}(a_{j_2 j_1 s_1} - a_{j_1 j_1 1})^2 > 0$ and $a_{j_2 j_2 0}(a_{j_2 j_1 s_1} - a_{j_2 j_1 s_2})^2 < 0$,

$$(a_{j_1 j_2 1}, a_{j_2 j_1 s_1}) = \underbrace{(\text{DP}_-, \text{DP}_+)}_{\text{CW center}}.$$

The equilibrium of $(x_{j_2}^*, x_{j_1}^*) = (a_{j_1 j_2 1}, a_{j_2 j_1 s_1})$ is a (DP$_-$,DP$_+$)-clockwise center.
- For $a_{j_1 j_1 0}(a_{j_2 j_1 s_1} - a_{j_1 j_1 1})^2 < 0$ and $a_{j_2 j_2 0}(a_{j_2 j_1 s_1} - a_{j_2 j_1 s_2})^2 < 0$,

$$(a_{j_1 j_2 1}, a_{j_2 j_1 s_1}) = \underbrace{(\text{UP}_-, \text{UP}_-)}_{\text{negative saddle}}.$$

The equilibrium of $(x_{j_2}^*, x_{j_1}^*) = (a_{j_1 j_2 1}, a_{j_2 j_1 s_1})$ is a (UP$_-$,UP$_-$)-negative saddle.

(v_{1b}) In phase space, at $x_{j_1}^* = a_{j_2 j_1 s_2}$ and $\bar{x}_{j_2} \neq a_{j_1 j_2 1}$,

$$\frac{dx_{j_2}}{dx_{j_1}}\bigg|_{x_{j_1}^* = a_{j_2 j_1 s_2}} = \frac{a_{j_2 j_2 0}}{a_{j_1 j_1 0}} \frac{(x_{j_1} - a_{j_2 j_1 s_1})(x_{j_1} - a_{j_2 j_1 s_2})^2}{(x_{j_1} - a_{j_1 j_1 1})^2 (\bar{x}_{j_2} - a_{j_1 j_2 1})}\bigg|_{x_{j_1}^* = a_{j_2 j_1 s_2}} = 0,$$

$$\frac{d^2 x_{j_2}}{dx_{j_1}^2}\bigg|_{x_{j_1}^* = a_{j_2 j_1 s_2}} = 0.$$

If

$$\frac{d^3 x_{j_2}}{dx_{j_1}^3}\bigg|_{x_{j_1}^* = a_{j_2 j_1 s_2}} = \frac{a_{j_2 j_2 0}}{a_{j_1 j_1 0}} \frac{2(a_{j_2 j_1 s_2} - a_{j_2 j_1 s_1})}{(x_{j_1} - a_{j_1 j_1 1})^2 (\overline{x}_{j_2} - a_{j_1 j_2 1})} > 0,$$

there is an increasing-inflection flow. If

$$\frac{d^3 x_{j_2}}{dx_{j_1}^3}\bigg|_{x_{j_1}^* = a_{j_2 j_1 s_2}} = \frac{a_{j_2 j_2 0}}{a_{j_1 j_1 0}} \frac{2(a_{j_2 j_1 s_2} - a_{j_2 j_1 s_1})}{(x_{j_1} - a_{j_1 j_1 1})^2 (\overline{x}_{j_2} - a_{j_1 j_2 1})} < 0,$$

there is a decreasing-inflection flow. Let

$$\dot{x}_{j_1} = a_{j_1 j_1 0}(a_{j_2 j_1 s_2} - a_{j_1 j_1 1})^2 (\overline{x}_{j_2} - a_{j_1 j_2 1}).$$

The inflection flows at $x_{j_1}^* = a_{j_2 j_1 1}$ are positive and negative for $\dot{x}_{j_1} > 0$ and $\dot{x}_{j_1} < 0$, respectively.

Thus, the equilibrium of $x_{j_1}^* = a_{j_2 j_1 s_2}$ has the following properties:

- For $a_{j_1 j_1 0}(a_{j_2 j_1 s_2} - a_{j_1 j_1 1})^2 (\overline{x}_{j_2} - a_{j_1 j_2 1}) > 0$ and $a_{j_2 j_2 0}(a_{j_2 j_1 s_2} - a_{j_2 j_1 s_1}) > 0$,

$$(\dot{x}_{j_1}, a_{j_2 j_1 s_2}) = \underbrace{(\mathrm{pF, II})}_{\text{increasing-infelction flow } (+)}.$$

- For $a_{j_1 j_1 0}(a_{j_2 j_1 s_2} - a_{j_1 j_1 1})^2 (\overline{x}_{j_2} - a_{j_1 j_2 1}) < 0$ and $a_{j_2 j_2 0}(a_{j_2 j_1 s_2} - a_{j_2 j_1 s_1}) > 0$,

$$(\dot{x}_{j_1}, a_{j_2 j_1 s_2}) = \underbrace{(\mathrm{nF, DI})}_{\text{decreasing-infelction flow } (-)}.$$

- For $a_{j_1 j_1 0}(a_{j_2 j_1 s_2} - a_{j_1 j_1 1})^2 (\overline{x}_{j_2} - a_{j_1 j_2 1}) > 0$ and $a_{j_2 j_2 0}(a_{j_2 j_1 s_2} - a_{j_2 j_1 s_1}) < 0$,

$$(\dot{x}_{j_1}, a_{j_2 j_1 s_2}) = \underbrace{(\mathrm{pF, DI})}_{\text{decreasing-infelction flow } (+)}.$$

- For $a_{j_1 j_1 0}(a_{j_2 j_1 s_2} - a_{j_1 j_1 1})^2 (\overline{x}_{j_2} - a_{j_1 j_2 1}) < 0$ and $a_{j_2 j_2 0}(a_{j_2 j_1 s_2} - a_{j_2 j_1 s_1}) < 0$,

$$(\dot{x}_{j_1}, a_{j_2 j_1 s_2}) = \underbrace{(\mathrm{nF, II})}_{\text{increasing-infelction flow } (-)}.$$

Therefore, from the above analysis with $\overline{x}_{j_2} = a_{j_1 j_2 1}$ and case (II) of the proof of case (v_{1a}) with $\overline{x}_{j_2} = a_{j_1 j_2 s_2}$, the equilibrium of $(x_{j_2}^*, x_{j_1}^*) = (a_{j_1 j_2 s_2}, a_{j_2 j_1 1})$ with $a_{j_1 j_1 1} \neq a_{j_2 j_1 s_2}, a_{j_2 j_1 s_1}$ has the following properties as in Eqs. (1.61)–(1.64):

- For $a_{j_1 j_1 0}(a_{j_2 j_1 s_2} - a_{j_1 j_1 1})^2 > 0$ and $a_{j_2 j_2 0}(a_{j_2 j_1 s_2} - a_{j_2 j_1 s_1}) > 0$,

$$(a_{j_1 j_2 1}, a_{j_2 j_1 s_2}) = \underbrace{\text{(UP, US)}}_{\text{up-parabola upper-saddle}} .$$

The equilibrium of $(x_{j_2}^*, x_{j_1}^*) = (a_{j_1 j_2 1}, a_{j_2 j_1 s_2})$ is a (UP,US)-up-parabola upper-saddle.

- For $a_{j_1 j_1 0}(a_{j_2 j_1 s_2} - a_{j_1 j_1 1})^2 < 0$ and $a_{j_2 j_2 0}(a_{j_2 j_1 s_2} - a_{j_2 j_1 s_1}) > 0$,

$$(a_{j_1 j_2 1}, a_{j_2 j_1 s_2}) = \underbrace{\text{(DP, US)}}_{\text{down-parabola upper-saddle}} .$$

The equilibrium of $(x_{j_2}^*, x_{j_1}^*) = (a_{j_1 j_2 1}, a_{j_2 j_1 s_2})$ is a (DP,US)-down-parabola upper-saddle.

- For $a_{j_1 j_1 0}(a_{j_2 j_1 s_2} - a_{j_1 j_1 1})^2 > 0$ and $a_{j_2 j_2 0}(a_{j_2 j_1 s_2} - a_{j_2 j_1 s_1}) < 0$,

$$(a_{j_1 j_2 1}, a_{j_2 j_1 s_2}) = \underbrace{\text{(DP, LS)}}_{\text{down-parabola lower-saddle}} .$$

The equilibrium of $(x_{j_2}^*, x_{j_1}^*) = (a_{j_1 j_2 1}, a_{j_2 j_1 s_2})$ is a (DP,LS)-down-parabola lower-saddle.

- For $a_{j_1 j_1 0}(a_{j_2 j_1 s_2} - a_{j_1 j_1 1})^2 < 0$ and $a_{j_2 j_2 0}(a_{j_2 j_1 s_2} - a_{j_2 j_1 s_1}) < 0$,

$$(a_{j_1 j_2 1}, a_{j_2 j_1 s_2}) = \underbrace{\text{(UP, LS)}}_{\text{up-parabola lower-saddle}} .$$

The equilibrium of $(x_{j_2}^*, x_{j_1}^*) = (a_{j_1 j_2 1}, a_{j_2 j_1 s_2})$ is a (UP,LS)-up-parabola lower-saddle.

- The parabola-saddles are the appearing bifurcations of saddle and center.

(v_{1c}) The variational equation at $x_{j_1}^* = a_{j_1 j_1 1}$ with $\bar{x}_{j_2} \neq a_{j_1 j_2 1}$ is

$$\Delta \dot{x}_{j_1} = a_{j_1 j_1 0}(\bar{x}_{j_2} - a_{j_1 j_2 1})(\Delta x_{j_1})^2$$

and

$$\dot{x}_{j_2} = a_{j_2 j_2 0}(a_{j_1 j_1 1} - a_{j_2 j_1 s_1})(a_{j_1 j_1 1} - a_{j_2 j_1 s_2})^2.$$

The equilibrium of $x_{j_1}^* = a_{j_1 j_1 1}$ with $\bar{x}_{j_2} \neq a_{j_1 j_2 1}$ is upper-saddle and lower-saddle for $\dot{x}_{j_2} > 0$ and $\dot{x}_{j_2} < 0$, respectively.

Thus, the equilibrium of $x_{j_1}^* = a_{j_1 j_1 1}$ with $\bar{x}_{j_2} \neq a_{j_1 j_2 1}$ has the following properties:

- For $a_{j_1 j_1 0}(\bar{x}_{j_2} - a_{j_1 j_2 1}) > 0$ and $a_{j_2 j_2 0}(a_{j_1 j_1 1} - a_{j_2 j_1 s_1})(a_{j_1 j_1 1} - a_{j_2 j_1 s_2})^2 > 0$,

$$(a_{j_1 j_1 1}, \dot{x}_{j_2}) = \underbrace{(\text{US}, \text{pF})}_{\text{upp-saddle flow } (+)} .$$

- For $a_{j_1 j_1 0}(\bar{x}_{j_2} - a_{j_1 j_2 1})^2 < 0$ and $a_{j_2 j_2 0}(a_{j_1 j_1 1} - a_{j_2 j_1 s_1})(a_{j_1 j_1 1} - a_{j_2 j_1 s_2})^2 > 0$,

$$(a_{j_1 j_1 1}, \dot{x}_{j_2}) = \underbrace{(\text{LS}, \text{pF})}_{\text{lower-saddle flow } (+)} .$$

- For $a_{j_1 j_1 0}(\bar{x}_{j_2} - a_{j_1 j_2 1})^2 > 0$ and $a_{j_2 j_2 0}(a_{j_1 j_1 1} - a_{j_2 j_1 s_1})(a_{j_1 j_1 1} - a_{j_2 j_1 s_2})^2 < 0$,

$$(a_{j_1 j_1 1}, \dot{x}_{j_2}) = \underbrace{(\text{US}, \text{nF})}_{\text{upper-saddle flow } (-)} .$$

- For $a_{j_1 j_1 0}(\bar{x}_{j_2} - a_{j_1 j_2 1})^2 < 0$ and $a_{j_2 j_2 0}(a_{j_1 j_1 1} - a_{j_2 j_1 s_1})(a_{j_1 j_1 1} - a_{j_2 j_1 s_2})^2 < 0$,

$$(a_{j_1 j_1 1}, \dot{x}_{j_2}) = \underbrace{(\text{LS}, \text{nF})}_{\text{lower-saddle flow } (-)} .$$

Therefore, from the above analysis at $\bar{x}_{j_2} = a_{j_1 j_2 s_1}$ and the case (II) of the proof of case (v_{1a}) with $\bar{x}_{j_1} = a_{j_1 j_1 1}$, the equilibrium of $(x_{j_1}^*, x_{j_2}^*) = (a_{j_1 j_1 1}, a_{j_1 j_2 s_1})$ with $a_{j_2 j_2 1} \neq a_{j_1 j_2 1}$ has the following properties as in Eqs. (1.65)–(1.68):

- For $a_{j_1 j_1 0} > 0$ and $a_{j_2 j_2 0}(a_{j_1 j_1 1} - a_{j_2 j_1 s_1})(a_{j_1 j_1 1} - a_{j_2 j_1 s_2})^2 > 0$,

$$(a_{j_1 j_1 1}, a_{j_1 j_2 1}) = \underbrace{(\text{UP:UP}, \text{pF})}_{\text{hyperbolic-secant-to-hyperbolic flow } (+)} .$$

The equilibrium of $(x_{j_1}^*, x_{j_2}^*) = (a_{j_1 j_1 1}, a_{j_1 j_2 1})$ is a (UP:UP, pF)-positive hyperbolic-secant-to-hyperbolic flow.

- For $a_{j_1 j_1 0} < 0$ and $a_{j_2 j_2 0}(a_{j_1 j_1 1} - a_{j_2 j_1 s_1})(a_{j_1 j_1 1} - a_{j_2 j_1 s_2})^2 > 0$,

$$(a_{j_1 j_1 1}, a_{j_1 j_2 1}) = \underbrace{(\text{DP:DP}, \text{pF})}_{\text{hyperbolic-to-hyperbolic-secant flow } (+)} .$$

The equilibrium of $(x_{j_1}^*, x_{j_2}^*) = (a_{j_1 j_1 1}, a_{j_1 j_2 1})$ is a (DP:DP,pF)-positive hyperbolic-to-hyperbolic-secant flow.

- For $a_{j_1 j_1 0} > 0$ and $a_{j_2 j_2 0}(a_{j_1 j_1 1} - a_{j_2 j_1 s_1})(a_{j_1 j_1 1} - a_{j_2 j_1 s_2})^2 < 0$,

$$(a_{j_1 j_1 1}, a_{j_1 j_2 1}) = \underbrace{(\text{DP:DP}, \text{nF})}_{\text{hyperbolic-to-hyperbolic-secant flow } (-)} .$$

The equilibrium of $(x^*_{j_1}, x^*_{j_2}) = (a_{j_1 j_1 1}, a_{j_1 j_2 1})$ is a (DP:DP, nF)-negative hyperbolic-secant-to-hyperbolic flow.

- For $a_{j_1 j_1 0} < 0$ and $a_{j_2 j_2 0}(a_{j_1 j_1 1} - a_{j_2 j_1 s_1})(a_{j_1 j_1 1} - a_{j_2 j_1 s_2})^2 < 0$,

$$(a_{j_1 j_1 1}, a_{j_1 j_2 1}) = \underbrace{(\text{UP:UP, nF})}_{\text{hyperbolic-secant-to-hyperbolic flow } (-)}.$$

The equilibrium of $(x^*_{j_1}, x^*_{j_2}) = (a_{j_1 j_1 1}, a_{j_1 j_2 1})$ is a (UP:UP,nF)-negative hyperbolic-secant-to-hyperbolic flow.

(v_2) For $a_{j_2 j_1 s_1} = a_{j_2 j_1 s_2} = a_{j_2 j_1 1}$, the standard form is

$$\dot{x}_{j_2} = a_{j_2 j_2 0}(x_{j_1} - a_{j_2 j_1 1})^3,$$
$$\dot{x}_{j_1} = a_{j_1 j_1 0}(x_{j_1} - a_{j_1 j_1 1})^2(x_{j_2} - a_{j_1 j_2 1}).$$

For $a_{j_2 j_1 1} \neq a_{j_1 j_1 1}$, in phase space,

$$\frac{dx_{j_1}}{dx_{j_2}} = \frac{a_{j_1 j_1 0}}{a_{j_2 j_2 0}} \frac{(x_{j_1} - a_{j_1 j_1 1})^2(x_{j_2} - a_{j_1 j_2 1})}{(x_{j_1} - a_{j_2 j_1 1})^3},$$

and the deformation of the foregoing equation is

$$[(x_{j_1} - a_{j_1 j_1 1}) + 3(a_{j_1 j_1 1} - a_{j_2 j_1 1}) + 3(a_{j_1 j_1 1} - a_{j_2 j_1 1})\frac{1}{x_{j_1} - a_{j_1 j_1 1}}$$
$$+ (a_{j_1 j_1 1} - a_{j_2 j_1 1})^3 \frac{1}{(x_{j_1} - a_{j_1 j_1 1})^2}]dx_{j_1} = \frac{a_{j_1 j_1 0}}{a_{j_2 j_2 0}}(x_{j_2} - a_{j_1 j_2 1})^2 dx_{j_2}.$$

With an initial condition of $(x_{j_1 0}, x_{j_2 0})$ at $t = t_0$, the integration of the above equation gives the first integral manifold for $a_{j_1 j_1 1} \neq a_{j_2 j_1 1}$ as

$$\frac{1}{2}\left[(x_{j_1} - a_{j_1 j_1 1})^2 - (x_{j_1 0} - a_{j_1 j_1 1})^2\right] + 3(a_{j_1 j_1 1} - a_{j_2 j_1 1})(x_{j_1} - x_{j_1 0})$$
$$+ 3(a_{j_1 j_1 1} - a_{j_2 j_1 1})^2 \ln \frac{|x_{j_1} - a_{j_1 j_1 1}|}{|x_{j_1 0} - a_{j_1 j_1 1}|} - (a_{j_1 j_1 1} - a_{j_2 j_1 1})^3(\frac{1}{x_{j_1} - a_{j_1 j_1 1}} - \frac{1}{x_{j_1 0} - a_{j_1 j_1 1}})$$
$$= \frac{1}{3}\frac{a_{j_1 j_1 0}}{a_{j_2 j_2 0}}\left[(x_{j_2} - a_{j_1 j_2 1})^3 - (x_{j_2 0} - a_{j_1 j_2 1})^3\right].$$

(v_{2a}) Consider two cases: (I) $x^*_{j_1} = a_{j_2 j_1 1}$ and (II) $x^*_{j_2} = a_{j_1 j_2 1}$.
Case (I): At $x^*_{j_1} = a_{j_2 j_1 1}$ and $\bar{x}_{j_2} \neq a_{j_1 j_2 1}$,

$$\frac{dx_{j_2}}{dx_{j_1}}\Big|_{x^*_{j_1} = a_{j_2 j_1 1}} = \frac{a_{j_2 j_2 0}}{a_{j_1 j_1 0}} \frac{(x_{j_1} - a_{j_2 j_1 1})^3}{(x_{j_1} - a_{j_1 j_1 1})^2(\bar{x}_{j_2} - a_{j_1 j_2 1})}\Big|_{x^*_{j_1} = a_{j_2 j_1 1}} = 0,$$
$$\frac{d^2 x_{j_2}}{dx_{j_1}^2}\Big|_{x^*_{j_1} = a_{j_2 j_1 1}} = \frac{d^3 x_{j_2}}{dx_{j_1}^3}\Big|_{x^*_{j_1} = a_{j_2 j_1 1}} = 0,$$

If

$$\frac{d^4 x_{j_2}}{dx_{j_1}^4}\bigg|_{x_{j_1}^* = a_{j_2 j_1 1}} = \frac{a_{j_2 j_2 0}}{a_{j_1 j_1 0}}\frac{6}{(a_{j_2 j_1 1} - a_{j_1 j_1 1})^2 (\overline{x}_{j_2} - a_{j_1 j_2 1})} > 0,$$

there is a third-order up-parabola flow at $x_{j_1}^* = a_{j_2 j_1 1}$. If

$$\frac{d^4 x_{j_2}}{dx_{j_1}^4}\bigg|_{x_{j_1}^* = a_{j_2 j_1 1}} = \frac{a_{j_2 j_2 0}}{a_{j_1 j_1 0}}\frac{6}{(a_{j_2 j_1 1} - a_{j_1 j_1 1})^2 (\overline{x}_{j_2} - a_{j_1 j_2 1})} < 0,$$

there is a third-order down-parabola flow at $x_{j_1}^* = a_{j_2 j_1 1}$. Let

$$\dot{x}_{j_1} = a_{j_1 j_1 0}(a_{j_2 j_1 1} - a_{j_1 j_1 1})^2 (\overline{x}_{j_2} - a_{j_1 j_2 1}).$$

The third-order parabola flows at $x_{j_1}^* = a_{j_2 j_1 1}$ are positive and negative for $\dot{x}_{j_1} > 0$ and $\dot{x}_{j_1} < 0$.

Therefore, the equilibrium of $x_{j_1}^* = a_{j_2 j_1 1}$ with $\overline{x}_{j_2} \neq a_{j_1 j_2 1}$ has the following properties:

- For $a_{j_1 j_1 0}(a_{j_2 j_1 1} - a_{j_1 j_1 1})^2 (\overline{x}_{j_2} - a_{j_1 j_2 1}) > 0$ and $a_{j_2 j_2 0} > 0$,

$$(\dot{x}_{j_1}, a_{j_2 j_1 1}) = \underbrace{(\text{pF}, 3^{\text{rd}}\text{UP})}_{\text{third-order up-parabola flow } (+)}.$$

- For $a_{j_1 j_1 0}(a_{j_2 j_1 1} - a_{j_1 j_1 1})^2 (\overline{x}_{j_2} - a_{j_1 j_2 1}) < 0$ and $a_{j_2 j_2 0} > 0$,

$$(\dot{x}_{j_1}, a_{j_2 j_1 1}) = \underbrace{(\text{nF}, 3^{\text{rd}}\text{DP})}_{\text{third-order down-parabola flow } (-)}.$$

- For $a_{j_1 j_1 0}(a_{j_2 j_1 1} - a_{j_1 j_1 1})^2 (\overline{x}_{j_2} - a_{j_1 j_2 1}) > 0$ and $a_{j_2 j_2 0} < 0$,

$$(\dot{x}_{j_1}, a_{j_2 j_1 1}) = \underbrace{(\text{pF}, 3^{\text{rd}}\text{DP})}_{\text{third-order down-parabola flow } (+)}.$$

- For $a_{j_1 j_1 0}(a_{j_2 j_1 1} - a_{j_1 j_1 1})^2 (\overline{x}_{j_2} - a_{j_1 j_2 1}) < 0$ and $a_{j_2 j_2 0} < 0$,

$$(\dot{x}_{j_1}, a_{j_2 j_1 1}) = \underbrace{(\text{nF}, 3^{\text{rd}}\text{UP})}_{\text{third-order up-parabola flow } (-)}.$$

Case (II): In phase space, at $x_{j_2}^* = a_{j_1 j_2 1}$ with $\overline{x}_{j_1} \neq a_{j_1 j_1 1}, a_{j_2 j_1 1}$,

$$\frac{dx_{j_1}}{dx_{j_2}}\bigg|_{x_{j_2}^*=a_{j_1j_21}} = \frac{a_{j_1j_10}}{a_{j_2j_20}} \frac{(\bar{x}_{j_1}-a_{j_1j_11})^2(x_{j_2}-a_{j_1j_21})}{(\bar{x}_{j_1}-a_{j_2j_11})^3}\bigg|_{x_{j_2}^*=a_{j_1j_21}} = 0.$$

If

$$\frac{d^2x_{j_1}}{dx_{j_2}^2}\bigg|_{x_{j_2}^*=a_{j_1j_21}} = \frac{a_{j_1j_10}}{a_{j_2j_20}} \frac{(\bar{x}_{j_1}-a_{j_1j_11})^2}{(\bar{x}_{j_1}-a_{j_2j_11})^3} > 0,$$

there is an up-parabola flow at $x_{j_2}^* = a_{j_1j_21}$ with $\bar{x}_{j_1} \neq a_{j_1j_11}$. If

$$\frac{d^2x_{j_1}}{dx_{j_2}^2}\bigg|_{x_{j_2}^*=a_{j_1j_21}} = \frac{a_{j_1j_10}}{a_{j_2j_20}} \frac{(\bar{x}_{j_1}-a_{j_1j_11})^2}{(\bar{x}_{j_1}-a_{j_2j_11})^3} < 0,$$

there is a down-parabola flow at $x_{j_2}^* = a_{j_1j_21}$ with $\bar{x}_{j_1} \neq a_{j_1j_11}$. Let

$$\dot{x}_{j_2} = a_{j_2j_20}(\bar{x}_{j_1}-a_{j_2j_11})^3.$$

The parabola flows at $x_{j_2}^* = a_{j_1j_21}$ with $\bar{x}_{j_1} \neq a_{j_1j_11}$ are positive and negative for $\dot{x}_{j_2} > 0$ and $\dot{x}_{j_2} < 0$, respectively.

Thus, the equilibrium of $x_{j_2}^* = a_{j_1j_21}$ with $\bar{x}_{j_1} \neq a_{j_1j_11}$ has the following properties:

- For $a_{j_1j_10}(\bar{x}_{j_1}-a_{j_1j_11})^2 > 0$ and $a_{j_2j_20}(\bar{x}_{j_1}-a_{j_2j_11})^3 > 0$,

$$(a_{j_1j_21}, \dot{x}_{j_2}) = \underbrace{(\text{UP}, \text{pF})}_{\text{up-paraobla flow } (+)}.$$

- For $a_{j_1j_10}(\bar{x}_{j_1}-a_{j_1j_11})^2 < 0$ and $a_{j_2j_20}(\bar{x}_{j_1}-a_{j_2j_11})^3 > 0$,

$$(a_{j_1j_21}, \dot{x}_{j_2}) = \underbrace{(\text{DP}, \text{pF})}_{\text{down-paraobla flow } (+)}.$$

- For $a_{j_1j_10}(\bar{x}_{j_1}-a_{j_1j_11})^2 > 0$ and $a_{j_2j_20}(\bar{x}_{j_1}-a_{j_2j_11})^3 < 0$,

$$(a_{j_1j_21}, \dot{x}_{j_2}) = \underbrace{(\text{DP}, \text{nF})}_{\text{down-paraobla flow } (-)}.$$

- For $a_{j_1j_10}(\bar{x}_{j_1}-a_{j_1j_11})^2 < 0$ and $a_{j_2j_20}(\bar{x}_{j_1}-a_{j_2j_11})^3 < 0$,

$$(a_{j_1j_21}, \dot{x}_{j_2}) = \underbrace{(\text{UP}, \text{nF})}_{\text{up-paraobla flow } (-)}.$$

Therefore, for case (I) with $\bar{x}_{j_2} = a_{j_1j_21}$ and case (II) with $\bar{x}_{j_1} = a_{j_2j_11}$, the equilibrium of $(x_{j_2}^*, x_{j_1}^*) = (a_{j_1j_21}, a_{j_2j_11})$ has the following properties as in Eqs. (1.71)–(1.74):

- For $a_{j_1j_10}(a_{j_2j_11} - a_{j_1j_11})^2 > 0$ and $a_{j_2j_20} > 0$,

$$(a_{j_1j_21}, a_{j_2j_11}) = \underbrace{(\text{UP}_+, 3^{\text{rd}}\text{UP}_+)}_{\text{third-order positive saddle}}.$$

The equilibrium of $(x_{j_2}^*, x_{j_1}^*) = (a_{j_1j_21}, a_{j_2j_11})$ is a $(\text{UP}_+,3^{\text{rd}}\text{UP}_+)$-third-order positive saddle.

- For $a_{j_1j_10}(a_{j_2j_11} - a_{j_1j_11})^2 < 0$ and $a_{j_2j_20} > 0$,

$$(a_{j_1j_21}, a_{j_2j_11}) = \underbrace{(\text{DP}_+, 3^{\text{rd}}\text{DP}_-)}_{\text{third-order CCW center}}.$$

The equilibrium of $(x_{j_2}^*, x_{j_1}^*) = (a_{j_1j_21}, a_{j_2j_11})$ is a $(\text{DP}_+,3^{\text{rd}}\text{DP}_-)$-third-order counter-clockwise center.

- For $a_{j_1j_10}(a_{j_2j_11} - a_{j_1j_11})^2 > 0$ and $a_{j_2j_20} < 0$,

$$(a_{j_1j_21}, a_{j_2j_11}) = \underbrace{(\text{DP}_-, 3^{\text{rd}}\text{DP}_+)}_{\text{third-order CW center}}.$$

The equilibrium of $(x_{j_2}^*, x_{j_1}^*) = (a_{j_1j_21}, a_{j_2j_11})$ is a $(\text{DP}_-,3^{\text{rd}}\text{DP}_+)$-third-order clockwise center.

- For $a_{j_1j_10}(a_{j_2j_11} - a_{j_1j_11})^2 < 0$ and $a_{j_2j_20} < 0$,

$$(a_{j_1j_21}, a_{j_2j_11}) = \underbrace{(\text{UP}_-, 3^{\text{rd}}\text{UP}_-)}_{\text{third-order negative saddle}}.$$

The equilibrium of $(x_{j_2}^*, x_{j_1}^*) = (a_{j_1j_21}, a_{j_2j_11})$ is a $(\text{UP}_-,3^{\text{rd}}\text{UP}_-)$-third-order negative saddle.

- The third-order saddles are the appearing and switching bifurcations of saddle, center, and saddle.
- The third-order centers are the appearing and switching bifurcations of center, saddle, and center.

(v_{2b}) At $x_{j_1}^* = a_{j_1j_11}$ with $\bar{x}_{j_2} \neq a_{j_1j_21}$, the variational equation is

$$\Delta \dot{x}_{j_1} = a_{j_1j_10}(\bar{x}_{j_2} - a_{j_1j_21})(\Delta x_{j_1})^2.$$

The flows at $x_{j_1}^* = a_{j_1j_11}$ for $a_{j_1j_10}(\bar{x}_{j_2} - a_{j_1j_21}) > 0$ and $a_{j_1j_10}(\bar{x}_{j_2} - a_{j_1j_21}) < 0$ are upper-saddle and lower-saddle, respectively. From case (II) of the proof of case (v_{2a}), let $\bar{x}_{j_1} = a_{j_1j_11}$, the equilibrium of $(x_{j_1}^*, x_{j_2}^*) = (a_{j_1j_11}, a_{j_1j_21})$ with $a_{j_1j_11} \neq a_{j_1j_2s_1}, a_{j_1j_2s_2}$ has the following properties as in Eqs. (1.75)–(1.78):

- For $a_{j_1j_10} > 0$ and $a_{j_2j_20}(a_{j_1j_11} - a_{j_2j_11})^3 > 0$,

$$(a_{j_1j_11}, a_{j_1j_21}) = \underbrace{\text{(UP:UP, pF)}}_{\text{hyperbolic-secant-to-hyperbolic flow } (+)} .$$

The equilibrium of $(x_{j_1}^*, x_{j_2}^*) = (a_{j_1j_11}, a_{j_1j_21})$ is a (UP:UP, pF)-positive hyperbolic-secant-to-hyperbolic flow.
- For $a_{j_1j_10} < 0$ and $a_{j_2j_20}(a_{j_1j_11} - a_{j_2j_11})^3 > 0$,

$$(a_{j_1j_11}, a_{j_1j_21}) = \underbrace{\text{(DP:DP, pF)}}_{\text{hyperbolic-to-hyperbolic-secant flow } (+)} .$$

The equilibrium of $(x_{j_1}^*, x_{j_2}^*) = (a_{j_1j_11}, a_{j_1j_21})$ is a (DP:DP,pF)-positive hyperbolic-to-hyperbolic-secant flow.
- For $a_{j_1j_10} > 0$ and $a_{j_2j_20}(a_{j_1j_11} - a_{j_2j_11})^3 < 0$,

$$(a_{j_1j_11}, a_{j_1j_21}) = \underbrace{\text{(DP:DP, nF)}}_{\text{hyperbolic-to-hyperbolic-secant flow } (-)} .$$

The equilibrium of $(x_{j_1}^*, x_{j_2}^*) = (a_{j_1j_11}, a_{j_1j_21})$ is a (DP:DP, nF)-negative hyperbolic-secant to hyperbolic flow.
- For $a_{j_1j_10} < 0$ and $a_{j_2j_20}(a_{j_1j_11} - a_{j_2j_11})^3 < 0$,

$$(a_{j_1j_11}, a_{j_1j_21}) = \underbrace{\text{(UP:UP, nF)}}_{\text{hyperbolic-secant-to-hyperbolic flow } (-)} .$$

The equilibrium of $(x_{j_1}^*, x_{j_2}^*) = (a_{j_1j_11}, a_{j_1j_21})$ is a (UP:UP,nF)-negative hyperbolic-secant-to-hyperbolic flow.

(v$_3$) For $a_{j_2j_1s_1} = a_{j_1j_11}$, in phase space,

$$\frac{dx_{j_1}}{dx_{j_2}} = \frac{a_{j_1j_10}}{a_{j_2j_20}} \frac{(x_{j_2} - a_{j_1j_21})(x_{j_1} - a_{j_1j_11})}{(x_{j_1} - a_{j_2j_1s_2})^2},$$

and the deformation of the foregoing equation is

$$\left[(x_{j_1} - a_{j_1j_11}) + 2(a_{j_1j_11} - a_{j_2j_1s_2}) + \frac{(a_{j_1j_11} - a_{j_2j_1s_2})^2}{x_{j_1} - a_{j_1j_11}}\right]dx_{j_1}$$
$$= \frac{a_{j_1j_10}}{a_{j_2j_20}}(x_{j_2} - a_{j_1j_2})dx_{j_2}.$$

With an initial condition of (x_{j_10}, x_{j_20}) at $t = t_0$, the integration of the above equation gives the first integral manifold for $a_{j_2j_1s_1} = a_{j_1j_11}$ as

$$\frac{1}{2}\left[(x_{j_1} - a_{j_1 j_1 1})^2 - (x_{j_1 0} - a_{j_1 j_1 1})^2\right] + 2(a_{j_1 j_1 1} - a_{j_2 j_1 1})(x_{j_1} - x_{j_1 0})$$

$$+(a_{j_1 j_1 1} - a_{j_2 j_1 1})^2 \ln \frac{|\, x_{j_1} - a_{j_1 j_1 1}\,|}{|\, x_{j_1} 0 - a_{j_1 j_1 1}\,|}$$

$$= \frac{1}{2}\frac{a_{j_1 j_1 0}}{a_{j_2 j_2 0}}\left[(x_{j_2} - a_{j_1 j_2 1})^2 - (x_{j_2 0} - a_{j_1 j_2 1})^2\right].$$

(v$_{3a}$) In phase space, at $x_{j_1}^* = a_{j_2 j_1 s_1} = a_{j_1 j_1 1}$ with $\bar{x}_{j_2} \neq a_{j_1 j_2 1}$,

$$\frac{dx_{j_2}}{dx_{j_1}}\bigg|_{x_{j_1} = a_{j_1 j_1 1} \pm \varepsilon} = \pm \frac{a_{j_2 j_2 0}}{a_{j_1 j_1 0}}\frac{(a_{j_2 j_1 1} - a_{j_1 j_2 s_2})^2}{(\bar{x}_{j_2} - a_{j_1 j_2 1})\varepsilon}.$$

If

$$\frac{a_{j_2 j_2 0}(a_{j_2 j_1 1} - a_{j_1 j_2 s_2})^2}{a_{j_1 j_1 0}(\bar{x}_{j_2} - a_{j_1 j_2 1})} > 0,$$

there is a down-up asymptotic flow. If

$$\frac{a_{j_2 j_2 0}(a_{j_2 j_1 1} - a_{j_1 j_2 s_2})^2}{a_{j_1 j_1 0}(\bar{x}_{j_2} - a_{j_1 j_2 1})} < 0,$$

there is an up-down asymptotic flow. The variational equation at $x_{j_1}^* = a_{j_1 j_1 1} = a_{j_2 j_1 1}$ with $\bar{x}_{j_2} \neq a_{j_1 j_2 1}$ is

$$\Delta \dot{x}_{j_1} = a_{j_1 j_1 0}(\bar{x}_{j_2} - a_{j_1 j_2 1})(\Delta x_{j_1})^2.$$

Thus, the up-down and down-up flows at $x_{j_1}^* = a_{j_1 j_1 1} = a_{j_2 j_1 1}$ with $\bar{x}_{j_2} \neq a_{j_1 j_2 1}$ are upper-saddle and lower-saddle for $a_{j_1 j_1 0}(\bar{x}_{j_2} - a_{j_1 j_2 1}) > 0$ and $a_{j_1 j_1 0}(\bar{x}_{j_2} - a_{j_1 j_2 1}) < 0$, respectively.

Therefore, the infinite-equilibrium of $x_{j_1}^* = a_{j_2 j_1 s_1} = a_{j_1 j_1 1}$ with $\bar{x}_{j_2} \neq a_{j_1 j_2 1}$ has the following properties as in Eqs. (1.80)–(1.83):

- For $a_{j_1 j_1 0}(\bar{x}_{j_2} - a_{j_1 j_2 1}) > 0$ and $a_{j_2 j_2 0}(a_{j_2 j_1 s_1} - a_{j_2 j_1 s_2})^2 > 0$,

$$(a_{j_1 j_1 1}, \bar{x}_{j_2}) = \underbrace{(\text{US}, \text{DU})}_{\text{down-up upper-saddle}}.$$

The infinite-equilibrium of $x_{j_1}^* = a_{j_2 j_1 s_1} = a_{j_1 j_1 1}$ is a (US,DU)-down-up asymptotic upper-saddle.

- For $a_{j_1 j_1 0}(\bar{x}_{j_2} - a_{j_1 j_2 1}) < 0$ and $a_{j_2 j_2 0}(a_{j_2 j_1 s_1} - a_{j_2 j_1 s_2})^2 > 0$,

$$(a_{j_1j_11}, \bar{x}_{j_2}) = \underbrace{(\text{LS}, \text{UD})}_{\text{up-down lower-saddle}} .$$

The infinite-equilibrium of $x_{j_1}^* = a_{j_2j_1s_1} = a_{j_1j_11}$ is an (LS,UD)-up-down asymptotic lower-saddle.

- For $a_{j_1j_10}(\bar{x}_{j_2} - a_{j_1j_21}) > 0$ and $a_{j_2j_20}(a_{j_2j_1s_1} - a_{j_2j_1s_2})^2 < 0$,

$$(a_{j_1j_11}, \bar{x}_{j_2}) = \underbrace{(\text{US}, \text{UD})}_{\text{up-down upper-saddle}} .$$

The infinite-equilibrium of $x_{j_1}^* = a_{j_2j_1s_1} = a_{j_1j_11}$ is a (US,UD)-up-down asymptotic upper-saddle.

- For $a_{j_1j_10}(\bar{x}_{j_2} - a_{j_1j_21}) < 0$ and $a_{j_2j_20}(a_{j_2j_1s_1} - a_{j_2j_1s_2})^2 < 0$,

$$(a_{j_1j_11}, \bar{x}_{j_2}) = \underbrace{(\text{LS}, \text{DU})}_{\text{down-up lower-saddle}} .$$

The infinite-equilibrium of $x_{j_1}^* = a_{j_2j_1s_1} = a_{j_1j_11}$ is an (LS,DU)-down-up asymptotic lower-saddle.

(v3b) From case (v3a) with $\bar{x}_{j_2} = a_{j_1j_21}$, thus, the equilibrium of $(x_{j_2}^*, x_{j_1}^*) = (a_{j_1j_21}, a_{j_2j_1s_1})$ with $a_{j_2j_1s_1} = a_{j_1j_11}$ has the following properties as in Eqs. (1.84)–(1.87):

- For $a_{j_1j_10} > 0$ and $a_{j_2j_20}(a_{j_2j_1s_1} - a_{j_2j_1s_2})^2 > 0$,

$$(a_{j_1j_21}, a_{j_2j_1s_1}) = \underbrace{(_{\text{UD}}\text{LS}:_{\text{DU}}\text{US}, \text{DP}_-:\text{UP}_+)}_{\text{hyperbolic lower-to-upper saddle}} .$$

The equilibrium of $(x_{j_2}^*, x_{j_1}^*) = (a_{j_1j_21}, a_{j_2j_1s_1})$ for $a_{j_1j_11} = a_{j_2j_1s_1}$ is a ($_{\text{UD}}\text{LS}:_{\text{DU}}\text{US}$, DP$_-$:UP$_+$)-hyperbolic lower-to-upper saddle.

- For $a_{j_1j_10} < 0$ and $a_{j_2j_20}(a_{j_2j_1s_1} - a_{j_2j_1s_2})^2 > 0$,

$$(a_{j_1j_21}, a_{j_2j_1s_1}) = \underbrace{(_{\text{DU}}\text{US}:_{\text{UD}}\text{LS}, \text{UP}_-:\text{DP}_+)}_{\text{hyperbolic-secant upper-to-lower saddle}} .$$

The equilibrium of $(x_{j_2}^*, x_{j_1}^*) = (a_{j_1j_21}, a_{j_2j_1s_1})$ for $a_{j_1j_11} = a_{j_2j_1s_1}$ is a ($_{\text{DU}}\text{US}:_{\text{UD}}\text{LS}$, UP$_-$:DP$_+$)-hyperbolic-secant upper-to-lower saddle.

- For $a_{j_1j_10} > 0$ and $a_{j_2j_20}(a_{j_2j_1s_1} - a_{j_2j_1s_2})^2 < 0$,

$$(a_{j_1j_21}, a_{j_2j_1s_1}) = \underbrace{(_{\text{DU}}\text{LS}:_{\text{UD}}\text{US}, \text{UP}_+:\text{DP}_-)}_{\text{hyperbolic-secant lower-to-upper saddle}} .$$

The equilibrium of $(x_{j_2}^*, x_{j_1}^*) = (a_{j_1j_21}, a_{j_2j_1s_1})$ for $a_{j_1j_11} = a_{j_2j_1s_1}$ is a ($_{\text{DU}}\text{LS}:_{\text{UD}}\text{US}$, UP$_+$:DP$_-$)-hyperbolic-secant lower-to-upper saddle.

- For $a_{j_1j_10} < 0$ and $a_{j_2j_20}(a_{j_2j_1s_1} - a_{j_2j_1s_2})^2 < 0$,

$$(a_{j_1j_21}, a_{j_2j_1s_1}) = (\underbrace{_{UD}US:\ _{DU}LS, DP_+ :UP_-)}_{\text{hyperbolic upper-to-lower saddle}}.$$

The equilibrium of $(x_{j_2}^*, x_{j_1}^*) = (a_{j_1j_21}, a_{j_2j_1s_1})$ for $a_{j_1j_11} = a_{j_2j_1s_1}$ is a $(_{UD}US:\ _{DU}LS,$ $DP_+ :UP_-)$-hyperbolic upper-to-lower saddle.

(v_4) At $a_{j_1j_11} = a_{j_2j_1s_2}$, in phase space,

$$\frac{dx_{j_1}}{dx_{j_2}} = \frac{a_{j_1j_10}}{a_{j_2j_20}}\frac{(x_{j_2} - a_{j_1j_21})}{(x_{j_1} - a_{j_2j_1s_1})},$$

and the deformation of the foregoing equation is

$$(x_{j_1} - a_{j_2j_1s_1})dx_{j_1} = \frac{a_{j_1j_10}}{a_{j_2j_20}}(x_{j_2} - a_{j_1j_21})dx_{j_2}.$$

With an initial condition of (x_{j_10}, x_{j_20}) at $t = t_0$, the integration of the above equation gives the first integral manifold for $a_{j_1j_11} = a_{j_2j_11}$ as

$$\frac{1}{2}\left[(x_{j_1} - a_{j_1j_11})^2 - (x_{j_10} - a_{j_1j_11})^2\right] = \frac{1}{2}\frac{a_{j_1j_10}}{a_{j_2j_20}}\left[(x_{j_2} - a_{j_1j_21})^2 - (x_{j_20} - a_{j_1j_21})^2\right].$$

(v_{4a}) At $x_{j_1}^* = a_{j_1j_11} = a_{j_2j_1s_2}$ and $\bar{x}_{j_2} \neq a_{j_1j_21}$, if

$$\frac{dx_{j_2}}{dx_{j_1}}\bigg|_{x_{j_1}^* = a_{j_2j_1s_2}} = \frac{a_{j_2j_20}}{a_{j_1j_10}}\frac{(a_{j_2j_1s_2} - a_{j_2j_1s_1})}{(\bar{x}_{j_2} - a_{j_1j_21})} > 0,$$

there is an increasing-inflection flow. If

$$\frac{dx_{j_2}}{dx_{j_1}}\bigg|_{x_{j_1}^* = a_{j_2j_1s_2}} = \frac{a_{j_2j_20}}{a_{j_1j_10}}\frac{(a_{j_2j_1s_2} - a_{j_2j_1s_1})}{(\bar{x}_{j_2} - a_{j_1j_21})} < 0,$$

there is a decreasing-inflection flow. The variational equation is

$$\Delta\dot{x}_{j_1} = a_{j_1j_10}(\bar{x}_{j_2} - a_{j_1j_21})(\Delta x_{j_1})^2.$$

The inflection flows at $x_{j_1}^* = a_{j_2j_11}$ are upper-saddle and lower-saddle in the x_{j_1}-direction for $a_{j_1j_10}(\bar{x}_{j_2} - a_{j_1j_21}) > 0$ and $a_{j_1j_10}(\bar{x}_{j_2} - a_{j_1j_21}) < 0$, respectively.

Therefore, the infinite equilibrium of $x_{j_1}^* = a_{j_1j_11} = a_{j_2j_1s_2}$ with $\bar{x}_{j_2} \neq a_{j_1j_21}$ has the following properties as in Eqs. (1.89)–(1.92):

- For $a_{j_1 j_1 0}(\bar{x}_{j_2} - a_{j_1 j_2 1}) > 0$ and $a_{j_2 j_2 0}(a_{j_2 j_1 s_2} - a_{j_2 j_1 s_1}) > 0$,

$$(a_{j_1 j_1 1}, \bar{x}_{j_2}) = \underbrace{(\text{US, II})}_{\text{increasing-inflection upper-saddle}}.$$

The infinite-equilibrium of $x_{j_1}^* = a_{j_1 j_1 1} = a_{j_2 j_1 s_2}$ is a (US,II)-increasing-inflection upper-saddle.

- For $a_{j_1 j_1 0}(\bar{x}_{j_2} - a_{j_1 j_2 1}) < 0$ and $a_{j_2 j_2 0}(a_{j_2 j_1 s_2} - a_{j_2 j_1 s_1}) > 0$,

$$(a_{j_1 j_1 1}, \bar{x}_{j_2}) = \underbrace{(\text{LS, DI})}_{\text{decreasing-inflection lower-saddle}}.$$

The infinite-equilibrium of $x_{j_1}^* = a_{j_1 j_1 1} = a_{j_2 j_1 s_2}$ is a (LS,DI)-decreasing-inflection lower-saddle.

- For $a_{j_1 j_1 0}(\bar{x}_{j_2} - a_{j_1 j_2 1}) > 0$ and $a_{j_2 j_2 0}(a_{j_2 j_1 s_2} - a_{j_2 j_1 s_1}) < 0$,

$$(a_{j_1 j_1 1}, \bar{x}_{j_2}) = \underbrace{(\text{US, DI})}_{\text{decreasing-inflection upper-saddle}}.$$

The infinite-equilibrium of $x_{j_1}^* = a_{j_1 j_1 1} = a_{j_2 j_1 s_2}$ is a (US,DI)-decreasing-inflection upper-saddle.

- For $a_{j_1 j_1 0}(\bar{x}_{j_2} - a_{j_1 j_2 1}) < 0$ and $a_{j_2 j_2 0}(a_{j_2 j_1 s_2} - a_{j_2 j_1 s_1}) < 0$,

$$(a_{j_1 j_1 1}, \bar{x}_{j_2}) = \underbrace{(\text{LS, II})}_{\text{increasing-inflection lower-saddle}}.$$

The infinite-equilibrium of $x_{j_1}^* = a_{j_1 j_1 1} = a_{j_2 j_1 s_2}$ is a (LS,II)-increasing-inflection lower-saddle.

(v_{4b}) From case (v_{4a}) with $\bar{x}_{j_2} = a_{j_1 j_2 1}$, the equilibrium of $(x_{j_2}^*, x_{j_1}^*) = (a_{j_1 j_2 1}, a_{j_2 j_1 s_2})$ with $a_{j_1 j_1 1} = a_{j_2 j_1 s_2}$ has the following properties as in Eqs. (1.93)–(1.96):

- For $a_{j_1 j_1 0} > 0$ and $a_{j_2 j_2 0}(a_{j_2 j_1 s_2} - a_{j_2 j_1 s_1}) > 0$,

$$(a_{j_1 j_2 1}, a_{j_2 j_1 s_2}) = \underbrace{({}_{\text{DI:II}}\text{UP}, {}_{\text{LS:US}}\text{US})}_{\text{up-parabola upper-saddle}}.$$

The equilibrium of $(x_{j_2}^*, x_{j_1}^*) = (a_{j_1 j_2 1}, a_{j_2 j_1 s_2})$ with $a_{j_1 j_1 1} = a_{j_2 j_1 s_2}$ is a (${}_{\text{DI:II}}\text{UP}$, ${}_{\text{LS:US}}\text{US}$)-up-parabola upper-saddle.

- For $a_{j_1 j_1 0} < 0$ and $a_{j_2 j_2 0}(a_{j_2 j_1 s_2} - a_{j_2 j_1 s_1}) > 0$,

$$(a_{j_1j_21}, a_{j_2j_1s_2}) = \underbrace{(_{\text{II:DI}}DP, _{\text{US:LS}}US)}_{\text{down-parabola upper-saddle}} .$$

The equilibrium of $(x_{j_2}^*, x_{j_1}^*) = (a_{j_1j_21}, a_{j_2j_1s_2})$ with $a_{j_1j_11} = a_{j_2j_1s_2}$ is an $(_{\text{II:DI}}DP,$ $_{\text{US:LS}}US)$-down-parabola upper-saddle.
- For $a_{j_1j_10} > 0$ and $a_{j_2j_20}(a_{j_2j_1s_2} - a_{j_2j_1s_1}) < 0$,

$$(a_{j_1j_21}, a_{j_2j_1s_2}) = \underbrace{(_{\text{II:DI}}DP, _{\text{LS:US}}LS)}_{\text{down-parabola lower-saddle}} .$$

The equilibrium of $(x_{j_2}^*, x_{j_1}^*) = (a_{j_1j_21}, a_{j_2j_1s_2})$ with $a_{j_1j_11} = a_{j_2j_1s_2}$ is an $(_{\text{II:DI}}DP,$ $_{\text{LS:US}}LS)$-down-parabola lower-saddle.
- For $a_{j_1j_10} < 0$ and $a_{j_2j_20}(a_{j_2j_1s_2} - a_{j_2j_1s_1}) < 0$,

$$(a_{j_1j_21}, a_{j_2j_1s_2}) = \underbrace{(_{\text{DI:II}}UP, _{\text{US:LS}}LS)}_{\text{up-parabola lower-saddle}}.$$

The equilibrium of $(x_{j_2}^*, x_{j_1}^*) = (a_{j_1j_21}, a_{j_2j_1s_2})$ with $a_{j_1j_11} = a_{j_2j_1s_2}$ is a $(_{\text{DI:II}}UP,$ $_{\text{US:LS}}LS)$-up-parabola lower-saddle.

(v_5) For $a_{j_2j_1s_1} = a_{j_2j_1s_2} = a_{j_2j_11} = a_{j_1j_11}$, in phase place,

$$\frac{dx_{j_1}}{dx_{j_2}} = \frac{a_{j_1j_10}}{a_{j_2j_20}} \frac{(x_{j_2} - a_{j_1j_21})}{(x_{j_1} - a_{j_2j_11})},$$

and the deformation of the foregoing equation is

$$(x_{j_1} - a_{j_2j_11})dx_{j_1} = \frac{a_{j_1j_10}}{a_{j_2j_20}}(x_{j_2} - a_{j_1j_21})dx_{j_2}.$$

With an initial condition of (x_{j_10}, x_{j_20}) at $t = t_0$, the integration of the above equation gives the first integral manifold for $a_{j_1j_11} \neq a_{j_2j_11}$ as

$$\frac{1}{2}\left[(x_{j_1} - a_{j_2j_11})^2 - (x_{j_10} - a_{j_2j_11})^2\right] = \frac{1}{2}\frac{a_{j_1j_10}}{a_{j_2j_20}}\left[(x_{j_2} - a_{j_1j_2})^2 - (x_{j_20} - a_{j_1j_2})^2\right].$$

At $x_{j_1}^* = a_{j_2j_11} = a_{j_1j_11}$ and $\bar{x}_{j_2} \neq a_{j_1j_21}$,

$$\frac{dx_{j_2}}{dx_{j_1}}\bigg|_{x_{j_1}^* = a_{j_2j_11}} = \frac{a_{j_2j_20}}{a_{j_1j_10}} \frac{(x_{j_1} - a_{j_2j_11})}{(\bar{x}_{j_2} - a_{j_1j_21})}\bigg|_{x_{j_1}^* = a_{j_2j_11}} = 0.$$

If

$$\frac{d^2 x_{j_2}}{dx_{j_1}^2}\bigg|_{x_{j_1}^* = a_{j_2 j_1 1}} = \frac{a_{j_2 j_2 0}}{a_{j_1 j_1 0}} \frac{1}{(\bar{x}_{j_2} - a_{j_1 j_2 1})} > 0,$$

there is an up-parabola flow at $x_{j_1}^* = a_{j_2 j_1 1}$. If

$$\frac{d^2 x_{j_2}}{dx_{j_1}^2}\bigg|_{x_{j_1}^* = a_{j_2 j_1 1}} = \frac{a_{j_2 j_2 0}}{a_{j_1 j_1 0}} \frac{1}{(\bar{x}_{j_2} - a_{j_1 j_2 1})} < 0,$$

there is a decreasing-inflection flow at $x_{j_1}^* = a_{j_2 j_1 1}$. Let

$$\Delta \dot{x}_{j_1} = a_{j_1 j_1 0}(\bar{x}_{j_2} - a_{j_1 j_2 1})(\Delta x_{j_1})^2.$$

The third-order parabola flows at $x_{j_1}^* = a_{j_2 j_1 1}$ are upper-saddle and lower-saddle for $a_{j_1 j_1 0}(\bar{x}_{j_2} - a_{j_1 j_2 1}) > 0$ and $a_{j_1 j_1 0}(\bar{x}_{j_2} - a_{j_1 j_2 1}) < 0$.

Therefore, the infinite equilibrium of $x_{j_1}^* = a_{j_1 j_1 1} = a_{j_2 j_1 1}$ has the following properties as in Eqs. (1.98)–(1.101):

- For $a_{j_1 j_1 0}(\bar{x}_{j_2} - a_{j_1 j_2 1}) > 0$ and $a_{j_2 j_2 0} > 0$,

$$(a_{j_1 j_1 1}, \bar{x}_{j_2}) = \underbrace{\text{(US, UP)}}_{\text{up-parabola upper-saddle}}.$$

  The infinite-equilibrium of $x_{j_1}^* = a_{j_2 j_1 1} = a_{j_1 j_1 1}$ is a (US,US)-up-parabola upper-saddle.

- For $a_{j_1 j_1 0}(\bar{x}_{j_2} - a_{j_1 j_2 1}) < 0$ and $a_{j_2 j_2 0} > 0$,

$$(a_{j_1 j_1 1}, \bar{x}_{j_2}) = \underbrace{\text{(LS, DP)}}_{\text{down-parabola lower-saddle}}.$$

  The infinite-equilibrium of $x_{j_1}^* = a_{j_2 j_1 1} = a_{j_1 j_1 1}$ is an (LS,DP)-down-parabola lower-saddle.

- For $a_{j_1 j_1 0}(\bar{x}_{j_2} - a_{j_1 j_2 1}) > 0$ and $a_{j_2 j_2 0} < 0$,

$$(a_{j_1 j_1 1}, \bar{x}_{j_2}) = \underbrace{\text{(US, DP)}}_{\text{down-parabola upper-saddle}}.$$

  The infinite-equilibrium of $x_{j_1}^* = a_{j_2 j_1 1} = a_{j_1 j_1 1}$ is a (US,DP)-down-parabola upper-saddle.

- For $a_{j_1j_10}(\bar{x}_{j_2} - a_{j_1j_21}) < 0$ and $a_{j_2j_20} < 0$,

$$(a_{j_1j_11}, \bar{x}_{j_2}) = \underbrace{\text{(LS, UP)}}_{\text{up-parabola lower-saddle}} \quad .$$

The infinite-equilibrium of $x^*_{j_1} = a_{j_2j_11} = a_{j_1j_11}$ is an (LS,UP)-up-parabola lower-saddle.

(v_{5b}) From case (v_{5a}), for $\bar{x}_{j_2} = a_{j_1j_21}$, the equilibrium of $(x^*_{j_1}, x^*_{j_2}) = (a_{j_1j_11}, a_{j_1j_21})$ with $a_{j_2j_11} = a_{j_1j_11}$ has the following properties as in Eqs. (1.102)–(1.105):

- For $a_{j_1j_10} > 0$ and $a_{j_2j_20} > 0$,

$$(a_{j_1j_21}, a_{j_2j_11}) = \underbrace{(_{\mathrm{DP}}\mathrm{LS} : {}_{\mathrm{UP}}\mathrm{US}, \mathrm{DP}_- : \mathrm{UP}_+)}_{\text{positive hyperbolic lower-to-upper saddle}} \quad .$$

The equilibrium of $(x^*_{j_2}, x^*_{j_1}) = (a_{j_1j_21}, a_{j_2j_11})$ with $a_{j_2j_11} = a_{j_1j_11}$ is a $(_{\mathrm{DP}}\mathrm{LS}:_{\mathrm{UP}}\mathrm{US},$ $\mathrm{DP}_-:\mathrm{UP}_+)$-positive hyperbolic lower-to-upper saddle.
- For $a_{j_1j_10} < 0$ and $a_{j_2j_20} > 0$,

$$(a_{j_1j_21}, a_{j_2j_11}) = \underbrace{(_{\mathrm{UP}}\mathrm{US} : {}_{\mathrm{DP}}\mathrm{LS}, \mathrm{UP}_- : \mathrm{DP}_+)}_{\text{CCW circular upper-to-lower saddle}}.$$

The equilibrium of $(x^*_{j_2}, x^*_{j_1}) = (a_{j_1j_21}, a_{j_2j_11})$ with $a_{j_2j_11} = a_{j_1j_11}$ is a $(_{\mathrm{UP}}\mathrm{US}:_{\mathrm{DP}}\mathrm{LS},$ $\mathrm{UP}_-:\mathrm{DP}_+)$-counter-clockwise circular upper-to-lower saddle.
- For $a_{j_1j_10} > 0$ and $a_{j_2j_20} < 0$,

$$(a_{j_1j_21}, a_{j_2j_11}) = \underbrace{(_{\mathrm{UP}}\mathrm{LS} : {}_{\mathrm{DP}}\mathrm{US}, \mathrm{UP}_+ : \mathrm{DP}_-)}_{\text{CW circular lower-to-upper saddle}}.$$

The equilibrium of $(x^*_{j_2}, x^*_{j_1}) = (a_{j_1j_21}, a_{j_2j_11})$ with $a_{j_2j_11} = a_{j_1j_11}$ is a $(_{\mathrm{UP}}\mathrm{LS}:_{\mathrm{DP}}\mathrm{US},$ $\mathrm{UP}_+:\mathrm{DP}_-)$-clockwise circular lower-to-upper saddle.
- For $a_{j_1j_10} < 0$ and $a_{j_2j_20} < 0$,

$$(a_{j_1j_21}, a_{j_2j_11}) = \underbrace{(_{\mathrm{DP}}\mathrm{US} : {}_{\mathrm{UP}}\mathrm{LS}, \mathrm{DP}_+ : \mathrm{UP}_-)}_{\text{negative hyperbolic upper-to-lower saddle}} \quad .$$

The equilibrium of $(x^*_{j_2}, x^*_{j_1}) = (a_{j_1j_21}, a_{j_2j_11})$ with $a_{j_2j_11} = a_{j_1j_11}$ is a $(_{\mathrm{DP}}\mathrm{US}:_{\mathrm{UP}}\mathrm{LS},$ $\mathrm{DP}_+:\mathrm{UP}_-)$-negative hyperbolic upper-to-lower saddle.

(vi) For $\Delta_{j_1j_1} = 0$ and $\Delta_{j_2j_1} > 0$, the standard form is

$$\dot{x}_{j_2} = a_{j_2 j_2 0}(x_{j_1} - a_{j_2 j_1 1})(x_{j_1} - a_{j_2 j_1 2})(x_{j_1} - a_{j_2 j_1 3}),$$

$$\dot{x}_{j_1} = a_{j_1 j_1 0}(x_{j_1} - a_{j_1 j_1 1})^2(x_{j_2} - a_{j_1 j_2 1})$$

where

$$a_{j_1 j_1 1} = -\frac{1}{2}B_{j_1 j_1},$$

$$b_{j_2 j_1 2}, b_{j_2 j_1 3} = -\frac{1}{2}(B_{j_2 j_1} \pm \sqrt{\Delta_{j_2 j_1}}),$$

$$\{a_{j_2 j_1 1}, a_{j_2 j_1 2}, a_{j_2 j_1 3}\} = \text{sort}\{b_{j_2 j_1 1}, b_{j_2 j_1 2}, b_{j_2 j_1 3}\},$$

$$a_{j_2 j_1 s_1} < a_{j_2 j_1 s_2}, s_1, s_2 \in \{1, 2, 3\}, s_1 < s_2.$$

(vi$_1$) For $a_{j_2 j_1 s_1} \neq a_{j_1 j_1 1}$, in phase space,

$$\frac{dx_{j_1}}{dx_{j_2}} = \frac{a_{j_1 j_1 0}}{a_{j_2 j_2 0}} \frac{(x_{j_1} - a_{j_1 j_1 1})(x_{j_2} - a_{j_1 j_2 1})^2}{\prod_{s_1=1}^{3}(x_{j_1} - a_{j_2 j_1 s_1})},$$

and the deformation of the foregoing equation is

$$[(x_{j_1} - a_{j_1 j_1 1}) + \sum_{s_1=1}^{3}(a_{j_1 j_1 1} - a_{j_2 j_1 s_1})$$

$$+ \sum_{s_1=1}^{3}\prod_{s_2=1, s_2 \neq s_1}^{3}(a_{j_1 j_1 1} - a_{j_2 j_1 s_1})\frac{1}{x_{j_1} - a_{j_1 j_1 1}}$$

$$+ \prod_{s_1=1}^{3}(a_{j_1 j_1 1} - a_{j_2 j_1 s_1})\frac{1}{(x_{j_1} - a_{j_1 j_1 1})^2}]dx_{j_1}$$

$$= \frac{a_{j_1 j_1 0}}{a_{j_2 j_2 0}}(x_{j_2} - a_{j_1 j_2 1})dx_{j_2}.$$

With an initial condition of $(x_{j_1 0}, x_{j_2 0})$ at $t = t_0$, the integration of the above equation gives the first integral manifold ($a_{j_2 j_1 s_1} \neq a_{j_1 j_1 1}$):

$$\frac{1}{2}\left[(x_{j_1} - a_{j_1 j_1 1})^2 - (x_{j_1 0} - a_{j_1 j_1 1})^2\right] + \sum_{s_1=1}^{3}(a_{j_1 j_1 1} - a_{j_2 j_1 s_1})(x_{j_1} - x_{j_1 0})$$

$$+ \sum_{s_1=1}^{3}\prod_{s_2=1, s_2 \neq s_1}^{3}(a_{j_1 j_1 1} - a_{j_2 j_1 s_2})\ln\frac{|x_{j_1} - a_{j_1 j_1 1}|}{|x_{j_1 0} - a_{j_1 j_1 1}|}$$

$$- \prod_{s_3=1}^{3}(a_{j_1 j_1 1} - a_{j_2 j_1 s_3})(\frac{1}{x_{j_1} - a_{j_1 j_1 1}} - \frac{1}{x_{j_1 0} - a_{j_1 j_1 1}})$$

$$= \frac{1}{2}\frac{a_{j_1 j_1 0}}{a_{j_2 j_2 0}}\left[(x_{j_2} - a_{j_1 j_2 1})^2 - (x_{j_2 0} - a_{j_1 j_2 1})^2\right].$$

(vi_{1a}) Consider two cases: (I) $x_{j_1}^* = a_{j_2 j_1 s_1}$ and (II) $x_{j_2}^* = a_{j_1 j_2 1}$ for this proof.

(I) In phase space, at $x_{j_1}^* = a_{j_2 j_1 s_1} \neq a_{j_1 j_1 1}$ ($s_1, s_2 \in \{1, 2, 3\}$; $s_1 \neq s_2$) with $\overline{x}_{j_2} \neq a_{j_1 j_2 1}$,

$$\frac{dx_{j_2}}{dx_{j_1}}\bigg|_{x_{j_1}^* = a_{j_2 j_1 s_1}} = \frac{a_{j_2 j_2 0}}{a_{j_1 j_1 0}} \frac{(x_{j_1} - a_{j_2 j_1 s_1}) \prod_{s_2 = 1, s_2 \neq s_1}^{3} (x_{j_1} - a_{j_2 j_1 s_2})}{(x_{j_1} - a_{j_1 j_1 1})^2 (\overline{x}_{j_2} - a_{j_1 j_2 1})}\bigg|_{x_{j_1}^* = a_{j_2 j_1 s_1}} = 0.$$

If

$$\frac{d^2 x_{j_2}}{dx_{j_1}^2}\bigg|_{x_{j_1}^* = a_{j_2 j_1 s_1}} = \frac{a_{j_2 j_2 0}}{a_{j_1 j_1 0}} \frac{\prod_{s_2 = 1, s_2 \neq s_1}^{3} (a_{j_2 j_1 s_1} - a_{j_2 j_1 s_2})}{(a_{j_2 j_1 s_1} - a_{j_1 j_1 1})^2 (\overline{x}_{j_2} - a_{j_1 j_2 1})} > 0,$$

there is an up-parabola flow. If

$$\frac{d^2 x_{j_2}}{dx_{j_1}^2}\bigg|_{x_{j_1}^* = a_{j_2 j_1 s_1}} = \frac{a_{j_2 j_2 0}}{a_{j_1 j_1 0}} \frac{\prod_{s_2 = 1, s_2 \neq s_1}^{3} (a_{j_2 j_1 s_1} - a_{j_2 j_1 s_2})}{(a_{j_2 j_1 s_1} - a_{j_1 j_1 1})^2 (\overline{x}_{j_2} - a_{j_1 j_2 1})} < 0,$$

there is a down-parabola flow. Let

$$\dot{x}_{j_1} = a_{j_1 j_1 0}(a_{j_2 j_1 s_1} - a_{j_1 j_1 1})^2 (\overline{x}_{j_2} - a_{j_1 j_2 1}).$$

The parabola flows for $\dot{x}_{j_1} > 0$ and $\dot{x}_{j_1} < 0$ at $x_{j_1}^* = a_{j_2 j_1 s_1}$ are positive and negative, respectively.

Thus, the equilibrium of $x_{j_1}^* = a_{j_2 j_1 s_1}$ has the following properties:

- For $a_{j_1 j_1 0}(a_{j_2 j_1 s_1} - a_{j_1 j_1 1})^2 (\overline{x}_{j_2} - a_{j_1 j_2 1}) > 0$ and $a_{j_2 j_2 0} \prod_{s_2 = 1, s_2 \neq s_1}^{3} (a_{j_2 j_1 s_1} - a_{j_2 j_1 s_2}) > 0$,

$$(\dot{x}_{j_1}, a_{j_2 j_1 s_1}) = \underbrace{(\mathrm{pF}, \mathrm{UP})}_{\text{up-parabola flow } (+)}.$$

- For $a_{j_1 j_1 0}(a_{j_2 j_1 s_1} - a_{j_1 j_1 1})^2 (\overline{x}_{j_2} - a_{j_1 j_2 1}) < 0$ and $a_{j_2 j_2 0} \prod_{s_2 = 1, s_2 \neq s_1}^{3} (a_{j_2 j_1 s_1} - a_{j_2 j_1 s_2}) > 0$,

$$(\dot{x}_{j_1}, a_{j_2 j_1 s_1}) = \underbrace{(\mathrm{nF}, \mathrm{DP})}_{\text{down-parabola flow } (-)}.$$

- For $a_{j_1 j_1 0}(a_{j_2 j_1 s_1} - a_{j_1 j_1 1})^2 (\overline{x}_{j_2} - a_{j_1 j_2 1}) > 0$ and $a_{j_2 j_2 0} \prod_{s_2 = 1, s_2 \neq s_1}^{3} (a_{j_2 j_1 s_1} - a_{j_2 j_1 s_2}) < 0$,

$$(\dot{x}_{j_1}, a_{j_2 j_1 s_1}) = \underbrace{(\mathrm{pF}, \mathrm{DP})}_{\text{down-parabola flow } (+)}.$$

- For $a_{j_1 j_1 0}(a_{j_2 j_1 s_1} - a_{j_1 j_1 1})^2 (\overline{x}_{j_2} - a_{j_1 j_2 1}) < 0$ and $a_{j_2 j_2 0} \prod_{s_2 = 1, s_2 \neq s_1}^{3} (a_{j_2 j_1 s_1} - a_{j_2 j_1 s_2}) < 0$,

$$(\dot{x}_{j_1}, a_{j_2 j_1 s_1}) = \underbrace{(nF, UP)}_{\text{up-parabola flow }(-)}.$$

(II) In phase space, at $x_{j_2}^* = a_{j_1 j_2 1}$ with $\bar{x}_{j_1} \neq a_{j_2 j_1 1}, a_{j_2 j_1 2}, a_{j_2 j_1 3}, a_{j_1 j_1 1}$,

$$\frac{dx_{j_1}}{dx_{j_2}}\Big|_{x_{j_2}^* = a_{j_1 j_2 1}} = \frac{a_{j_1 j_1 0}}{a_{j_2 j_2 0}} \frac{(\bar{x}_{j_1} - a_{j_1 j_1 1})^2 (x_{j_2} - a_{j_1 j_2 1})}{\prod_{s_1 = 1}^{3}(\bar{x}_{j_1} - a_{j_2 j_1 s_1})}\Big|_{x_{j_2}^* = a_{j_1 j_2 1}} = 0.$$

If

$$\frac{d^2 x_{j_1}}{dx_{j_2}^2}\Big|_{x_{j_2}^* = a_{j_1 j_2 1}} = \frac{a_{j_1 j_1 0}}{a_{j_2 j_2 0}} \frac{(\bar{x}_{j_1} - a_{j_1 j_1 1})^2}{\prod_{s_1 = 1}^{3}(\bar{x}_{j_1} - a_{j_2 j_1 s_1})} > 0,$$

there is an up-parabola flow. If

$$\frac{d^2 x_{j_1}}{dx_{j_2}^2}\Big|_{x_{j_2}^* = a_{j_1 j_2 1}} = \frac{a_{j_1 j_1 0}}{a_{j_2 j_2 0}} \frac{(\bar{x}_{j_1} - a_{j_1 j_1 1})^2}{\prod_{s_1 = 1}^{3}(\bar{x}_{j_1} - a_{j_2 j_1 s_1})} < 0,$$

there is a down-parabola flow. Let

$$\dot{x}_{j_2} = a_{j_2 j_2 0} \prod_{s_1 = 1}^{3}(\bar{x}_{j_1} - a_{j_2 j_1 s_1}).$$

The parabola flows for $\dot{x}_{j_2} > 0$ and $\dot{x}_{j_2} < 0$ at $x_{j_2}^* = a_{j_1 j_2 1}$ are positive and negative, respectively.

Therefore, the equilibrium of $x_{j_2}^* = a_{j_1 j_2 1}$ has the following properties:

- For $a_{j_1 j_1 0}(\bar{x}_{j_1} - a_{j_1 j_1 1})^2 > 0$ and $a_{j_2 j_2 0}\prod_{s_1 = 1}^{3}(\bar{x}_{j_1} - a_{j_2 j_1 s_1}) > 0$,

$$(a_{j_1 j_2 1}, \dot{x}_{j_2}) = \underbrace{(UP, pF)}_{\text{up-parabola flow }(+)}.$$

- For $a_{j_1 j_1 0}(\bar{x}_{j_1} - a_{j_1 j_1 1})^2 < 0$ and $a_{j_2 j_2 0}\prod_{s_1 = 1}^{3}(\bar{x}_{j_1} - a_{j_2 j_1 s_1}) > 0$,

$$(a_{j_1 j_2 1}, \dot{x}_{j_2}) = \underbrace{(DP, pF)}_{\text{down-parabola flow }(+)}.$$

- For $a_{j_1 j_1 0}(\bar{x}_{j_1} - a_{j_1 j_1 1})^2 > 0$ and $a_{j_2 j_2 0}\prod_{s_1 = 1}^{3}(\bar{x}_{j_1} - a_{j_2 j_1 s_1}) < 0,$

$$(a_{j_1j_21}, \dot{x}_{j_2}) = \underbrace{(\text{DP}, \text{nF})}_{\text{down-parabola flow } (-)} .$$

- For $a_{j_1j_10}(\overline{x}_{j_1} - a_{j_1j_11})^2 < 0$ and $a_{j_2j_20}\prod_{s_1=1}^{3}(\overline{x}_{j_1} - a_{j_2j_1s_1}) < 0$,

$$(a_{j_1j_21}, \dot{x}_{j_2}) = \underbrace{(\text{UP}, \text{nF})}_{\text{up-parabola flow } (-)} .$$

Therefore, from the above analysis for case (I) and (II), the equilibrium of $(x_{j_1}^*, x_{j_1}^*) = (a_{j_1j_21}, a_{j_2j_1s_1})$ $(s_1, s_2 \in \{1, 2\}, s_1 \neq s_2)$ has the following properties as in Eqs. (1.109)–(1.112):

- For $a_{j_1j_10}(a_{j_2j_1s_1} - a_{j_1j_11})^2 > 0$ and $a_{j_2j_20}\prod_{s_2=1, s_2 \neq s_1}^{3}(a_{j_2j_1s_1} - a_{j_2j_1s_2}) > 0$,

$$(a_{j_1j_21}, a_{j_2j_1s_1}) = \underbrace{(\text{UP}_+, \text{UP}_+)}_{\text{positive saddle}} .$$

The equilibrium of $(x_{j_2}^*, x_{j_1}^*) = (a_{j_1j_21}, a_{j_2j_1s_1})$ is a (UP$_+$,UP$_+$)-positive saddle.
- For $a_{j_1j_10}(a_{j_2j_1s_1} - a_{j_1j_11})^2 < 0$ and $a_{j_2j_20}\prod_{s_2=1, s_2 \neq s_1}^{3}(a_{j_2j_1s_1} - a_{j_2j_1s_2}) > 0$,

$$(a_{j_1j_21}, a_{j_2j_1s_1}) = \underbrace{(\text{DP}_+, \text{DP}_-)}_{\text{CCW center}} .$$

The equilibrium of $(x_{j_2}^*, x_{j_1}^*) = (a_{j_1j_21}, a_{j_2j_1s_1})$ is a (DP$_+$,DP$_-$)-counter-clockwise center.
- For $a_{j_1j_10}(a_{j_2j_1s_1} - a_{j_1j_11})^2 > 0$ and $a_{j_2j_20}\prod_{s_2=1, s_2 \neq s_1}^{3}(a_{j_2j_1s_1} - a_{j_2j_1s_2}) < 0$,

$$(a_{j_1j_21}, a_{j_2j_1s_1}) = \underbrace{(\text{DP}_-, \text{DP}_+)}_{\text{CW center}} .$$

The equilibrium of $(x_{j_2}^*, x_{j_1}^*) = (a_{j_1j_21}, a_{j_2j_1s_1})$ is a (DP$_-$,DP$_+$)-clockwise center.
- For $a_{j_1j_10}(a_{j_2j_1s_1} - a_{j_1j_11})^2 < 0$ and $a_{j_2j_20}\prod_{s_2=1, s_2 \neq s_1}^{3}(a_{j_2j_1s_1} - a_{j_2j_1s_2}) < 0$,

$$(a_{j_1j_21}, a_{j_2j_1s_1}) = \underbrace{(\text{UP}_-, \text{UP}_-)}_{\text{negative saddle}} .$$

The equilibrium of $(x_{j_2}^*, x_{j_1}^*) = (a_{j_1j_21}, a_{j_2j_1s_1})$ is a (UP$_-$,UP$_-$)-negative saddle.

(vi$_{1b}$) Similar to case (v$_{1b}$), the equilibrium of $(x_{j_1}^*, x_{j_2}^*) = (a_{j_1j_11}, a_{j_1j_21})$ with $a_{j_1j_21} \neq a_{j_2j_2s_1}$ $(s_1 = 1, 2, 3)$ has the following properties as in Eqs. (1.113)–(1.116):

- For $a_{j_1j_10} > 0$ and $a_{j_2j_20}\prod_{s_1=1}^{3}(a_{j_1j_11} - a_{j_2j_1s_1}) > 0$,

$$(a_{j_1 j_1 1}, a_{j_1 j_2 1}) = \underbrace{(\text{UP:UP, pF})}_{\text{hyperbolic-secant-to-hyperbolic flow } (+)} \quad .$$

The equilibrium of $(x_{j_1}^*, x_{j_2}^*) = (a_{j_1 j_1 1}, a_{j_1 j_2 1})$ is a (UP:UP, pF)-positive hyperbolic-secant-to-hyperbolic flow.

- For $a_{j_1 j_1 0} < 0$ and $a_{j_2 j_2 0} \prod_{s_1 = 1}^{3} (a_{j_1 j_1 1} - a_{j_2 j_1 s_1}) > 0$,

$$(a_{j_1 j_1 1}, a_{j_1 j_2 1}) = \underbrace{(\text{DP:DP, pF})}_{\text{hyperbolic-to-hyperbolic-secant flow } (+)} \quad .$$

The equilibrium of $(x_{j_1}^*, x_{j_2}^*) = (a_{j_1 j_1 1}, a_{j_1 j_2 1})$ is a (DP:DP,pF)-positive hyperbolic-to-hyperbolic-secant flow.

- For $a_{j_1 j_1 0} > 0$ and $a_{j_2 j_2 0} \prod_{s_1 = 1}^{3} (a_{j_1 j_1 1} - a_{j_2 j_1 s_1}) < 0$,

$$(a_{j_1 j_1 1}, a_{j_1 j_2 1}) = \underbrace{(\text{DP:DP, nF})}_{\text{hyperbolic-to-hyperbolic-secant flow } (-)} \quad .$$

The equilibrium of $(x_{j_1}^*, x_{j_2}^*) = (a_{j_1 j_1 1}, a_{j_1 j_2 1})$ is a (DP:DP, nF)-negative hyperbolic-secant-to-hyperbolic flow.

- For $a_{j_1 j_1 0} < 0$ and $a_{j_2 j_2 0} \prod_{s_1 = 1}^{3} (a_{j_1 j_1 1} - a_{j_2 j_1 s_1}) < 0$,

$$(a_{j_1 j_1 1}, a_{j_1 j_2 1}) = \underbrace{(\text{UP:UP, nF})}_{\text{hyperbolic-secant-to-hyperbolic flow } (-)} \quad .$$

The equilibrium of $(x_{j_1}^*, x_{j_2}^*) = (a_{j_1 j_1 1}, a_{j_1 j_2 1})$ is a (UP:UP,nF)-negative hyperbolic-secant-to-hyperbolic flow.

(vi_2) For $a_{j_2 j_1 s_1} = a_{j_1 j_1 1}$, in phase space,

$$\frac{dx_{j_1}}{dx_{j_2}} = \frac{a_{j_1 j_1 0}}{a_{j_2 j_2 0}} \frac{(x_{j_2} - a_{j_1 j_2 1})(x_{j_1} - a_{j_1 j_1 1})}{(x_{j_1} - a_{j_2 j_1 s_2})(x_{j_1} - a_{j_2 j_1 s_3})},$$

and the deformation of the foregoing equation is

$$[(x_{j_1} - a_{j_1 j_1 1}) + (2a_{j_1 j_1 1} - a_{j_2 j_1 s_2} - a_{j_2 j_1 s_3})$$
$$+ (a_{j_1 j_1 1} - a_{j_2 j_1 s_2})(a_{j_1 j_1 1} - a_{j_2 j_1 s_3}) \frac{1}{(x_{j_1} - a_{j_1 j_1 1})}] dx_{j_1}$$
$$= \frac{a_{j_1 j_1 0}}{a_{j_2 j_2 0}} (x_{j_2} - a_{j_1 j_2 1}) dx_{j_2}.$$

With an initial condition of $(x_{j_1 0}, x_{j_2 0})$ at $t = t_0$, the integration of the above equation gives the first integral manifold as

$$\frac{1}{2}\left[(x_{j_1} - a_{j_1 j_1 1})^2 - (x_{j_1 0} - a_{j_1 j_1 1})^2\right] + (2a_{j_1 j_1 1} - a_{j_2 j_1 s_2} - a_{j_2 j_1 s_3})(x_{j_1} - x_{j_1 0})$$

$$+(a_{j_1 j_1 1} - a_{j_2 j_1 s_2})(a_{j_1 j_1 1} - a_{j_2 j_1 s_3}) \ln \frac{|x_{j_1} - a_{j_1 j_1 1}|}{|x_{j_1 0} - a_{j_1 j_1 1}|}$$

$$= \frac{1}{2}\frac{a_{j_1 j_1 0}}{a_{j_2 j_2 0}}\left[(x_{j_2} - a_{j_1 j_2 1})^2 - (x_{j_2 0} - a_{j_1 j_2 1})^2\right].$$

(vi$_{2a}$) For $x^*_{j_1} = a_{j_1 j_1 1} = a_{j_2 j_1 s_1}$ with $\bar{x}_{j_2} \neq a_{j_1 j_2 1}$,

$$\frac{dx_{j_2}}{dx_{j_1}}\bigg|_{x_{j_1} = a_{j_2 j_1 s_1} \pm \varepsilon} = \pm \frac{a_{j_2 j_2 0}}{a_{j_1 j_1 0}}\frac{\prod_{s_2 = 1, s_2 \neq s_1}^{3}(a_{j_2 j_1 s_1} - a_{j_2 j_1 s_2})}{(\bar{x}_{j_2} - a_{j_1 j_2 1})\varepsilon}.$$

If

$$\frac{a_{j_2 j_2 0}\prod_{s_2 = 1, s_2 \neq s_1}^{3}(a_{j_2 j_1 s_1} - a_{j_2 j_1 s_2})}{a_{j_1 j_1 0}(\bar{x}_{j_2} - a_{j_1 j_2 1})} > 0,$$

there is a down-up asymptotic flow, and if

$$\frac{a_{j_2 j_2 0}\prod_{s_2 = 1, s_2 \neq s_1}^{3}(a_{j_2 j_1 s_1} - a_{j_2 j_1 s_2})}{a_{j_1 j_1 0}(\bar{x}_{j_2} - a_{j_1 j_2 1})} < 0,$$

there is an up-down asymptotic flow. The variational equation is

$$\Delta \dot{x}_{j_1} = a_{j_1 j_1 0}(\bar{x}_{j_2} - a_{j_1 j_2 1})(\Delta x_{j_1})^2.$$

The up-down and down-up asymptotic flows are upper-saddle and lower-saddle for $a_{j_1 j_1 0}(\bar{x}_{j_2} - a_{j_1 j_2 1}) > 0$ and $a_{j_1 j_1 0}(\bar{x}_{j_2} - a_{j_1 j_2 1}) < 0$ at $x^*_{j_1} = a_{j_2 j_1 1}$, respectively.

Therefore, the infinite-equilibrium of $x^*_{j_1} = a_{j_1 j_1 1} = a_{j_2 j_1 s_1}$ with $\bar{x}_{j_2} \neq a_{j_1 j_2 1}$ has the following properties as in Eqs. (1.118)–(1.121):

- For $a_{j_1 j_1 0}(\bar{x}_{j_2} - a_{j_1 j_2 1}) > 0$ and $a_{j_2 j_2 0}\prod_{s_2 = 1, s_2 \neq s_1}^{3}(a_{j_2 j_1 s_1} - a_{j_2 j_1 s_2}) > 0$,

$$(a_{j_1 j_1 1}, \bar{x}_{j_2}) = \underbrace{(\text{US}, \text{DU})}_{\text{down-up upper-saddle}}.$$

The infinite-equilibrium of $x^*_{j_1} = a_{j_1 j_1 1} = a_{j_2 j_1 s_1}$ is a (US,DU)-down-up asymptotic upper-saddle.

- For $a_{j_1 j_1 0}(\bar{x}_{j_2} - a_{j_1 j_2 1}) < 0$ and $a_{j_2 j_2 0}\prod_{s_2 = 1, s_2 \neq s_1}^{3}(a_{j_2 j_1 s_1} - a_{j_2 j_1 s_2}) > 0$,

$$(a_{j_1 j_1}1, \bar{x}_{j_2}) = \underbrace{(\text{LS, UD})}_{\text{up-down lower-saddle}}.$$

The infinite-equilibrium of $x_{j_1}^* = a_{j_1 j_1}1 = a_{j_2 j_1 s_1}$ is an (LS,UD)-up-down asymptotic lower-saddle.

- For $a_{j_1 j_1 0}(\bar{x}_{j_2} - a_{j_1 j_2}1) > 0$ and $a_{j_2 j_2 0}\prod_{s_2=1, s_2 \neq s_1}^{3}(a_{j_2 j_1 s_1} - a_{j_2 j_1 s_2}) < 0$,

$$(a_{j_1 j_1}1, \bar{x}_{j_2}) = \underbrace{(\text{US, UD})}_{\text{down-parabola upper-saddle}}.$$

The infinite-equilibrium of $x_{j_1}^* = a_{j_1 j_1}1 = a_{j_2 j_1 s_1}$ is a (US,UD)-up-down asymptotic upper-saddle.

- For $a_{j_1 j_1 0}(\bar{x}_{j_2} - a_{j_1 j_2}1) < 0$ and $a_{j_2 j_2 0}\prod_{s_2=1, s_2 \neq s_1}^{3}(a_{j_2 j_1 s_1} - a_{j_2 j_1 s_2}) < 0$,

$$(a_{j_1 j_1}1, \bar{x}_{j_2}) = \underbrace{(\text{LS, DU})}_{\text{down-up lower-saddle}}.$$

The infinite-equilibrium of $x_{j_1}^* = a_{j_1 j_1}1 = a_{j_2 j_1 s_1}$ is an (LS,DU)-down-up asymptotic lower-saddle.

(vi$_{2b}$) From case (vi$_{2a}$), for $\bar{x}_{j_2} = a_{j_1 j_2}1$, the equilibrium of $(x_{j_2}^*, x_{j_1}^*) = (a_{j_1 j_2}1, a_{j_2 j_1 s_1})$ with $a_{j_2 j_1 s_1} = a_{j_1 j_1}1$ has the following properties as in Eqs. (1.122)–(1.125):

- For $a_{j_1 j_1 0} > 0$ and $a_{j_2 j_2 0}\prod_{s_2=1, s_2 \neq s_1}^{3}(a_{j_2 j_1 s_1} - a_{j_2 j_1 s_2}) > 0$,

$$(a_{j_1 j_2}1, a_{j_2 j_1 s_1}) = \underbrace{(_{\text{UD}}\text{LS:}_{\text{DU}}\text{US, DP}_- : \text{UP}_+)}_{\text{hyperbolic lower-to-upper saddle}}.$$

The equilibrium of $(x_{j_2}^*, x_{j_1}^*) = (a_{j_1 j_2}1, a_{j_2 j_1 s_1})$ for $a_{j_1 j_1}1 = a_{j_2 j_1 s_1}$ is a $(_{\text{UD}}\text{LS: }_{\text{DU}}\text{US}$, DP:UP)-hyperbolic lower-to-upper saddle.

- For $a_{j_1 j_1 0} < 0$ and $a_{j_2 j_2 0}\prod_{s_2=1, s_2 \neq s_1}^{3}(a_{j_2 j_1 s_1} - a_{j_2 j_1 s_2}) > 0$,

$$(a_{j_1 j_2}1, a_{j_2 j_1 s_1}) = \underbrace{(_{\text{DU}}\text{US:}_{\text{UD}}\text{LS, UP}_- : \text{DP}_+)}_{\text{hyperbolic-secant upper-to-lower saddle}}.$$

The equilibrium of $(x_{j_2}^*, x_{j_1}^*) = (a_{j_1 j_2}1, a_{j_2 j_1 s_1})$ for $a_{j_1 j_1}1 = a_{j_2 j_1 s_1}$ is a $(_{\text{DU}}\text{US: }_{\text{UD}}\text{LS}$, UP:DP)-hyperbolic-secant upper-to-lower saddle.

- For $a_{j_1 j_1 0} > 0$ and $a_{j_2 j_2 0}\prod_{s_2=1, s_2 \neq s_1}^{3}(a_{j_2 j_1 s_1} - a_{j_2 j_1 s_2}) < 0$,

$$(a_{j_1j_21}, a_{j_2j_1s_1}) = \underbrace{(_{\text{DU}}\text{LS:}_{\text{UD}}\text{US}, \text{UP}_+ :\text{DP}_-)}_{\text{hyperbolic-secant lower-to-upper saddle}} .$$

The equilibrium of $(x^*_{j_2}, x^*_{j_1}) = (a_{j_1j_21}, a_{j_2j_1s_1})$ for $a_{j_1j_11} = a_{j_2j_1s_1}$ is a $(_{\text{DU}}\text{LS:}_{\text{UD}}\text{US}, \text{UP:DP})$-hyperbolic-secant lower-to-upper saddle.

- For $a_{j_1j_10} < 0$ and $a_{j_2j_20}\prod^3_{s_2 = 1, s_2 \neq s_1}(a_{j_2j_1s_1} - a_{j_2j_1s_2}) < 0$,

$$(a_{j_1j_21}, a_{j_2j_1s_1}) = \underbrace{(_{\text{UD}}\text{US:}_{\text{DU}}\text{LS}, \text{DP}_+ :\text{UP}_-)}_{\text{hyperbolic upper-to-lower saddle}}.$$

The equilibrium of $(x^*_{j_2}, x^*_{j_1}) = (a_{j_1j_21}, a_{j_2j_1s_1})$ for $a_{j_1j_11} = a_{j_2j_1s_1}$ is a $(_{\text{UD}}\text{US:}_{\text{DU}}\text{LS}, \text{DP:UP})$-hyperbolic upper-to-lower saddle.

(vii) For $\Delta_{j_1j_2} > 0$ and $\Delta_{j_2j_1} < 0$, the standard form is

$$\dot{x}_{j_2} = a_{j_2j_20}(x_{j_1} - a_{j_2j_11})\left[(x_{j_1} - a_{j_2j_1})^2 + b_{j_2j_1}\right],$$
$$\dot{x}_{j_1} = a_{j_1j_10}(x_{j_1} - a_{j_1j_11})(x_{j_1} - a_{j_1j_12})(x_{j_2} - a_{j_1j_21})$$

where

$$b_{j_1j_11}, b_{j_1j_12} = -\frac{1}{2}(B_{j_1j_1} \pm \sqrt{\Delta_{j_1j_1}}),$$
$$\{a_{j_1j_11}, a_{j_1j_11}\} = \text{sort}\{b_{j_1j_11}, b_{j_1j_12}\}, a_{j_1j_11} < a_{j_1j_12};$$
$$a_{j_2j_11} = b_{j_2j_11}, \ a_{j_2j_1} = -\frac{1}{2}B_{j_2j_1}, b_{j_2j_1} = \frac{1}{4}(-\Delta_{j_2j_1}).$$

In phase space,

$$\frac{dx_{j_1}}{dx_{j_2}} = \frac{a_{j_1j_10}}{a_{j_2j_20}}\frac{(x_{j_1} - a_{j_1j_11})(x_{j_1} - a_{j_1j_12})(x_{j_2} - a_{j_1j_21})}{(x_{j_1} - a_{j_2j_11})\left[(x_{j_1} - a_{j_2j_1})^2 + b_{j_2j_1}\right]},$$

and the deformation of the foregoing equation is

$$\begin{aligned}
&\left\{(x_{j_1} - a_{j_2j_11}) + (a_{j_1j_11} + a_{j_1j_12} - 2a_{j_2j_1})\right.\\
&+ \frac{(a_{j_1j_11} - a_{j_2j_11})\left[(a_{j_1j_11} - a_{j_2j_1})^2 + b_{j_2j_1}\right]}{(x_{j_1} - a_{j_1j_11})(a_{j_1j_11} - a_{j_1j_12})}\\
&+ \left.\frac{(a_{j_1j_12} - a_{j_2j_11})\left[(a_{j_1j_12} - a_{j_2j_1})^2 + b_{j_2j_1}\right]}{(x_{j_1} - a_{j_1j_12})(a_{j_1j_12} - a_{j_1j_11})}\right\}dx_{j_1}\\
&= \frac{a_{j_1j_10}}{a_{j_2j_20}}(x_{j_2} - a_{j_1j_21})dx_{j_2}.
\end{aligned}$$

With an initial condition of $(x_{j_1 0}, x_{j_2 0})$ at $t = t_0$, the integration of the above equation gives

$$\frac{1}{2}\left[(x_{j_1} - a_{j_2 j_1 1})^2 - (x_{j_1 0} - a_{j_2 j_1 1})^2\right] + (a_{j_1 j_1 1} + a_{j_1 j_1 2} - 2a_{j_2 j_1})(x_{j_1} - x_{j_1 0})$$

$$+ \frac{(a_{j_1 j_1 1} - a_{j_2 j_1 1})\left[(a_{j_1 j_1 1} - a_{j_2 j_1})^2 + b_{j_2 j_1}\right]}{(a_{j_1 j_1 1} - a_{j_1 j_1 2})} \ln \frac{|x_{j_1} - a_{j_1 j_1 1}|}{|x_{j_1 0} - a_{j_1 j_1 1}|}$$

$$+ \frac{(a_{j_1 j_1 2} - a_{j_2 j_1 1})\left[(a_{j_1 j_1 2} - a_{j_2 j_1})^2 + b_{j_2 j_1}\right]}{(a_{j_1 j_1 2} - a_{j_1 j_1 1})} \ln \frac{|x_{j_1} - a_{j_1 j_1 2}|}{|x_{j_1 0} - a_{j_1 j_1 2}|}$$

$$= \frac{1}{2}\frac{a_{j_1 j_1 0}}{a_{j_2 j_2 0}}\left[(x_{j_2} - a_{j_1 j_2 1})^2 - (x_{j_2 0} - a_{j_1 j_2 1})^2\right].$$

In phase space, at $x_{j_2}^* = a_{j_1 j_2 1}$ with $\bar{x}_{j_1} \neq a_{j_1 j_1 l_1}, a_{j_2 j_1 1}, (l_1 = 1, 2)$,

$$\frac{dx_{j_1}}{dx_{j_2}}\bigg|_{x_{j_2}^* = a_{j_1 j_2 1}} = \frac{a_{j_1 j_1 0}}{a_{j_2 j_2 0}} \frac{(x_{j_2} - a_{j_1 j_2 1})\prod_{l_1=1}^2(\bar{x}_{j_1} - a_{j_1 j_1 l_1})}{(\bar{x}_{j_1} - a_{j_2 j_1 1})\left[(\bar{x}_{j_1} - a_{j_2 j_1})^2 + b_{j_2 j_1}\right]}\bigg|_{x_{j_2}^* = a_{j_1 j_2 1}} = 0.$$

If

$$\frac{d^2 x_{j_1}}{dx_{j_2}^2}\bigg|_{x_{j_2}^* = a_{j_1 j_2 1}} = \frac{a_{j_1 j_1 0}}{a_{j_2 j_2 0}} \frac{\prod_{l_1=1}^2(\bar{x}_{j_1} - a_{j_1 j_1 l_1})}{(\bar{x}_{j_1} - a_{j_2 j_1 1})\left[(\bar{x}_{j_1} - a_{j_2 j_1})^2 + b_{j_2 j_1}\right]} > 0,$$

there is an up-parabola flow at $x_{j_2}^* = a_{j_1 j_2 1}$ with $\bar{x}_{j_1} \neq a_{j_1 j_1 1}$. If

$$\frac{d^2 x_{j_1}}{dx_{j_2}^2}\bigg|_{x_{j_2}^* = a_{j_1 j_2 s_1}} = \frac{a_{j_1 j_1 0}}{a_{j_2 j_2 0}} \frac{\prod_{l_1=1}^2(\bar{x}_{j_1} - a_{j_1 j_1 l_1})}{(\bar{x}_{j_1} - a_{j_2 j_1 1})\left[(\bar{x}_{j_1} - a_{j_2 j_1})^2 + b_{j_2 j_1}\right]} < 0,$$

there is a down-parabola flow at $x_{j_2}^* = a_{j_1 j_2 1}$ with $\bar{x}_{j_1} \neq a_{j_1 j_1 1}$. Let

$$\dot{x}_{j_2} = a_{j_2 j_2 0}(\bar{x}_{j_1} - a_{j_2 j_1 1})\left[(\bar{x}_{j_1} - a_{j_2 j_1})^2 + b_{j_2 j_1}\right],$$

with

$$(\bar{x}_{j_1} - a_{j_2 j_1})^2 + b_{j_2 j_1} > 0.$$

The parabola flows at $x_{j_2}^* = a_{j_1 j_2 1}$ with $\bar{x}_{j_1} \neq a_{j_1 j_1 1}$ are positive and negative for $a_{j_2 j_2 0}(\bar{x}_{j_1} - a_{j_2 j_1 1}) > 0$ and $a_{j_2 j_2 0}(\bar{x}_{j_1} - a_{j_2 j_1 1}) < 0$, respectively.

Thus, the equilibrium of $x_{j_2}^* = a_{j_1 j_2 1}$ with $\bar{x}_{j_1} \neq a_{j_1 j_1 1}$ has the following properties:

- For $a_{j_1 j_1 0} \prod_{l_1=1}^{2} (\bar{x}_{j_1} - a_{j_1 j_1 l_1}) > 0$ and $a_{j_2 j_2 0} (\bar{x}_{j_1} - a_{j_2 j_1 1}) > 0$,

$$(a_{j_1 j_2 1}, \dot{x}_{j_2}) = \underbrace{(\mathrm{UP}, \mathrm{pF})}_{\text{up-parabola flow }(+)} .$$

- For $a_{j_1 j_1 0} \prod_{l_1=1}^{2} (\bar{x}_{j_1} - a_{j_1 j_1 l_1}) > 0$ and $a_{j_2 j_2 0} (\bar{x}_{j_1} - a_{j_2 j_1 1}) > 0$,

$$(a_{j_1 j_2 1}, \dot{x}_{j_2}) = \underbrace{(\mathrm{DP}, \mathrm{pF})}_{\text{down-parabola flow }(+)} .$$

- For $a_{j_1 j_1 0} \prod_{l_1=1}^{2} (\bar{x}_{j_1} - a_{j_1 j_1 l_1}) > 0$ and $a_{j_2 j_2 0} (\bar{x}_{j_1} - a_{j_2 j_1 1}) < 0$,

$$(a_{j_1 j_2 1}, \dot{x}_{j_2}) = \underbrace{(\mathrm{DP}, \mathrm{nF})}_{\text{down-parabola flow }(-)} .$$

- For $a_{j_1 j_1 0} \prod_{l_1=1}^{2} (\bar{x}_{j_1} - a_{j_1 j_1 l_1}) > 0$ and $a_{j_2 j_2 0} (\bar{x}_{j_1} - a_{j_2 j_1 1}) < 0$,

$$(a_{j_1 j_2 1}, \dot{x}_{j_2}) = \underbrace{(\mathrm{UP}, \mathrm{nF})}_{\text{up-parabola flow }(-)} .$$

The variational equation at $x_{j_1}^* = a_{j_1 j_1 l_1}$ with $\bar{x}_{j_2} \neq a_{j_1 j_2 1}$ is

$$\Delta \dot{x}_{j_1} = a_{j_1 j_1 0} (\bar{x}_{j_2} - a_{j_1 j_2 1})(a_{j_1 j_2 l_1} - a_{j_1 j_2 l_2}) \Delta x_{j_1}.$$

The flows at $x_{j_1}^* = a_{j_1 j_1 1}$ with $\bar{x}_{j_2} \neq a_{j_1 j_2 1}$ are source and sink in the x_{j_1}-direction for $a_{j_1 j_1 0}(\bar{x}_{j_2} - a_{j_1 j_2 1})(a_{j_1 j_2 l_1} - a_{j_1 j_2 l_2}) > 0$ and $a_{j_1 j_1 0}(\bar{x}_{j_2} - a_{j_1 j_2 1})(a_{j_1 j_1 l_1} - a_{j_1 j_1 l_2}) < 0$, respectively. Let

$$\dot{x}_{j_2} = a_{j_2 j_2 0}(\bar{x}_{j_1} - a_{j_2 j_1 1}) \left[(a_{j_1 j_1 1} - a_{j_2 j_1})^2 + b_{j_2 j_1} \right].$$

The sink and source flows are positive and negative, respectively.

From the above analysis, the equilibrium of $(x_{j_1}^*, x_{j_2}^*) = (a_{j_1 j_1 l_1}, a_{j_1 j_2 1})$ ($l_1, l_2 \in \{1, 2\}$, $l_1 \neq l_2$) has the following properties as in Eqs. (1.129)–(1.132):

- For $a_{j_1 j_1 0}(a_{j_1 j_1 l_1} - a_{j_1 j_1 l_2}) > 0$ and $a_{j_2 j_2 0}(a_{j_1 j_1 l_1} - a_{j_2 j_1 1}) > 0$,

$$(a_{j_1 j_1 l_1}, a_{j_1 j_2 1}) = \underbrace{(\mathrm{DP : UP}, \mathrm{pF})}_{\text{hyperbolic flow }(+)} .$$

The equilibrium of $(x_{j_1}^*, x_{j_2}^*) = (a_{j_1 j_1 l_1}, a_{j_1 j_2 1})$ is a (DP:UP,pF)-positive hyperbolic flow.

- For $a_{j_1 j_1 0}(a_{j_1 j_1 l_1} - a_{j_1 j_1 l_2}) < 0$ and $a_{j_2 j_2 0}(a_{j_1 j_1 l_1} - a_{j_2 j_1 1}) > 0,$

$$(a_{j_1 j_1 l_1}, a_{j_1 j_2 1}) = \underbrace{(\text{UP}:\text{DP}, \text{pF})}_{\text{hyperbolic-secant flow }(+)}.$$

The equilibrium of $(x_{j_1}^*, x_{j_2}^*) = (a_{j_1 j_1 l_1}, a_{j_1 j_2 1})$ is a (UP:DP,pF)-positive hyperbolic-secant flow.

- For $a_{j_1 j_1 0}(a_{j_1 j_1 l_1} - a_{j_1 j_1 l_2}) > 0$ and $a_{j_2 j_2 0}(a_{j_1 j_1 l_1} - a_{j_2 j_1 1}) < 0,$

$$(a_{j_1 j_1 l_1}, a_{j_1 j_2 1}) = \underbrace{(\text{UP}:\text{DP}, \text{nF})}_{\text{hyperbolic-secant flow }(-)}.$$

The equilibrium of $(x_{j_1}^*, x_{j_2}^*) = (a_{j_1 j_1 l_1}, a_{j_1 j_2 1})$ is a (UP:DP,nF)-negative hyperbolic-secant flow.

- For $a_{j_1 j_1 0}(a_{j_1 j_1 l_1} - a_{j_1 j_1 l_2}) < 0$ and $a_{j_2 j_2 0}(a_{j_1 j_1 l_1} - a_{j_2 j_1 1}) < 0,$

$$(a_{j_1 j_1 l_1}, a_{j_1 j_2 1}) = \underbrace{(\text{DP}:\text{UP}, \text{nF})}_{\text{hyperbolic flow }(-)}.$$

The equilibrium of $(x_{j_1}^*, x_{j_2}^*) = (a_{j_1 j_1 l_1}, a_{j_1 j_2 1})$ is a (DP:UP,nF)-negative hyperbolic flow.

(vii$_{1b}$) Consider two cases: (I) $x_{j_1}^* = a_{j_2 j_1 1}$ and (II) $x_{j_2}^* = a_{j_1 j_2 1}$.

Case (I): At $x_{j_1}^* = a_{j_2 j_1 1}$ and $\bar{x}_{j_2} \neq a_{j_1 j_2 1}$,

$$\frac{dx_{j_2}}{dx_{j_1}}\bigg|_{x_{j_1}^* = a_{j_2 j_1 1}} = \frac{a_{j_2 j_2 0}}{a_{j_1 j_1 0}} \frac{(x_{j_1} - a_{j_2 j_1 1})\left[(x_{j_1} - a_{j_2 j_1})^2 + b_{j_2 j_1}\right]}{\prod_{l_1=1}^{2}(x_{j_1} - a_{j_1 j_1 l_1})(\bar{x}_{j_2} - a_{j_1 j_2 1})}\bigg|_{x_{j_1}^* = a_{j_2 j_1 1}} = 0.$$

If

$$\frac{d^2 x_{j_2}}{dx_{j_1}^2}\bigg|_{x_{j_1}^* = a_{j_2 j_1 1}} = \frac{a_{j_2 j_2 0}}{a_{j_1 j_1 0}} \frac{(a_{j_2 j_1 1} - a_{j_2 j_1})^2 + b_{j_2 j_1}}{\prod_{l_1=1}^{2}(a_{j_2 j_1 1} - a_{j_1 j_1 l_1})(\bar{x}_{j_2} - a_{j_1 j_2 1})} > 0,$$

there is an up-parabola flow at $x_{j_1}^* = a_{j_2 j_1 1}$. If

$$\frac{d^2 x_{j_2}}{dx_{j_1}^2}\bigg|_{x_{j_1}^* = a_{j_2 j_1 1}} = \frac{a_{j_2 j_2 0}}{a_{j_1 j_1 0}} \frac{(a_{j_2 j_1 1} - a_{j_2 j_1})^2 + b_{j_2 j_1}}{\prod_{l_1=1}^{2}(a_{j_2 j_1 1} - a_{j_1 j_1 l_1})(\bar{x}_{j_2} - a_{j_1 j_2 1})} < 0,$$

there is a down-parabola flow at $x_{j_1}^* = a_{j_2 j_1 1}$. Let

$$\dot{x}_{j_1} = a_{j_1 j_1 0} \prod_{l_1=1}^{2}(a_{j_2 j_1 1} - a_{j_1 j_1 l_1})(\bar{x}_{j_2} - a_{j_1 j_2 1}).$$

Because of

$$(a_{j_2 j_1 1} - a_{j_2 j_1})^2 + b_{j_2 j_1} > 0,$$

the parabola flows at $x_{j_1}^* = a_{j_2 j_1 1}$ are positive and negative for $\dot{x}_{j_1} > 0$ and $\dot{x}_{j_1} < 0$. Thus, the flows at $x_{j_1}^* = a_{j_2 j_1 1}$ with $\bar{x}_{j_2} \neq a_{j_1 j_2 l_1}, a_{j_1 j_2 l_2}$, we have

- For $a_{j_1 j_1 0} \prod_{l_1=1}^{2}(a_{j_2 j_1 1} - a_{j_1 j_1 l_1})(\bar{x}_{j_2} - a_{j_1 j_2 1}) > 0$ and $a_{j_2 j_2 0} > 0$,

$$(\dot{x}_{j_1}, a_{j_2 j_1 1}) = \underbrace{(\text{pF}, \text{UP})}_{\text{up-parabola flow }(+)}.$$

- For $a_{j_1 j_1 0} \prod_{l_1=1}^{2}(a_{j_2 j_1 1} - a_{j_1 j_1 l_1})(\bar{x}_{j_2} - a_{j_1 j_2 1}) > 0 < 0$ and $a_{j_2 j_2 0} > 0$,

$$(\dot{x}_{j_1}, a_{j_2 j_1 1}) = \underbrace{(\text{nF}, \text{DP})}_{\text{down-parabola flow }(-)}.$$

- For $a_{j_1 j_1 0} \prod_{l_1=1}^{2}(a_{j_2 j_1 1} - a_{j_1 j_1 l_1})(\bar{x}_{j_2} - a_{j_1 j_2 1}) > 0$ and $a_{j_2 j_2 0} < 0$,

$$(\dot{x}_{j_1}, a_{j_2 j_1 1}) = \underbrace{(\text{pF}, \text{DP})}_{\text{down-parabola flow }(+)}.$$

- For $a_{j_1 j_1 0} \prod_{l_1=1}^{2}(a_{j_2 j_1 1} - a_{j_1 j_1 l_1})(\bar{x}_{j_2} - a_{j_1 j_2 1}) < 0$ and $a_{j_2 j_2 0} < 0$,

$$(\dot{x}_{j_1}, a_{j_2 j_1 1}) = \underbrace{(\text{nF}, \text{UP})}_{\text{up-parabola flow }(-)}.$$

Case (II): At $x_{j_2}^* = a_{j_1 j_2 1}$ and $\bar{x}_{j_1} \neq a_{j_2 j_1 1}, a_{j_1 j_1 1}, a_{j_1 j_1 2}$,

$$\frac{dx_{j_1}}{dx_{j_2}}\bigg|_{x_{j_2}^*=a_{j_1 j_2 1}} = \frac{a_{j_1 j_1 0}}{a_{j_2 j_2 0}} \frac{\prod_{l_1=1}^{2}(\bar{x}_{j_1} - a_{j_1 j_1 l_1})(x_{j_2} - a_{j_1 j_2 1})}{(\bar{x}_{j_1} - a_{j_2 j_1 1})[(\bar{x}_{j_1} - a_{j_2 j_1})^2 + b_{j_2 j_1}]}\bigg|_{x_{j_2}^*=a_{j_1 j_2 1}} = 0.$$

If

$$\frac{d^2 x_{j_1}}{dx_{j_2}^2}\bigg|_{x_{j_2}^*=a_{j_1 j_2 1}} = \frac{a_{j_1 j_1 0}}{a_{j_2 j_2 0}} \frac{\prod_{l_1=1}^{2}(\bar{x}_{j_1} - a_{j_1 j_1 l_1})}{(\bar{x}_{j_1} - a_{j_2 j_1 1})[(\bar{x}_{j_1} - a_{j_2 j_1})^2 + b_{j_2 j_1}]} > 0,$$

there is an up-parabola flow at $x_{j_2}^* = a_{j_1 j_2 1}$. If

$$\frac{d^2 x_{j_1}}{dx_{j_2}^2}\bigg|_{x_{j_2}^* = a_{j_1 j_2 1}} = \frac{a_{j_1 j_1 0}}{a_{j_2 j_2 0}} \frac{\prod_{l_1=1}^2 (\bar{x}_{j_1} - a_{j_1 j_1 l_1})}{(\bar{x}_{j_1} - a_{j_2 j_1 1})[(\bar{x}_{j_1} - a_{j_2 j_1})^2 + b_{j_2 j_1}]} < 0,$$

there is a down-parabola flow at $x_{j_2}^* = a_{j_1 j_2 1}$. Let

$$\dot{x}_{j_2} = a_{j_2 j_2 0}(\bar{x}_{j_1} - a_{j_2 j_1 1})[(\bar{x}_{j_1} - a_{j_2 j_1})^2 + b_{j_2 j_1}].$$

Because of

$$(\bar{x}_{j_1} - a_{j_2 j_1})^2 + b_{j_2 j_1} > 0,$$

the parabola flows at $x_{j_1}^* = a_{j_2 j_1 1}$ are positive and negative for $a_{j_2 j_2 0}(\bar{x}_{j_1} - a_{j_2 j_1 1}) > 0$ and $a_{j_2 j_2 0}(\bar{x}_{j_1} - a_{j_2 j_1 1}) < 0$.

Thus, the equilibrium of $x_{j_2}^* = a_{j_1 j_2 1}$ has the following properties:

- For $a_{j_1 j_1 0} \prod_{l_1=1}^2 (\bar{x}_{j_1} - a_{j_1 j_1 l_1}) > 0$ and $a_{j_2 j_2 0}(\bar{x}_{j_1} - a_{j_2 j_1 1}) > 0$,

$$(a_{j_1 j_2 1}, \dot{x}_{j_2}) = \underbrace{(\text{UP, pF})}_{\text{up-parabola flow } (+)}.$$

- For $a_{j_1 j_1 0} \prod_{l_1=1}^2 (\bar{x}_{j_1} - a_{j_1 j_1 l_1}) < 0$ and $a_{j_2 j_2 0}(\bar{x}_{j_1} - a_{j_2 j_1 1}) > 0$,

$$(a_{j_1 j_2 1}, \dot{x}_{j_2}) = \underbrace{(\text{DP, pF})}_{\text{down-parabola flow } (+)}.$$

- For $a_{j_1 j_1 0} \prod_{l_1=1}^2 (\bar{x}_{j_1} - a_{j_1 j_1 l_1}) > 0$ and $a_{j_2 j_2 0}(\bar{x}_{j_1} - a_{j_2 j_1 1}) < 0$,

$$(a_{j_1 j_2 1}, \dot{x}_{j_2}) = \underbrace{(\text{DP, nF})}_{\text{down-parabola flow } (-)}.$$

- For $a_{j_1 j_1 0} \prod_{l_1=1}^2 (\bar{x}_{j_1} - a_{j_1 j_1 l_1}) < 0$ and $a_{j_2 j_2 0}(\bar{x}_{j_1} - a_{j_2 j_1 1}) < 0$,

$$(a_{j_1 j_2 1}, \dot{x}_{j_2}) = \underbrace{(\text{UP, nF})}_{\text{up-parabola flow } (-)}.$$

Therefore, from case (I) for $\bar{x}_{j_2} = a_{j_1 j_2 1}$ and case (II) for $\bar{x}_{j_1} = a_{j_2 j_1 1}$, the equilibrium of $(x_{j_2}^*, x_{j_1}^*) = (a_{j_1 j_2 1}, a_{j_2 j_1 1})$ has the following properties as in Eqs. (1.133)–(1.136):

- For $a_{j_1 j_1 0} \prod_{l_1=1}^2 (a_{j_2 j_1 1} - a_{j_1 j_1 l_1}) > 0$ and $a_{j_2 j_2 0} > 0$,

$$(a_{j_1 j_2 1}, a_{j_2 j_1 1}) = \underbrace{(\text{UP}_+, \text{UP}_+)}_{\text{positive saddle}}.$$

The equilibrium of $(x^*_{j_2}, x^*_{j_1}) = (a_{j_1 j_2 1}, a_{j_2 j_1 1})$ is a $(\text{UP}_+, \text{UP}_+)$-positive saddle.

- For $a_{j_1 j_1 0} \prod^2_{l_1 = 1} (a_{j_2 j_1 1} - a_{j_1 j_1 l_1}) < 0$ and $a_{j_2 j_2 0} > 0$,

$$(a_{j_1 j_2 1}, a_{j_2 j_1 1}) = \underbrace{(\text{DP}_+, \text{DP}_-)}_{\text{CCW center}}.$$

The equilibrium of $(x^*_{j_2}, x^*_{j_1}) = (a_{j_1 j_2 1}, a_{j_2 j_1 1})$ is a $(\text{DP}_+, \text{DP}_-)$-counter-clockwise center.

- For $a_{j_1 j_1 0} \prod^2_{l_1 = 1} (a_{j_2 j_1 1} - a_{j_1 j_1 l_1}) > 0$ and $a_{j_2 j_2 0} < 0$,

$$(a_{j_1 j_2 1}, a_{j_2 j_1 1}) = \underbrace{(\text{DP}_-, \text{DP}_+)}_{\text{CW center}}.$$

The equilibrium of $(x^*_{j_2}, x^*_{j_1}) = (a_{j_1 j_2 1}, a_{j_2 j_1 1})$ is a $(\text{DP}_-, \text{DP}_+)$-clockwise center.

- For $a_{j_1 j_1 0} \prod^2_{l_1 = 1} (a_{j_2 j_1 1} - a_{j_1 j_1 l_1}) < 0$ and $a_{j_2 j_2 0} < 0$,

$$(a_{j_1 j_2 1}, a_{j_2 j_1 1}) = \underbrace{(\text{UP}_-, \text{UP}_-)}_{\text{negative saddle}}.$$

The equilibrium of $(x^*_{j_2}, x^*_{j_1}) = (a_{j_1 j_2 1}, a_{j_2 j_1 1})$ is a $(\text{UP}_-, \text{UP}_-)$-negative saddle.

(vii$_2$) For $a_{j_2 j_1 1} = a_{j_1 j_1 l_1}$, in phase space,

$$\frac{dx_{j_1}}{dx_{j_2}} = \frac{a_{j_1 j_1 0}}{a_{j_2 j_2 0}} \frac{(x_{j_1} - a_{j_1 j_1 l_2})(x_{j_2} - a_{j_1 j_2 1})}{(x_{j_1} - a_{j_2 j_1})^2 + b_{j_2 j_1}},$$

and the deformation of the foregoing equation is

$$\left[(x_{j_1} - a_{j_1 j_1 l_2}) + 2(a_{j_1 j_1 l_2} - a_{j_2 j_1}) + \frac{(a_{j_1 j_1 l_2} - a_{j_2 j_1})^2 + b_{j_2 j_1}}{x_{j_1} - a_{j_1 j_1 l_2}} \right] dx_{j_1}$$
$$= \frac{a_{j_1 j_1 0}}{a_{j_2 j_2 0}} (x_{j_2} - a_{j_1 j_2 1}) dx_{j_2}.$$

With an initial condition of $(x_{j_1 0}, x_{j_2 0})$ at $t = t_0$, the integration of the above equation gives the first integral manifold for $a_{j_1 j_1 1} \neq a_{j_2 j_1 1}$ as

$$\frac{1}{2}\left[(x_{j_1} - a_{j_1 j_1 l_2})^2 - (x_{j_1 0} - a_{j_1 j_1 l_2})^2\right] + 2(a_{j_1 j_1 l_2} - a_{j_2 j_1})(x_{j_1} - x_{j_1 0})$$

$$+ \left[(a_{j_1 j_1 l_2} - a_{j_2 j_1})^2 + b_{j_2 j_1}\right] \ln \frac{|x_{j_1} - a_{j_1 j_1 l_2}|}{|x_{j_1 0} - a_{j_1 j_1 l_2}|}$$

$$= \frac{1}{2}\frac{a_{j_1 j_1 0}}{a_{j_2 j_2 0}}\left[(x_{j_2} - a_{j_1 j_2 1})^2 - (x_{j_2 0} - a_{j_1 j_2 1})^2\right].$$

(vii$_{2a}$) Similar to case (iv$_{2a}$), in phase space, at $x_{j_1}^* = a_{j_2 j_1 1} = a_{j_1 j_1 1}$ with $\bar{x}_{j_2} \neq a_{j_1 j_2 1}$, if

$$\frac{dx_{j_2}}{dx_{j_1}}\bigg|_{x_{j_1}^* = a_{j_2 j_1 1}} = \frac{a_{j_2 j_2 0}}{a_{j_1 j_1 0}}\frac{(a_{j_2 j_1 1} - a_{j_2 j_1})^2 + b_{j_2 j_1}}{(a_{j_2 j_1 1} - a_{j_1 j_1 l_2})(\bar{x}_{j_2} - a_{j_1 j_2 1})} > 0,$$

there is an increasing-inflection flow. If

$$\frac{dx_{j_2}}{dx_{j_1}}\bigg|_{x_{j_1}^* = a_{j_2 j_1 1}} = \frac{a_{j_2 j_2 0}}{a_{j_1 j_1 0}}\frac{(a_{j_2 j_1 1} - a_{j_2 j_1})^2 + b_{j_2 j_1}}{(a_{j_2 j_1 1} - a_{j_1 j_1 l_2})(\bar{x}_{j_2} - a_{j_1 j_2 1})} < 0,$$

there is a decreasing-inflection flow at $x_{j_1}^* = a_{j_2 j_1 1} = a_{j_1 j_1 1}$ with $\bar{x}_{j_2} \neq a_{j_1 j_2 1}$. The variational equation is

$$\Delta \dot{x}_{j_1} = a_{j_1 j_1 0}(a_{j_2 j_1 1} - a_{j_1 j_1 l_2})(\bar{x}_{j_2} - a_{j_1 j_2 1})\Delta x_{j_1},$$

with

$$(a_{j_2 j_1 1} - a_{j_2 j_1})^2 + b_{j_2 j_1} > 0.$$

The inflection flows at $x_{j_1}^* = a_{j_2 j_1 1} = a_{j_1 j_1 1}$ with $\bar{x}_{j_2} \neq a_{j_1 j_2 1}$ are source and sink, for $a_{j_1 j_1 0}$ $(a_{j_2 j_1 1} - a_{j_1 j_1 l_2})(\bar{x}_{j_2} - a_{j_1 j_2 1}) > 0$ and $a_{j_1 j_1 0}(a_{j_2 j_1 1} - a_{j_1 j_1 l_2})(\bar{x}_{j_2} - a_{j_1 j_2 1}) < 0$, respectively.

Therefore, the infinite-equilibrium of $x_{j_1}^* = a_{j_2 j_1 1} = a_{j_1 j_1 l_1}$ with $\bar{x}_{j_2} \neq a_{j_1 j_2 1}$ has the following properties as in Eqs. (1.138)–(1.141):

- For $a_{j_1 j_1 0}(a_{j_1 j_1 l_1} - a_{j_1 j_1 l_2})(\bar{x}_{j_2} - a_{j_1 j_2 1}) > 0$ and $a_{j_2 j_2 0} > 0$,

$$(a_{j_1 j_1 l_1}, \bar{x}_{j_2}) = \underbrace{(\text{SO, II})}_{\text{increasing-inflection source}}.$$

The infinite-equilibrium of $x_{j_1}^* = a_{j_2 j_1 1} = a_{j_1 j_1 l_1}$ is an (SO,II)-increasing-inflection source.

- For $a_{j_1j_10}(a_{j_1j_1l_1} - a_{j_1j_1l_2})(\bar{x}_{j_2} - a_{j_1j_21}) < 0$ and $a_{j_2j_20} > 0$,

$$(a_{j_1j_1l_1}, \bar{x}_{j_2}) = \underbrace{(\text{SI, DI})}_{\text{decreasing-inflection sink}}.$$

The infinite-equilibrium of $x^*_{j_1} = a_{j_2j_11} = a_{j_1j_1l_1}$ is an (SI,DI)-decreasing-inflection sink.

- For $a_{j_1j_10}(a_{j_1j_1l_1} - a_{j_1j_1l_2})(\bar{x}_{j_2} - a_{j_1j_21}) > 0$ and $a_{j_2j_20} < 0$,

$$(a_{j_1j_1l_1}, \bar{x}_{j_2}) = \underbrace{(\text{SO, DI})}_{\text{decreasing-inflection source}}.$$

The infinite-equilibrium of $x^*_{j_1} = a_{j_2j_11} = a_{j_1j_1l_1}$ is an (SO,DI)-decreasing-inflection source.

- For $a_{j_1j_10}(a_{j_1j_1l_1} - a_{j_1j_1l_2})(\bar{x}_{j_2} - a_{j_1j_21}) < 0$ and $a_{j_2j_20} < 0$,

$$(a_{j_1j_1l_1}, \bar{x}_{j_2}) = \underbrace{(\text{SI, II})}_{\text{increasing-inflection sink}}.$$

The infinite-equilibrium of $x^*_{j_1} = a_{j_2j_11} = a_{j_1j_1l_1}$ is an (SI,II)-increasing-inflection sink.

(vii$_{2b}$) For $x^*_{j_2} = a_{j_1j_21}$ and $\bar{x}_{j_1} \neq a_{j_2j_11} = a_{j_1j_1l_1}, a_{j_1j_1l_2}$, in phase space,

$$\frac{dx_{j_1}}{dx_{j_2}}\Big|_{x^*_{j_2} = a_{j_1j_21}} = \frac{a_{j_1j_10}}{a_{j_2j_20}} \frac{(\bar{x}_{j_1} - a_{j_1j_1l_2})(x_{j_2} - a_{j_1j_21})}{(\bar{x}_{j_1} - a_{j_2j_1})^2 + b_{j_2j_1}}\Big|_{x^*_{j_2} = a_{j_1j_21}} = 0.$$

If

$$\frac{d^2x_{j_1}}{dx^2_{j_2}}\Big|_{x^*_{j_2} = a_{j_1j_21}} = \frac{a_{j_1j_10}}{a_{j_2j_20}} \frac{(\bar{x}_{j_1} - a_{j_1j_1l_2})}{(\bar{x}_{j_1} - a_{j_2j_1})^2 + b_{j_2j_1}} > 0,$$

there is an up-parabola flow. If

$$\frac{d^2x_{j_1}}{dx^2_{j_2}}\Big|_{x^*_{j_2} = a_{j_1j_21}} = \frac{a_{j_1j_10}}{a_{j_2j_20}} \frac{(\bar{x}_{j_1} - a_{j_1j_1l_2})}{(\bar{x}_{j_1} - a_{j_2j_1})^2 + b_{j_2j_1}} < 0,$$

there is a down-parabola flow at $x^*_{j_2} = a_{j_1j_21}$ and $\bar{x}_{j_1} \neq a_{j_2j_11} = a_{j_1j_1l_1}, a_{j_1j_1l_2}$. The variational equation is

$$\dot{x}_{j_2} = a_{j_2 j_2 0}(\bar{x}_{j_1} - a_{j_2 j_1 1})\left[(\bar{x}_{j_1} - a_{j_2 j_1})^2 + b_{j_2 j_1}\right],$$

with

$$(a_{j_2 j_1 1} - a_{j_2 j_1})^2 + b_{j_2 j_1} > 0.$$

The parabola flows at $x_{j_2}^* = a_{j_1 j_2 1}$ and $\bar{x}_{j_1} \neq a_{j_2 j_1 1} = a_{j_1 j_1 l_1}$ are positive and negative for $a_{j_2 j_2 0}(\bar{x}_{j_1} - a_{j_2 j_1 1}) > 0$ and $a_{j_2 j_2 0}(\bar{x}_{j_1} - a_{j_2 j_1 1}) < 0$, respectively.

From case (vii$_{2a}$) with $\bar{x}_{j_2} = a_{j_1 j_2 1}$ and the above analysis, the equilibrium of $(x_{j_2}^*, x_{j_1}^*) = (a_{j_1 j_2 1}, a_{j_2 j_1 1})$ with $a_{j_2 j_1 1} = a_{j_1 j_1 1}$ has the following properties as in Eqs. (1.142)–(1.145):

- For $a_{j_1 j_1 0}(a_{j_1 j_1 l_1} - a_{j_1 j_1 l_2}) > 0$ and $a_{j_2 j_2 0} > 0$,

$$(a_{j_1 j_2 1}, a_{j_2 j_1 1}) = \underbrace{(_{\text{DI:II}}\text{UP}, _{\text{SI:SO}}\text{US})}_{\text{up-parabola upper-saddle}}.$$

The equilibrium of $(x_{j_2}^*, x_{j_1}^*) = (a_{j_1 j_2 1}, a_{j_2 j_1 1})$ is a ($_{\text{DI:II}}$UP,$_{\text{SI:SO}}$US)-up-parabola upper-saddle.
- For $a_{j_1 j_1 0}(a_{j_1 j_1 l_1} - a_{j_1 j_1 l_2}) < 0$ and $a_{j_2 j_2 0} > 0$,

$$(a_{j_1 j_2 1}, a_{j_2 j_1 1}) = \underbrace{(_{\text{II:DI}}\text{DP}, _{\text{SO:SI}}\text{LS})}_{\text{down-parabola lower-saddle}}.$$

The equilibrium of $(x_{j_2}^*, x_{j_1}^*) = (a_{j_1 j_2 1}, a_{j_2 j_1 1})$ is a ($_{\text{II:DI}}$DP,$_{\text{SO:SI}}$LS)-down-parabola lower-saddle.
- For $a_{j_1 j_1 0}(a_{j_1 j_1 l_1} - a_{j_1 j_1 l_2}) > 0$ and $a_{j_2 j_2 0} < 0$,

$$(a_{j_1 j_2 1}, a_{j_2 j_1 1}) = \underbrace{(_{\text{II:DI}}\text{DP}, _{\text{SI:SO}}\text{US})}_{\text{down-parabola upper-saddle}}.$$

The equilibrium of $(x_{j_2}^*, x_{j_1}^*) = (a_{j_1 j_2 1}, a_{j_2 j_1 1})$ is a ($_{\text{II:DI}}$DP,$_{\text{SI:SO}}$US)-down-parabola upper-saddle.
- For $a_{j_1 j_1 0}(a_{j_1 j_1 l_1} - a_{j_1 j_1 l_2}) < 0$ and $a_{j_2 j_2 0} < 0$,

$$(a_{j_1 j_2 1}, a_{j_2 j_1 1}) = \underbrace{(_{\text{DI:II}}\text{UP}, _{\text{SO:SI}}\text{LS})}_{\text{up-parabola lower-saddle}}.$$

The equilibrium of $(x_{j_2}^*, x_{j_1}^*) = (a_{j_1 j_2 1}, a_{j_2 j_1 1})$ is a ($_{\text{DI:II}}$UP,$_{\text{SO:SI}}$LS)-up-parabola lower-saddle.

(viii) For $\Delta_{j_1 j_2} > 0$ and $\Delta_{j_2 j_1} = 0$, the standard form is

$$\dot{x}_{j_2} = a_{j_2 j_2 0}(x_{j_1} - a_{j_2 j_1 s_1})(x_{j_1} - a_{j_2 j_1 s_2})^2,$$

$$\dot{x}_{j_1} = a_{j_1 j_1 0}(x_{j_1} - a_{j_1 j_1 1})(x_{j_1} - a_{j_1 j_1 2})(x_{j_2} - a_{j_1 j_2 1})$$

where

$$b_{j_1 j_1 1}, b_{j_1 j_1 2} = -\frac{1}{2}(B_{j_1 j_1} \pm \sqrt{\Delta_{j_1 j_1}}),$$

$$\{a_{j_1 j_1 1}, a_{j_1 j_1 2}\} = \text{sort}\{b_{j_1 j_1 1}, b_{j_1 j_1 2}\}, a_{j_1 j_1 1} < a_{j_1 j_1 2};$$

$$a_{j_2 j_1 s_2} = b_{j_2 j_1 1}, a_{j_2 j_1 s_2} = -\frac{1}{2}B_{j_2 j_1}.$$

(viii$_1$) For $a_{j_1 j_1 1}, a_{j_1 j_1 2} \neq a_{j_2 j_1 s_1}, a_{j_2 j_1 s_2}$, in phase space,

$$\frac{dx_{j_1}}{dx_{j_2}} = \frac{a_{j_1 j_1 0}}{a_{j_2 j_2 0}} \frac{(x_{j_1} - a_{j_1 j_1 l_1})(x_{j_1} - a_{j_1 j_1 l_2})(x_{j_2} - a_{j_1 j_2 1})}{(x_{j_1} - a_{j_2 j_1 s_1})(x_{j_1} - a_{j_2 j_1 s_2})^2},$$

and the deformation of the foregoing equation is

$$\left[(x_{j_1} - a_{j_2 j_1 s_1}) + (a_{j_1 j_1 1} + a_{j_1 j_1 2} - 2a_{j_2 j_1 s_2})\right.$$
$$\left. + \frac{(a_{j_1 j_1 1} - a_{j_2 j_1 s_1})(a_{j_1 j_1 1} - a_{j_2 j_1 s_2})^2}{(x_{j_1} - a_{j_1 j_1 1})(a_{j_1 j_1 1} - a_{j_1 j_1 2})} + \frac{(a_{j_1 j_1 2} - a_{j_2 j_1 s_1})(a_{j_1 j_1 2} - a_{j_2 j_1 s_2})^2}{(x_{j_1} - a_{j_1 j_1 2})(a_{j_1 j_1 2} - a_{j_1 j_1 1})}\right] dx_{j_1}$$
$$= \frac{a_{j_1 j_1 0}}{a_{j_2 j_2 0}}(x_{j_2} - a_{j_1 j_2 1})dx_{j_2}.$$

With an initial condition of $(x_{j_1 0}, x_{j_2 0})$ at $t = t_0$, the integration of the above equation gives

$$\frac{1}{2}\left[(x_{j_1} - a_{j_2 j_1 s_1})^2 - (x_{j_1 0} - a_{j_2 j_1 s_1})^2\right] + (a_{j_1 j_1 1} + a_{j_1 j_1 2} - 2a_{j_2 j_1 s_2})(x_{j_1} - x_{j_1 0})$$

$$+ \frac{(a_{j_1 j_1 1} - a_{j_2 j_1 s_1})(a_{j_1 j_1 1} - a_{j_2 j_1 s_2})^2}{(a_{j_1 j_1 1} - a_{j_1 j_1 2})} \ln \frac{|x_{j_1} - a_{j_1 j_1 1}|}{|x_{j_1 0} - a_{j_1 j_1 1}|}$$

$$+ \frac{(a_{j_1 j_1 2} - a_{j_2 j_1 s_1})(a_{j_1 j_1 2} - a_{j_2 j_1 s_2})^2}{(a_{j_1 j_1 2} - a_{j_1 j_1 1})} \ln \frac{|x_{j_1} - a_{j_1 j_1 2}|}{|x_{j_1 0} - a_{j_1 j_1 2}|}$$

$$= \frac{1}{2}\frac{a_{j_1 j_1 0}}{a_{j_2 j_2 0}}\left[(x_{j_2} - a_{j_1 j_2 1})^2 - (x_{j_2 0} - a_{j_1 j_2 1})^2\right].$$

(viii$_{1a}$) Consider two cases: (I) $x_{j_1}^* = a_{j_2 j_1 s_1}$ and (II)$x_{j_2}^* = a_{j_1 j_2 1}$.

Case (I): At $x_{j_1}^* = a_{j_2 j_1 s_1}$ ($s_1 \in \{1, 2\}$ and $\bar{x}_{j_2} \neq a_{j_1 j_2 1}$

$$\frac{dx_{j_2}}{dx_{j_1}}\Big|_{x_{j_1}^* = a_{j_2 j_1 s_1}} = \frac{a_{j_2 j_2 0}}{a_{j_1 j_1 0}} \frac{(x_{j_1} - a_{j_2 j_1 s_1})(x_{j_1} - a_{j_2 j_1 s_2})^2}{(\overline{x}_{j_2} - a_{j_1 j_2 1})\prod_{l_1 = 1}^{2}(x_{j_1} - a_{j_1 j_1 l_1})}\Big|_{x_{j_1}^* = a_{j_2 j_1 s_1}} = 0.$$

If

$$\frac{d^2 x_{j_2}}{dx_{j_1}^2}\Big|_{x_{j_1}^* = a_{j_2 j_1 s_1}} = \frac{a_{j_2 j_2 0}}{a_{j_1 j_1 0}} \frac{(a_{j_2 j_1 s_1} - a_{j_2 j_1 s_2})^2}{(\overline{x}_{j_2} - a_{j_1 j_2 1})\prod_{l_1 = 1}^{2}(a_{j_2 j_1 s_1} - a_{j_1 j_1 l_1})} > 0,$$

there is an up-parabola flow at $x_{j_1}^* = a_{j_2 j_1 s_1}$. If

$$\frac{d^2 x_{j_2}}{dx_{j_1}^2}\Big|_{x_{j_1}^* = a_{j_2 j_1 s_1}} = \frac{a_{j_2 j_2 0}}{a_{j_1 j_1 0}} \frac{(a_{j_2 j_1 s_1} - a_{j_2 j_1 s_2})^2}{(\overline{x}_{j_2} - a_{j_1 j_2 1})\prod_{l_1 = 1}^{2}(a_{j_2 j_1 s_1} - a_{j_1 j_1 l_1})} < 0,$$

there is a down-parabola flow at $x_{j_1}^* = a_{j_2 j_1 s_1}$. Let

$$\dot{x}_{j_1} = a_{j_1 j_1 0}(\overline{x}_{j_2} - a_{j_1 j_2 1})\prod_{l_1 = 1}^{2}(a_{j_2 j_1 s_1} - a_{j_1 j_1 l_1}).$$

The parabola flows at $x_{j_1}^* = a_{j_2 j_1 s_1}$ are positive and negative for $\dot{x}_{j_1} > 0$ and $\dot{x}_{j_1} < 0$.
Thus, the flows at $x_{j_1}^* = a_{j_2 j_1 s_1}$ with $\overline{x}_{j_2} \neq a_{j_1 j_2 1}$ have the following properties:

- For $a_{j_1 j_1 0}(\overline{x}_{j_2} - a_{j_1 j_2 1})\prod_{l_1 = 1}^{2}(a_{j_2 j_1 s_1} - a_{j_1 j_1 l_1}) > 0$ and $a_{j_2 j_2 0}(a_{j_2 j_1 s_1} - a_{j_2 j_1 s_2})^2 > 0$,

$$(\dot{x}_{j_1}, a_{j_2 j_1 s_1}) = \underbrace{(\text{pF}, \text{UP})}_{\text{up-parabola flow } (+)}.$$

- For $a_{j_1 j_1 0}(\overline{x}_{j_2} - a_{j_1 j_2 1})\prod_{l_1 = 1}^{2}(a_{j_2 j_1 s_1} - a_{j_1 j_1 l_1}) < 0$ and $a_{j_2 j_2 0}(a_{j_2 j_1 s_1} - a_{j_2 j_1 s_2})^2 > 0$,

$$(\dot{x}_{j_1}, a_{j_2 j_1 s_1}) = \underbrace{(\text{nF}, \text{DP})}_{\text{down-parabola flow } (-)}.$$

- For $a_{j_1 j_1 0}(\overline{x}_{j_2} - a_{j_1 j_2 1})\prod_{l_1 = 1}^{2}(a_{j_2 j_1 s_1} - a_{j_1 j_1 l_1}) > 0$ and $a_{j_2 j_2 0}(a_{j_2 j_1 s_1} - a_{j_2 j_1 s_2})^2 < 0$,

$$(\dot{x}_{j_1}, a_{j_2 j_1 s_1}) = \underbrace{(\text{pF}, \text{DP})}_{\text{down-parabola flow } (+)}.$$

- For $a_{j_1 j_1 0}(\overline{x}_{j_2} - a_{j_1 j_2 1})\prod_{l_1 = 1}^{2}(a_{j_2 j_1 s_1} - a_{j_1 j_1 l_1}) < 0$ and $a_{j_2 j_2 0}(a_{j_2 j_1 s_1} - a_{j_2 j_1 s_2})^2 < 0$,

$$(\dot{x}_{j_1}, a_{j_2 j_1 s_1}) = \underbrace{(\text{nF}, \text{UP})}_{\text{up-parabola flow } (-)}.$$

Case (II): At $x_{j_2}^* = a_{j_1 j_2 1}$ and $\bar{x}_{j_1} \neq a_{j_1 j_1 l_1}, a_{j_2 j_1 s_1}, a_{j_2 j_1 s_2}$ $(l_1 = 1, 2)$,

$$\frac{dx_{j_1}}{dx_{j_2}}\bigg|_{x_{j_2}^* = a_{j_1 j_2 1}} = \frac{a_{j_1 j_1 0}}{a_{j_2 j_2 0}} \frac{(x_{j_2} - a_{j_1 j_2 1}) \prod_{l_1 = 1}^2 (\bar{x}_{j_1} - a_{j_1 j_1 l_1})}{(\bar{x}_{j_1} - a_{j_2 j_1 s_1})(\bar{x}_{j_1} - a_{j_2 j_1 s_2})^2}\bigg|_{x_{j_2}^* = a_{j_1 j_2 1}} = 0.$$

If

$$\frac{d^2 x_{j_1}}{dx_{j_2}^2}\bigg|_{x_{j_2}^* = a_{j_1 j_2 1}} = \frac{a_{j_1 j_1 0}}{a_{j_2 j_2 0}} \frac{\prod_{l_1 = 1}^2 (\bar{x}_{j_1} - a_{j_1 j_1 l_1})}{(\bar{x}_{j_1} - a_{j_2 j_1 s_1})(\bar{x}_{j_1} - a_{j_2 j_1 s_2})^2} > 0,$$

there is an up-parabola flow at $x_{j_2}^* = a_{j_1 j_2 1}$. If

$$\frac{d^2 x_{j_1}}{dx_{j_2}^2}\bigg|_{x_{j_2}^* = a_{j_1 j_2 1}} = \frac{a_{j_1 j_1 0}}{a_{j_2 j_2 0}} \frac{\prod_{l_1 = 1}^2 (\bar{x}_{j_1} - a_{j_1 j_1 l_1})}{(\bar{x}_{j_1} - a_{j_2 j_1 s_1})(\bar{x}_{j_1} - a_{j_2 j_1 s_2})^2} < 0,$$

there is a down-parabola flow at $x_{j_2}^* = a_{j_1 j_2 1}$. Let

$$\dot{x}_{j_2} = a_{j_2 j_2 0}(\bar{x}_{j_1} - a_{j_2 j_1 s_1})(\bar{x}_{j_1} - a_{j_2 j_1 s_2})^2.$$

The inflection flows at $x_{j_2}^* = a_{j_1 j_2 1}$ are positive and negative for $\dot{x}_{j_2} > 0$ and $\dot{x}_{j_2} < 0$.

Thus, the equilibrium of $x_{j_2}^* = a_{j_1 j_2 1}$ has the following properties:

- For $a_{j_1 j_1 0} \prod_{l_1 = 1}^2 (\bar{x}_{j_1} - a_{j_1 j_1 l_1}) > 0$ and $a_{j_2 j_2 0}(\bar{x}_{j_1} - a_{j_2 j_1 s_1})(\bar{x}_{j_1} - a_{j_2 j_1 s_2})^2 > 0$,

$$(a_{j_1 j_2 1}, \dot{x}_{j_2}) = \underbrace{\text{(UP, pF)}}_{\text{up-parabola flow } (+)}.$$

- For $a_{j_1 j_1 0} \prod_{l_1 = 1}^2 (\bar{x}_{j_1} - a_{j_1 j_1 l_1}) < 0$ and $a_{j_2 j_2 0}(\bar{x}_{j_1} - a_{j_2 j_1 s_1})(\bar{x}_{j_1} - a_{j_2 j_1 s_2})^2 > 0$,

$$(a_{j_1 j_2 1}, \dot{x}_{j_2}) = \underbrace{\text{(DP, pF)}}_{\text{down-parabola flow } (+)}.$$

- For $a_{j_1 j_1 0} \prod_{l_1 = 1}^2 (\bar{x}_{j_1} - a_{j_1 j_1 l_1}) > 0$ and $a_{j_2 j_2 0}(\bar{x}_{j_1} - a_{j_2 j_1 s_1})(\bar{x}_{j_1} - a_{j_2 j_1 s_2})^2 < 0$,

$$(a_{j_1 j_2 1}, \dot{x}_{j_2}) = \underbrace{\text{(DP, nF)}}_{\text{down-parabola flow } (-)}.$$

- For $a_{j_1 j_1 0} \prod_{l_1 = 1}^2 (\bar{x}_{j_1} - a_{j_1 j_1 l_1}) < 0$ and $a_{j_2 j_2 0}(\bar{x}_{j_1} - a_{j_2 j_1 s_1})(\bar{x}_{j_1} - a_{j_2 j_1 s_2})^2 < 0$,

$$(a_{j_1 j_2 1}, \dot{x}_{j_2}) = \underbrace{(\text{UP}, \text{nF})}_{\text{up-parabola flow } (-)} .$$

Therefore, from case (I) and (II), the equilibrium of $(x_{j_2}^*, x_{j_1}^*) = (a_{j_1 j_2 1}, a_{j_2 j_1 s_1})$ has the following properties as in Eqs. (1.149)–(1.152):

- For $a_{j_1 j_1 0} \prod_{l_1 = 1}^{2} (a_{j_2 j_1 s_1} - a_{j_1 j_1 l_1}) > 0$ and $a_{j_2 j_2 0}(a_{j_2 j_1 s_1} - a_{j_2 j_1 s_2})^2 > 0$,

$$(a_{j_1 j_2 1}, a_{j_2 j_1 s_1}) = \underbrace{(\text{UP}_+, \text{UP}_+)}_{\text{positive saddle}}.$$

The equilibrium of $(x_{j_2}^*, x_{j_1}^*) = (a_{j_1 j_2 1}, a_{j_2 j_1 s_1})$ is a $(\text{UP}_+, \text{UP}_+)$-positive saddle.
- For $a_{j_1 j_1 0} \prod_{l_1 = 1}^{2} (a_{j_2 j_1 s_1} - a_{j_1 j_1 l_1}) < 0$ and $a_{j_2 j_2 0}(a_{j_2 j_1 s_1} - a_{j_2 j_1 s_2})^2 > 0$,

$$(a_{j_1 j_2 1}, a_{j_2 j_1 s_1}) = \underbrace{(\text{DP}_+, \text{DP}_-)}_{\text{CCW center}}.$$

The equilibrium of $(x_{j_2}^*, x_{j_1}^*) = (a_{j_1 j_2 1}, a_{j_2 j_1 s_1})$ is a $(\text{DP}_+, \text{DP}_-)$-counter-clockwise center.
- For $a_{j_1 j_1 0} \prod_{l_1 = 1}^{2} (a_{j_2 j_1 s_1} - a_{j_1 j_1 l_1}) > 0$ and $a_{j_2 j_2 0}(a_{j_2 j_1 s_1} - a_{j_2 j_1 s_2})^2 < 0$,

$$(a_{j_1 j_2 1}, a_{j_2 j_1 s_1}) = \underbrace{(\text{DP}_-, \text{DP}_+)}_{\text{CW center}}.$$

The equilibrium of $(x_{j_2}^*, x_{j_1}^*) = (a_{j_1 j_2 1}, a_{j_2 j_1 s_1})$ is a $(\text{DP}_+, \text{DP}_-)$-clockwise center.
- For $a_{j_1 j_1 0} \prod_{l_1 = 1}^{2} (a_{j_2 j_1 s_1} - a_{j_1 j_1 l_1}) < 0$ and $a_{j_2 j_2 0}(a_{j_2 j_1 s_1} - a_{j_2 j_1 s_2})^2 < 0$,

$$(a_{j_1 j_2 1}, a_{j_2 j_1 s_1}) = \underbrace{(\text{UP}_-, \text{UP}_-)}_{\text{negative saddle}}.$$

The equilibrium of $(x_{j_2}^*, x_{j_1}^*) = (a_{j_1 j_2 1}, a_{j_2 j_1 s_1})$ is a $(\text{UP}_-, \text{UP}_-)$-negative saddle.

(viii$_{1b}$) For $x_{j_1}^* = a_{j_2 j_1 s_2} \neq a_{j_1 j_1 l_1}, a_{j_2 j_1 s_1}$ and $\bar{x}_{j_2} \neq a_{j_1 j_2 1}$ ($l_1 = 1, 2$), in phase space,

$$\frac{dx_{j_2}}{dx_{j_1}} \bigg|_{x_{j_1}^* = a_{j_2 j_1 s_2}} = \frac{a_{j_2 j_2 0}}{a_{j_1 j_1 0}} \frac{(x_{j_1} - a_{j_2 j_1 s_1})(x_{j_1} - a_{j_2 j_1 s_2})^2}{(\bar{x}_{j_2} - a_{j_1 j_2 1}) \prod_{l_1 = 1}^{2} (x_{j_1} - a_{j_1 j_1 l_1})} = 0,$$

$$\frac{d^2 x_{j_2}}{dx_{j_1}^2} \bigg|_{x_{j_1}^* = a_{j_2 j_1 s_2}} = 0.$$

If

$$\left.\frac{d^3 x_{j_2}}{dx_{j_1}^3}\right|_{x_{j_1}^* = a_{j_2 j_1 s_2}} = \frac{a_{j_2 j_2 0}}{a_{j_1 j_1 0}} \frac{2(a_{j_2 j_1 s_2} - a_{j_2 j_1 s_1})}{(\bar{x}_{j_2} - a_{j_1 j_2 1})\prod_{l_1=1}^{2}(a_{j_2 j_1 s_2} - a_{j_1 j_1 l_1})} > 0,$$

there is a second-order increasing-inflection flow. If

$$\left.\frac{d^3 x_{j_2}}{dx_{j_1}^3}\right|_{x_{j_1}^* = a_{j_2 j_1 s_2}} = \frac{a_{j_2 j_2 0}}{a_{j_1 j_1 0}} \frac{2(a_{j_2 j_1 s_2} - a_{j_2 j_1 s_1})}{(\bar{x}_{j_2} - a_{j_1 j_2 1})\prod_{l_1=1}^{2}(a_{j_2 j_1 s_2} - a_{j_1 j_1 l_1})} < 0,$$

there is a second-order decreasing-inflection flow. Let

$$\dot{x}_{j_1} = a_{j_1 j_1 0}(\bar{x}_{j_2} - a_{j_1 j_2 1})\prod_{l_1=1}^{2}(a_{j_2 j_1 s_2} - a_{j_1 j_1 l_1}).$$

Thus, the inflection flow at $x_{j_1}^* = a_{j_2 j_1 s_2} \neq a_{j_1 j_1 1}, a_{j_2 j_1 s_1}$ with $\bar{x}_{j_2} \neq a_{j_1 j_2 1}$ has the following properties:

- For $a_{j_1 j_1 0}(\bar{x}_{j_2} - a_{j_1 j_2 1})\prod_{l_1=1}^{2}(a_{j_2 j_1 s_2} - a_{j_1 j_1 l_1}) > 0$ and $a_{j_2 j_2 0}(a_{j_2 j_1 s_2} - a_{j_2 j_1 s_1}) > 0$,

$$(\dot{x}_{j_1}, a_{j_2 j_1 s_2}) = \underbrace{(\text{pF, II})}_{\text{increasing-inflection flow } (+)}.$$

- For $a_{j_1 j_1 0}(\bar{x}_{j_2} - a_{j_1 j_2 1})\prod_{l_1=1}^{2}(a_{j_2 j_1 s_2} - a_{j_1 j_1 l_1}) < 0$ and $a_{j_2 j_2 0}(a_{j_2 j_1 s_2} - a_{j_2 j_1 s_1}) > 0$,

$$(\dot{x}_{j_1}, a_{j_2 j_1 s_2}) = \underbrace{(\text{nF, DI})}_{\text{decreasing-inflection flow } (-)}.$$

- For $a_{j_1 j_1 0}(\bar{x}_{j_2} - a_{j_1 j_2 1})\prod_{l_1=1}^{2}(a_{j_2 j_1 s_2} - a_{j_1 j_1 l_1}) > 0$ and $a_{j_2 j_2 0}(a_{j_2 j_1 s_2} - a_{j_2 j_1 s_1}) < 0$,

$$(\dot{x}_{j_1}, a_{j_2 j_1 s_2}) = \underbrace{(\text{pF, DI})}_{\text{decreasing-inflection flow } (+)}.$$

- For $a_{j_1 j_1 0}(\bar{x}_{j_2} - a_{j_1 j_2 1})\prod_{l_1=1}^{2}(a_{j_2 j_1 s_2} - a_{j_1 j_1 l_1}) < 0$ and $a_{j_2 j_2 0}(a_{j_2 j_1 s_2} - a_{j_2 j_1 s_1}) < 0$,

$$(\dot{x}_{j_1}, a_{j_2 j_1 s_2}) = \underbrace{(\text{nF, II})}_{\text{increasing-inflection flow } (-)}.$$

Therefore, from the above analysis and the proof (II) of (viii$_{1a}$) with $\bar{x}_{j_1} = a_{j_2 j_1 s_2}$, the equilibrium of $(x_{j_2}^*, x_{j_1}^*) = (a_{j_1 j_2 1}, a_{j_2 j_1 s_2})$ has the following properties as in Eqs. (1.153)–(1.156):

- For $a_{j_1j_10}(a_{j_2j_1s_2} - a_{j_2j_11})(a_{j_1j_2l_1} - a_{j_1j_2l_2}) > 0$ and $a_{j_2j_20}(a_{j_2j_1s_2} - a_{j_2j_1s_1}) > 0$,

$$(a_{j_1j_21}, a_{j_2j_1s_2}) = \underbrace{(\text{UP, US})}_{\text{up-parabola upper-saddle}}.$$

The equilibrium of $(x_{j_2}^*, x_{j_1}^*) = (a_{j_1j_21}, a_{j_2j_1s_2})$ is a (UP,US)-up-parabola upper-saddle.

- For $a_{j_1j_10}(a_{j_2j_1s_2} - a_{j_1j_11})(a_{j_1j_2l_1} - a_{j_1j_2l_2}) < 0$ and $a_{j_2j_20}(a_{j_2j_1s_2} - a_{j_2j_1s_1}) > 0$,

$$(a_{j_1j_21}, a_{j_2j_1s_2}) = \underbrace{(\text{DP, US})}_{\text{down-parabola upper-saddle}}.$$

The equilibrium of $(x_{j_2}^*, x_{j_1}^*) = (a_{j_1j_21}, a_{j_2j_1s_2})$ is a (DP,US)-down-parabola upper-saddle.

- For $a_{j_1j_10}(a_{j_2j_1s_2} - a_{j_1j_11})(a_{j_1j_2l_1} - a_{j_1j_2l_2}) > 0$ and $a_{j_2j_20}(a_{j_2j_1s_2} - a_{j_2j_1s_1}) < 0$,

$$(a_{j_1j_21}, a_{j_2j_1s_2}) = \underbrace{(\text{DP, LS})}_{\text{down-parabola upper-saddle}}.$$

The equilibrium of $(x_{j_2}^*, x_{j_1}^*) = (a_{j_1j_21}, a_{j_2j_1s_2})$ is a (DP,LS)-down-parabola lower-saddle.

- For $a_{j_1j_10}(a_{j_2j_1s_2} - a_{j_1j_11})(a_{j_1j_2l_1} - a_{j_1j_2l_2}) < 0$ and $a_{j_2j_20}(a_{j_2j_1s_2} - a_{j_2j_1s_1}) < 0$,

$$(a_{j_1j_21}, a_{j_2j_1s_2}) = \underbrace{(\text{UP, LS})}_{\text{up-parabola lower-saddle}}.$$

The equilibrium of $(x_{j_2}^*, x_{j_1}^*) = (a_{j_1j_21}, a_{j_2j_1s_2})$ is a (UP,LS)-up-parabola lower-saddle.

- The parabola upper-saddle and lower-saddle are the appearing bifurcations of positive and negative saddles with clockwise and counter-clockwise centers.

(viii$_{1c}$) From case (I) of (viii$_{1a}$) for $\bar{x}_{j_1} = a_{j_1j_1l_1}$, the equilibrium of $(x_{j_1}^*, x_{j_2}^*) = (a_{j_1j_1l_1}, a_{j_1j_21})$ ($l_1, l_2 \in \{1, 2\}$, $l_1 \neq l_2$) with $a_{j_2j_1s_1} \neq a_{j_2j_1s_2}$ ($s_1, s_2 \in \{1, 2\}$, $s_1 \neq s_2$) has the following properties as in Eqs. (1.157)–(1.160):

- For $a_{j_1j_10}(a_{j_1j_1l_1} - a_{j_1j_1l_2}) > 0$ and $a_{j_2j_20}(a_{j_1j_1l_1} - a_{j_2j_1s_1})(a_{j_1j_1l_1} - a_{j_2j_1s_2})^2 > 0$,

$$(a_{j_1j_1l_1}, a_{j_1j_21}) = \underbrace{(\text{DP:UP, pF})}_{\text{hyperbolic flow }(+)}.$$

The equilibrium of $(x_{j_1}^*, x_{j_2}^*) = (a_{j_1j_1l_1}, a_{j_1j_21})$ is a (DP:UP, pF)-positive hyperbolic flow.

- For $a_{j_1j_10}(a_{j_1j_1l_1} - a_{j_1j_1l_2}) < 0$ and $a_{j_2j_20}(a_{j_1j_1l_1} - a_{j_2j_1s_1})(a_{j_1j_1l_1} - a_{j_2j_1s_2})^2 > 0$,

$$(a_{j_1j_1l_1}, a_{j_1j_21}) = \underbrace{\text{(UP:DP, pF)}}_{\text{hyperbolic-secant flow }(+)} .$$

The equilibrium of $(x_{j_1}^*, x_{j_2}^*) = (a_{j_1j_1l_1}, a_{j_1j_21})$ is a (UP:DP, pF)-positive hyperbolic-secant flow.

- For $a_{j_1j_10}(a_{j_1j_1l_1} - a_{j_1j_1l_2}) > 0$ and $a_{j_2j_20}(a_{j_1j_1l_1} - a_{j_2j_1s_1})(a_{j_1j_1l_1} - a_{j_2j_1s_2})^2 < 0$,

$$(a_{j_1j_1l_1}, a_{j_1j_21}) = \underbrace{\text{(UP:DP, nF)}}_{\text{hyperbolic-secant flow }(-)} .$$

The equilibrium of $(x_{j_1}^*, x_{j_2}^*) = (a_{j_1j_1l_1}, a_{j_1j_21})$ is a (UP:DP,nF)-negative hyperbolic-secant flow.

- For $a_{j_1j_10}(a_{j_1j_1l_1} - a_{j_1j_1l_2}) < 0$ and $a_{j_2j_20}(a_{j_1j_1l_1} - a_{j_2j_1s_1})(a_{j_1j_1l_1} - a_{j_2j_1s_2})^2 < 0$,

$$(a_{j_1j_1l_1}, a_{j_1j_21}) = \underbrace{\text{(DP:UP, nF)}}_{\text{hyperbolic flow }(-)} .$$

The equilibrium of $(x_{j_1}^*, x_{j_2}^*) = (a_{j_1j_1l_1}, a_{j_1j_21})$ is a (DP:UP,nF)-negative hyperbolic flow.

(viii$_2$) For $a_{j_2j_1s_1} = a_{j_2j_1s_2} = a_{j_2j_11}$, the standard form is

$$\dot{x}_{j_2} = a_{j_2j_20}(x_{j_1} - a_{j_2j_11})^3,$$
$$\dot{x}_{j_1} = a_{j_1j_10}(x_{j_1} - a_{j_1j_1l_1})(x_{j_1} - a_{j_1j_1l_2})(x_{j_2} - a_{j_1j_21}).$$

For $a_{j_2j_11} \neq a_{j_1j_11}$, in phase space,

$$\frac{dx_{j_1}}{dx_{j_2}} = \frac{a_{j_1j_10}}{a_{j_2j_20}} \frac{(x_{j_2} - a_{j_1j_21})(x_{j_1} - a_{j_1j_1l_1})(x_{j_1} - a_{j_1j_1l_2})}{(x_{j_1} - a_{j_2j_11})^3},$$

and the deformation of the foregoing equation is

$$[(x_{j_1} - a_{j_2j_11}) + (a_{j_1j_11} + a_{j_1j_12} - 2a_{j_2j_11}) + \frac{(a_{j_1j_11} - a_{j_2j_11})^3}{(x_{j_1} - a_{j_1j_11})(a_{j_1j_11} - a_{j_1j_12})}$$
$$+ \frac{(a_{j_1j_12} - a_{j_2j_11})^3}{(x_{j_1} - a_{j_1j_12})(a_{j_1j_12} - a_{j_1j_11})}]dx_{j_1} = \frac{a_{j_1j_10}}{a_{j_2j_20}}(x_{j_2} - a_{j_1j_21})dx_{j_2}.$$

With an initial condition of $(x_{j_1 0}, x_{j_2 0})$ at $t = t_0$, the integration of the above equation gives

$$\frac{1}{2}\left[(x_{j_1} - a_{j_2 j_1 1})^2 - (x_{j_1 0} - a_{j_2 j_1 1})^2\right] + (a_{j_1 j_1 1} + a_{j_1 j_1 2} - 2a_{j_2 j_1 1})(x_{j_1} - x_{j_1 0})$$

$$+ \frac{(a_{j_1 j_1 1} - a_{j_2 j_1 1})^3}{(a_{j_1 j_1 1} - a_{j_1 j_1 2})} \ln\frac{|x_{j_1} - a_{j_1 j_1 1}|}{|x_{j_1 0} - a_{j_1 j_1 1}|} + \frac{(a_{j_1 j_1 2} - a_{j_2 j_1 1})^3}{(a_{j_1 j_1 2} - a_{j_1 j_1 1})} \ln\frac{|x_{j_1} - a_{j_1 j_1 2}|}{|x_{j_1 0} - a_{j_1 j_1 2}|}$$

$$= \frac{1}{2}\frac{a_{j_1 j_1 0}}{a_{j_2 j_2 0}}\left[(x_{j_2} - a_{j_1 j_2 1})^2 - (x_{j_2 0} - a_{j_1 j_2 1})^2\right].$$

(viii$_{2a}$) Consider two cases: (I) $x_{j_1}^* = a_{j_2 j_1 1}$ and (II) $x_{j_2}^* = a_{j_1 j_2 1}$.

(I) At $x_{j_1}^* = a_{j_2 j_1 1}$ and $\bar{x}_{j_2} \neq a_{j_1 j_2 1}$,

$$\frac{dx_{j_2}}{dx_{j_1}}\bigg|_{x_{j_1}^* = a_{j_2 j_1 1}} = \frac{a_{j_2 j_2 0}}{a_{j_1 j_1 0}}\frac{(x_{j_1} - a_{j_2 j_1 1})^3}{(\bar{x}_{j_2} - a_{j_1 j_2 1})\prod_{l_1 = 1}^{2}(x_{j_1} - a_{j_1 j_1 l_1})}\bigg|_{x_{j_1}^* = a_{j_2 j_1 1}} = 0,$$

$$\frac{d^2 x_{j_2}}{dx_{j_1}^2}\bigg|_{x_{j_1}^* = a_{j_2 j_1 1}} = \frac{d^3 x_{j_2}}{dx_{j_1}^3}\bigg|_{x_{j_1}^* = a_{j_2 j_1 1}} = 0.$$

If

$$\frac{d^4 x_{j_2}}{dx_{j_1}^4}\bigg|_{x_{j_1}^* = a_{j_2 j_1 1}} = \frac{a_{j_2 j_2 0}}{a_{j_1 j_1 0}}\frac{6}{(\bar{x}_{j_2} - a_{j_1 j_2 1})\prod_{l_1 = 1}^{2}(a_{j_2 j_1 1} - a_{j_1 j_1 l_1})} > 0,$$

there is a third-order up-parabola flow at $x_{j_1}^* = a_{j_2 j_1 1}$. If

$$\frac{d^4 x_{j_2}}{dx_{j_1}^4}\bigg|_{x_{j_1}^* = a_{j_2 j_1 1}} = \frac{a_{j_2 j_2 0}}{a_{j_1 j_1 0}}\frac{6}{(\bar{x}_{j_2} - a_{j_1 j_2 1})\prod_{l_1 = 1}^{2}(a_{j_2 j_1 1} - a_{j_1 j_1 l_1})} < 0,$$

there is a third-order down-parabola flow at $x_{j_1}^* = a_{j_2 j_1 1}$. Let

$$\dot{x}_{j_1} = a_{j_1 j_1 0}(\bar{x}_{j_2} - a_{j_1 j_2 1})\prod_{l_1 = 1}^{2}(a_{j_2 j_1 1} - a_{j_1 j_1 l_1}).$$

The third-order parabola flows at $x_{j_1}^* = a_{j_2 j_1 1}$ are positive and negative for $\dot{x}_{j_1} > 0$ and $\dot{x}_{j_1} < 0$.

Therefore, the equilibrium of $x_{j_1}^* = a_{j_2 j_1 1}$ has the following properties:

- For $a_{j_1j_10}(\bar{x}_{j_2} - a_{j_1j_21})\prod_{l_1=1}^{2}(a_{j_2j_11} - a_{j_1j_1l_1}) > 0$ and $a_{j_2j_20} > 0$,

$$(\dot{x}_{j_1}, a_{j_2j_11}) = \underbrace{(\text{pF}, 3^{\text{rd}}\text{UP})}_{\text{third-order up-parabola flow }(+)}.$$

The equilibrium of $x_{j_1}^* = a_{j_2j_11}$ is a (pF,3^{rd}UP)-positive third-order up-parabola flow.

- For $a_{j_1j_10}(\bar{x}_{j_2} - a_{j_1j_21})\prod_{l_1=1}^{2}(a_{j_2j_11} - a_{j_1j_1l_1}) < 0$ and $a_{j_2j_20} > 0$,

$$(\dot{x}_{j_1}, a_{j_2j_11}) = \underbrace{(\text{nF}, 3^{\text{rd}}\text{DP})}_{\text{third-order down-parabola flow }(-)}.$$

The equilibrium of $x_{j_1}^* = a_{j_2j_11}$ is an (nF,3^{rd}DP)-negative third-order down-parabola flow.

- For $(\bar{x}_{j_2} - a_{j_1j_21})\prod_{l_1=1}^{2}(a_{j_2j_11} - a_{j_1j_1l_1}) > 0$ and $a_{j_2j_20} < 0$,

$$(\dot{x}_{j_1}, a_{j_2j_11}) = \underbrace{(\text{pF}, 3^{\text{rd}}\text{DP})}_{\text{third-order down-parabola flow }(+)}.$$

The equilibrium of $x_{j_1}^* = a_{j_2j_11}$ is a (pF,3^{rd}DP)-positive third-order down-parabola flow.

- For $a_{j_1j_10}(\bar{x}_{j_2} - a_{j_1j_21})\prod_{l_1=1}^{2}(a_{j_2j_11} - a_{j_1j_1l_1}) < 0$ and $a_{j_2j_20} < 0$,

$$(\dot{x}_{j_1}, a_{j_2j_11}) = \underbrace{(\text{nF}, 3^{\text{rd}}\text{UP})}_{\text{third-order up-parabola flow }(-)}.$$

The equilibrium of $x_{j_1}^* = a_{j_2j_11}$ is an (nF,3^{rd}UP)-negative third-order up-parabola flow.

(II) In phase space, at $x_{j_2}^* = a_{j_1j_21}$ with $\bar{x}_{j_1} \neq a_{j_2j_11}, a_{j_1j_1l_1}$ ($l_1 = 1, 2$),

$$\frac{dx_{j_1}}{dx_{j_2}}\bigg|_{x_{j_2}^* = a_{j_1j_21}} = \frac{a_{j_1j_10}}{a_{j_2j_20}}\frac{(x_{j_2} - a_{j_1j_21})\prod_{l_1=1}^{2}(\bar{x}_{j_1} - a_{j_1j_1l_1})}{(\bar{x}_{j_1} - a_{j_2j_11})^3}\bigg|_{x_{j_2}^* = a_{j_1j_21}} = 0.$$

If

$$\frac{d^2x_{j_1}}{dx_{j_2}^2}\bigg|_{x_{j_2}^* = a_{j_1j_21}} = \frac{a_{j_1j_10}}{a_{j_2j_20}}\frac{\prod_{l_1=1}^{2}(\bar{x}_{j_1} - a_{j_1j_1l_1})}{(\bar{x}_{j_1} - a_{j_2j_11})^3} > 0,$$

there is an up-parabola flow at $x_{j_2}^* = a_{j_1j_21}$ with $\bar{x}_{j_1} = a_{j_2j_11}, a_{j_1j_1l_1}$ ($l_1 = 1, 2$). If

$$\left.\frac{d^2 x_{j_1}}{dx_{j_2}^2}\right|_{x_{j_2}^* = a_{j_1 j_2 1}} = \frac{a_{j_1 j_1 0}}{a_{j_2 j_2 0}} \frac{\prod_{l_1=1}^{2}(\bar{x}_{j_1} - a_{j_1 j_1 l_1})}{(\bar{x}_{j_1} - a_{j_2 j_1 1})^3} < 0,$$

there is a down-parabola flow at $x_{j_2}^* = a_{j_1 j_2 1}$ with $\bar{x}_{j_1} \neq a_{j_2 j_1 1}, a_{j_1 j_1 l_1}$ $(l_1 = 1, 2)$. Let

$$\dot{x}_{j_2} = a_{j_2 j_2 0}(\bar{x}_{j_1} - a_{j_2 j_1 1})^3.$$

The parabola flows at $x_{j_2}^* = a_{j_1 j_2 1}$ with $\bar{x}_{j_1} \neq a_{j_2 j_1 1}$ are positive and negative for $\dot{x}_{j_2} > 0$ and $\dot{x}_{j_2} < 0$, respectively.

Thus, the equilibrium of $x_{j_2}^* = a_{j_1 j_2 1}$ with $\bar{x}_{j_1} \neq a_{j_2 j_1 1}, a_{j_1 j_1 l_1}$ $(l_1 = 1, 2)$ has the following properties:

- For $a_{j_1 j_1 0} \prod_{l_1=1}^{2}(\bar{x}_{j_1} - a_{j_1 j_1 l_1}) > 0$ and $a_{j_2 j_2 0}(\bar{x}_{j_1} - a_{j_2 j_1 1})^3 > 0$,

$$(a_{j_1 j_2 1}, \dot{x}_{j_2}) = \underbrace{(\text{UP}, \text{pF})}_{\text{up-parabola flow } (+)}.$$

- For $a_{j_1 j_1 0} \prod_{l_1=1}^{2}(\bar{x}_{j_1} - a_{j_1 j_1 l_1}) < 0$ and $a_{j_2 j_2 0}(\bar{x}_{j_1} - a_{j_2 j_1 1})^3 > 0$,

$$(a_{j_1 j_2 1}, \dot{x}_{j_2}) = \underbrace{(\text{DP}, \text{pF})}_{\text{down-parabola flow } (+)}.$$

- For $a_{j_1 j_1 0} \prod_{l_1=1}^{2}(\bar{x}_{j_1} - a_{j_1 j_1 l_1}) > 0$ and $a_{j_2 j_2 0}(\bar{x}_{j_1} - a_{j_2 j_1 1})^3 < 0$,

$$(a_{j_1 j_2 1}, \dot{x}_{j_2}) = \underbrace{(\text{DP}, \text{nF})}_{\text{down-parabola flow } (-)}.$$

- For $a_{j_1 j_1 0} \prod_{l_1=1}^{2}(\bar{x}_{j_1} - a_{j_1 j_1 l_1}) < 0$ and $a_{j_2 j_2 0}(\bar{x}_{j_1} - a_{j_2 j_1 1})^3 < 0$,

$$(a_{j_1 j_2 1}, \dot{x}_{j_2}) = \underbrace{(\text{UP}, \text{nF})}_{\text{up-paraonbla flow } (-)}.$$

Therefore, for case (I) with $\bar{x}_{j_2} = a_{j_1 j_2 1}$ and case (II) with $\bar{x}_{j_1} = a_{j_2 j_1 1}$, the equilibrium of $(x_{j_2}^*, x_{j_1}^*) = (a_{j_1 j_2 1}, a_{j_2 j_1 1})$ has the following properties as in Eqs. (1.163)–(1.166):

- For $a_{j_1 j_1 0} \prod_{l_1=1}^{2}(a_{j_2 j_1 1} - a_{j_1 j_1 l_1}) > 0$ and $a_{j_2 j_2 0} > 0$,

$$(a_{j_1 j_2 1}, a_{j_2 j_1 1}) = \underbrace{(\text{UP}_+, 3^{\text{rd}}\text{UP}_+)}_{\text{third-order positive saddle}}.$$

The equilibrium of $(x_{j_2}^*, x_{j_1}^*) = (a_{j_1 j_2 1}, a_{j_2 j_1 1})$ is a $(\text{UP}_+, 3^{\text{rd}}\text{UP}_+)$-third-order positive saddle.

- For $a_{j_1j_10}\prod_{l_1=1}^{2}(a_{j_2j_11}-a_{j_1j_1l_1}) < 0$ and $a_{j_2j_20} > 0$,

$$(a_{j_1j_21}, a_{j_2j_11}) = \underbrace{(\mathrm{DP}_+, 3^{\mathrm{rd}}\mathrm{DP}_-)}_{\text{third-order CCW center}}.$$

The equilibrium of $(x^*_{j_2}, x^*_{j_1}) = (a_{j_1j_21}, a_{j_2j_11})$ is a $(\mathrm{DP}_+, 3^{\mathrm{rd}}\mathrm{DP}_-)$-third-order counter-clockwise center.

- For $a_{j_1j_10}\prod_{l_1=1}^{2}(a_{j_2j_11}-a_{j_1j_1l_1}) > 0$ and $a_{j_2j_20} < 0$,

$$(a_{j_1j_21}, a_{j_2j_11}) = \underbrace{(\mathrm{DP}_-, 3^{\mathrm{rd}}\mathrm{DP}_+)}_{\text{third-order CW center}}.$$

The equilibrium of $(x^*_{j_2}, x^*_{j_1}) = (a_{j_1j_21}, a_{j_2j_11})$ is a $(\mathrm{DP}_-, 3^{\mathrm{rd}}\mathrm{DP}_+)$-third-order clockwise center.

- For $a_{j_1j_10}\prod_{l_1=1}^{2}(a_{j_2j_11}-a_{j_1j_1l_1}) < 0$ and $a_{j_2j_20} < 0$,

$$(a_{j_1j_21}, a_{j_2j_11}) = \underbrace{(\mathrm{UP}_-, 3^{\mathrm{rd}}\mathrm{UP}_-)}_{\text{third-order negative saddle}}.$$

The equilibrium of $(x^*_{j_2}, x^*_{j_1}) = (a_{j_1j_21}, a_{j_2j_11})$ is a $(\mathrm{UP}_-, 3^{\mathrm{rd}}\mathrm{UP}_-)$-third-order negative saddle.

- The third-order saddles are the appearing and switching bifurcations of saddle, center, and saddle.
- The third-order centers are the appearing and switching bifurcations of center, saddle, and center.

(viii$_{2b}$) At $x^*_{j_1} = a_{j_1j_1l_1}$ with $\bar{x}_{j_2} \neq a_{j_1j_21}$, the variational equation is

$$\Delta\dot{x}_{j_1} = a_{j_1j_10}(\bar{x}_{j_2} - a_{j_1j_21})(a_{j_1j_1l_1} - a_{j_1j_1l_2})\Delta x_{j_1}.$$

The flows at $x^*_{j_1} = a_{j_1j_1l_1}$ are source and sink for $a_{j_1j_10}(\bar{x}_{j_2} - a_{j_1j_21})(a_{j_1j_1l_1} - a_{j_1j_1l_2}) > 0$ and $a_{j_1j_10}(\bar{x}_{j_2} - a_{j_1j_21})(a_{j_1j_1l_1} - a_{j_1j_1l_2}) < 0$, respectively. From the case (II) of the proof of (viii$_{2a}$), for $\bar{x}_{j_1} = a_{j_1j_1l_1}$, the equilibrium of $(x^*_{j_1}, x^*_{j_2}) = (a_{j_1j_1l_1}, a_{j_1j_21})$ with $a_{j_1j_1l_1} \neq a_{j_1j_1l_2}, a_{j_2j_11}$ has the following properties as in Eqs. (1.167)–(1.170):

- For $a_{j_1j_10}(a_{j_1j_1l_1} - a_{j_1j_1l_2}) > 0$ and $a_{j_2j_20}(a_{j_1j_1l_1} - a_{j_2j_11})^3 > 0$,

$$(a_{j_1j_1l_1}, a_{j_1j_21}) = \underbrace{(\mathrm{DP:UP, pF})}_{\text{hyperbolic flow } (+)}.$$

The equilibrium of $(x^*_{j_1}, x^*_{j_2}) = (a_{j_1j_11}, a_{j_1j_2l_1})$ is a (DP:UP, pF)-positive hyperbolic flow.

- For $a_{j_1 j_1 0}(a_{j_1 j_1 l_1} - a_{j_1 j_1 l_2}) < 0$ and $a_{j_2 j_2 0}(a_{j_1 j_1 l_1} - a_{j_2 j_1 1})^3 > 0$,

$$(a_{j_1 j_1 l_1}, a_{j_1 j_2 1}) = \underbrace{(\text{UP:DP, pF})}_{\text{hyperbolic-secant flow } (+)} .$$

The equilibrium of $(x_{j_1}^*, x_{j_2}^*) = (a_{j_1 j_1 1}, a_{j_1 j_2 l_1})$ is a (UP:DP, pF)-positive hyperbolic-secant flow.

- For $a_{j_1 j_1 0}(a_{j_1 j_1 l_1} - a_{j_1 j_1 l_2}) > 0$ and $a_{j_2 j_2 0}(a_{j_1 j_1 l_1} - a_{j_2 j_1 1})^3 < 0$,

$$(a_{j_1 j_1 l_1}, a_{j_1 j_2 1}) = \underbrace{(\text{UP:DP, nF})}_{\text{hyperbolic-secant flow } (-)} .$$

The equilibrium of $(x_{j_1}^*, x_{j_2}^*) = (a_{j_1 j_1 1}, a_{j_1 j_2 l_1})$ is a (UP:DP, nF)-negative hyperbolic-secant flow.

- For $a_{j_1 j_1 0}(a_{j_1 j_1 l_1} - a_{j_1 j_1 l_2}) < 0$ and $a_{j_2 j_2 0}(a_{j_1 j_1 l_1} - a_{j_2 j_1 1})^3 < 0$,

$$(a_{j_1 j_1 l_1}, a_{j_1 j_2 1}) = \underbrace{(\text{DP:UP, nF})}_{\text{hyperbolic flow } (-)} .$$

The equilibrium of $(x_{j_1}^*, x_{j_2}^*) = (a_{j_1 j_1 1}, a_{j_1 j_2 l_1})$ is a (DP:UP, nF)-negative hyperbolic flow.

(viii$_3$) For $a_{j_2 j_1 s_1} = a_{j_1 j_1 l_1}$, in phase space,

$$\frac{dx_{j_1}}{dx_{j_2}} = \frac{a_{j_1 j_1 0}}{a_{j_2 j_2 0}} \frac{(x_{j_1} - a_{j_1 j_1 l_2})(x_{j_2} - a_{j_1 j_2 1})}{(x_{j_1} - a_{j_2 j_1 s_2})^2},$$

and the deformation of the foregoing equation is

$$\left[(x_{j_1} - a_{j_1 j_1 l_2}) + 2(a_{j_1 j_1 l_2} - a_{j_2 j_1 s_2}) + \frac{(a_{j_1 j_1 l_2} - a_{j_2 j_1 s_2})^2}{x_{j_1} - a_{j_1 j_1 l_2}} \right] dx_{j_1}$$
$$= \frac{a_{j_1 j_1 0}}{a_{j_2 j_2 0}} (x_{j_2} - a_{j_1 j_2 1}) dx_{j_2}.$$

With an initial condition of $(x_{j_1 0}, x_{j_2 0})$ at $t = t_0$, the integration of the above equation gives the first integral manifold for $a_{j_2 j_1 s_1} = a_{j_1 j_1 1}$ is

$$\frac{1}{2}\left[(x_{j_1} - a_{j_1 j_1 l_2})^2 - (x_{j_1 0} - a_{j_2 j_1 l_2})^2\right] + 2(a_{j_1 j_1 l_2} - a_{j_2 j_1 s_2})(x_{j_1} - x_{j_1 0})$$

$$+(a_{j_1 j_1 l_2} - a_{j_2 j_1 s_2})^2 \ln \frac{|x_{j_1} - a_{j_1 j_1 l_2}|}{|x_{j_1 0} - a_{j_2 j_1 l_2}|}$$

$$= \frac{1}{2}\frac{a_{j_1 j_1 0}}{a_{j_2 j_2 0}}\left[(x_{j_2} - a_{j_1 j_2 1})^2 - (x_{j_2 0} - a_{j_1 j_2 1})^2\right].$$

(viii$_{3a}$) In phase space, at $x_{j_1}^* = a_{j_2 j_1 s_1} = a_{j_1 j_1 1}$ with $\bar{x}_{j_2} \neq a_{j_1 j_2 1}$, if

$$\frac{dx_{j_2}}{dx_{j_1}}\Big|_{x_{j_1}^* = a_{j_2 j_1 s_1}} = \frac{a_{j_2 j_2 0}}{a_{j_1 j_1 0}}\frac{(a_{j_2 j_1 s_1} - a_{j_2 j_1 s_2})^2}{(\bar{x}_{j_2} - a_{j_1 j_2 1})(a_{j_2 j_1 s_1} - a_{j_1 j_1 l_2})} > 0,$$

there is an increasing-inflection flow. If

$$\frac{dx_{j_2}}{dx_{j_1}}\Big|_{x_{j_1}^* = a_{j_2 j_1 s_1}} = \frac{a_{j_2 j_2 0}}{a_{j_1 j_1 0}}\frac{(a_{j_2 j_1 s_1} - a_{j_2 j_1 s_2})^2}{(\bar{x}_{j_2} - a_{j_1 j_2 1})(a_{j_2 j_1 s_1} - a_{j_1 j_1 l_2})} < 0,$$

there is a decreasing-inflection flow. The variational equation at $x_{j_1}^* = a_{j_1 j_1 1} = a_{j_2 j_1 s_1}$ with $\bar{x}_{j_2} \neq a_{j_1 j_2 1}$ is

$$\Delta \dot{x}_{j_1} = a_{j_1 j_1 0}(\bar{x}_{j_2} - a_{j_1 j_2 1})(a_{j_2 j_1 s_1} - a_{j_1 j_1 l_2})\Delta x_{j_1}.$$

The inflection flows at $x_{j_1}^* = a_{j_1 j_1 1} = a_{j_2 j_1 s_1}$ with $\bar{x}_{j_2} \neq a_{j_1 j_2 1}$ are source and sink in the x_{j_1}-direction for $a_{j_1 j_1 0}(\bar{x}_{j_2} - a_{j_1 j_2 1})(a_{j_2 j_1 s_1} - a_{j_1 j_1 l_2}) > 0$ and $a_{j_1 j_1 0}(\bar{x}_{j_2} - a_{j_1 j_2 1})(a_{j_2 j_1 s_1} - a_{j_1 j_1 l_2})$ < 0, respectively.

Therefore, the infinite-equilibrium of $x_{j_1}^* = a_{j_2 j_1 s_1} = a_{j_1 j_1 l_1}$ with $\bar{x}_{j_2} \neq a_{j_1 j_2 1}$ has the following properties as in Eqs. (1.172)–(1.175):

- For $a_{j_1 j_1 0}(\bar{x}_{j_2} - a_{j_1 j_2 1})(a_{j_1 j_1 l_1} - a_{j_1 j_1 l_2}) > 0$ and $a_{j_2 j_2 0}(a_{j_2 j_1 s_1} - a_{j_2 j_1 s_2})^2 > 0$,

$$(a_{j_1 j_1 l_1}, \bar{x}_{j_2}) = \underbrace{(SO, II)}_{\text{increasing-inflection source}}.$$

The infinite-equilibrium of $x_{j_1}^* = a_{j_2 j_1 s_1} = a_{j_1 j_1 l_1}$ is an (SO,II)-increasing-inflection source.

- For $a_{j_1 j_1 0}(\bar{x}_{j_2} - a_{j_1 j_2 1})(a_{j_1 j_1 l_1} - a_{j_1 j_1 l_2}) < 0$ and $a_{j_2 j_2 0}(a_{j_2 j_1 s_1} - a_{j_2 j_1 s_2})^2 > 0$,

$$(a_{j_1 j_1 l_1}, \bar{x}_{j_2}) = \underbrace{(SI, DI)}_{\text{decreasing-inflection sink}}.$$

The infinite-equilibrium of $x_{j_1}^* = a_{j_2 j_1 s_1} = a_{j_1 j_1 l_1}$ is an (SI,DI)-decreasing-inflection sink.

- For $a_{j_1j_10}(\bar{x}_{j_2} - a_{j_1j_21})(a_{j_1j_1l_1} - a_{j_1j_1l_2}) > 0$ and $a_{j_2j_20}(a_{j_2j_1s_1} - a_{j_2j_1s_2})^2 < 0$,

$$(a_{j_1j_1l_1}, \bar{x}_{j_2}) = \underbrace{(\text{SO, DI})}_{\text{decreasing-inflection source}} .$$

The infinite-equilibrium of $x^*_{j_1} = a_{j_2j_1s_1} = a_{j_1j_1l_1}$ is an (SO,DI)-decreasing-inflection source.

- For $a_{j_1j_10}(\bar{x}_{j_2} - a_{j_1j_21})(a_{j_1j_1l_1} - a_{j_1j_1l_2}) < 0$ and $a_{j_2j_20}(a_{j_2j_1s_1} - a_{j_2j_1s_2})^2 < 0$,

$$(a_{j_1j_1l_1}, \bar{x}_{j_2}) = \underbrace{(\text{SI, II})}_{\text{increasing-inflection sink}} .$$

The infinite-equilibrium of $x^*_{j_1} = a_{j_2j_1s_1} = a_{j_1j_1l_1}$ is an (SI,II)-increasing-inflection sink.

(viii$_{3b}$) For $x^*_{j_2} = a_{j_1j_21}$ and $\bar{x}_{j_1} \neq a_{j_1j_1l_1}, a_{j_1j_1j_2}a_{j_2j_1s_1}, a_{j_2j_1s_2}$, in phase space,

$$\frac{dx_{j_1}}{dx_{j_2}}\bigg|_{x^*_{j_2}=a_{j_1j_21}} = \frac{a_{j_1j_10}}{a_{j_2j_20}} \frac{(x_{j_2} - a_{j_1j_21})(\bar{x}_{j_1} - a_{j_1j_1l_2})}{(\bar{x}_{j_1} - a_{j_2j_1s_2})^2}\bigg|_{x^*_{j_2}=a_{j_1j_21}} = 0.$$

if

$$\frac{d^2x_{j_1}}{dx^2_{j_2}}\bigg|_{x^*_{j_2}=a_{j_1j_21}} = \frac{a_{j_1j_10}}{a_{j_2j_20}} \frac{(\bar{x}_{j_1} - a_{j_1j_1l_1})(\bar{x}_{j_1} - a_{j_1j_1l_2})}{(\bar{x}_{j_1} - a_{j_2j_1s_1})(\bar{x}_{j_1} - a_{j_2j_1s_2})^2} > 0,$$

there is an up-parabola flow. If

$$\frac{d^2x_{j_1}}{dx^2_{j_2}}\bigg|_{x^*_{j_2}=a_{j_1j_21}} = \frac{a_{j_1j_10}}{a_{j_2j_20}} \frac{(\bar{x}_{j_1} - a_{j_1j_1l_1})(\bar{x}_{j_1} - a_{j_1j_1l_2})}{(\bar{x}_{j_1} - a_{j_2j_1s_1})(\bar{x}_{j_1} - a_{j_2j_1s_2})^2} < 0,$$

there is a down-parabola flow at $x^*_{j_2} = a_{j_1j_21}$ and $\bar{x}_{j_1} \neq a_{j_1j_1l}, a_{j_1j_1l_2}, a_{j_2j_1s_1}, a_{j_2j_1s_2}$. The variational equation is

$$\dot{x}_{j_2} = a_{j_2j_20}(\bar{x}_{j_1} - a_{j_2j_1s_1})(\bar{x}_{j_1} - a_{j_2j_1s_2})^2.$$

The parabola flows at $x^*_{j_2} = a_{j_1j_21}$ and $\bar{x}_{j_1} \neq a_{j_1j_1l}, a_{j_1j_1l_2}, a_{j_2j_1s_1}, a_{j_2j_1s_2}$ are positive and negative, for $\dot{x}_{j_2} > 0$ and $\dot{x}_{j_2} < 0$, respectively.

From case (viii$_{3a}$) with $\bar{x}_{j_2} = a_{j_1j_21}$ and the above analysis, thus, the equilibrium of $(x^*_{j_2}, x^*_{j_1}) = (a_{j_1j_21}, a_{j_2j_1s_1})$ with $a_{j_2j_1s_1} = a_{j_1j_1l_1}$ has the following properties as in Eqs. (1.176)–(1.179):

- For $a_{j_1j_10}(a_{j_1j_1l_1} - a_{j_1j_1l_2}) > 0$ and $a_{j_2j_20}(a_{j_2j_1s_1} - a_{j_2j_1s_2})^2 > 0$,

$$(a_{j_1j_21}, a_{j_2j_1s_1}) = \underbrace{(_{\text{DI:II}}\text{UP}, _{\text{SI:SO}}\text{US})}_{\text{up-parabola upper-saddle}}.$$

The equilibrium of $(x_{j_2}^*, x_{j_1}^*) = (a_{j_1j_21}, a_{j_2j_1s_1})$ is a $(_{\text{DI:II}}\text{UP},_{\text{SI:SO}}\text{US})$-up-parabola upper-saddle.

- For $a_{j_1j_10}(a_{j_1j_1l_1} - a_{j_1j_1l_2}) < 0$ and $a_{j_2j_20}(a_{j_2j_1s_1} - a_{j_2j_1s_2})^2 > 0$,

$$(a_{j_1j_21}, a_{j_2j_1s_1}) = \underbrace{(_{\text{II:DI}}\text{DP}, _{\text{SO:SI}}\text{LS})}_{\text{down-parabola lower-saddle}}.$$

The equilibrium of $(x_{j_2}^*, x_{j_1}^*) = (a_{j_1j_21}, a_{j_2j_1s_1})$ is an $(_{\text{II:DI}}\text{DP},_{\text{SO:SI}}\text{LS})$-down-parabola lower-saddle.

- For $a_{j_1j_10}(a_{j_1j_1l_1} - a_{j_1j_1l_2}) > 0$ and $a_{j_2j_20}(a_{j_2j_1s_1} - a_{j_2j_1s_2})^2 < 0$,

$$(a_{j_1j_21}, a_{j_2j_1s_1}) = \underbrace{(_{\text{II:DI}}\text{DP}, _{\text{SI:SO}}\text{US})}_{\text{down-parabola upper-saddle}}.$$

The equilibrium of $(x_{j_2}^*, x_{j_1}^*) = (a_{j_1j_21}, a_{j_2j_1s_1})$ is an $(_{\text{II:DI}}\text{DP},_{\text{SI:SO}}\text{US})$-down-parabola upper-saddle.

- For $a_{j_1j_10}(a_{j_1j_1l_1} - a_{j_1j_1l_2}) < 0$ and $a_{j_2j_20}(a_{j_2j_1s_1} - a_{j_2j_1s_2})^2 < 0$,

$$(a_{j_1j_21}, a_{j_2j_1s_1}) = \underbrace{(_{\text{DI:II}}\text{UP}, _{\text{SO:SI}}\text{LS})}_{\text{up-parabola lower-saddle}}.$$

The equilibrium of $(x_{j_2}^*, x_{j_1}^*) = (a_{j_1j_21}, a_{j_2j_1s_1})$ is a $(_{\text{DI:II}}\text{UP},_{\text{SO:SI}}\text{LS})$-up-parabola lower-saddle.

(viii$_4$) For $a_{j_1j_1l_1} = a_{j_2j_1s_2}$ (s_1, s_2, l_1, $l_2 \in \{1, 2\}$, $s_1 \neq s_2$, $l_1 \neq l_2$), in phase space,

$$\frac{dx_{j_1}}{dx_{j_2}} = \frac{a_{j_1j_10}}{a_{j_2j_20}} \frac{(x_{j_2} - a_{j_1j_21})(x_{j_1} - a_{j_1j_1l_2})}{(x_{j_1} - a_{j_2j_1s_1})(x_{j_1} - a_{j_2j_1s_2})},$$

and the deformation of the foregoing equation is

$$[(x_{j_1} - a_{j_1j_1l_2}) + (2a_{j_1j_1l_2} - a_{j_2j_1s_1} - a_{j_2j_1s_2})$$
$$+ (a_{j_1j_1l_2} - a_{j_2j_1s_1})(a_{j_1j_1l_2} - a_{j_2j_1s_2})\frac{1}{x_{j_1} - a_{j_1j_1l_2}}]dx_{j_1}$$
$$= \frac{a_{j_1j_10}}{a_{j_2j_20}}(x_{j_2} - a_{j_1j_21})dx_{j_2}.$$

With an initial condition of $(x_{j_1 0}, x_{j_2 0})$ at $t = t_0$, the integration of the above equation gives

$$\frac{1}{2}\left[(x_{j_1} - a_{j_1 j_1 l_2})^2 - (x_{j_1 0} - a_{j_1 j_1 l_2})^2\right] + (2a_{j_1 j_1 l_2} - a_{j_2 j_1 s_1} - a_{j_2 j_1 s_2})(x_{j_1} - x_{j_1 0})$$

$$+(a_{j_1 j_1 l_2} - a_{j_2 j_1 s_1})(a_{j_1 j_1 l_2} - a_{j_2 j_1 s_2})\ln\frac{|x_{j_1} - a_{j_1 j_1 l_2}|}{|x_{j_1 0} - a_{j_1 j_1 l_2}|}$$

$$= \frac{1}{2}\frac{a_{j_1 j_1 0}}{a_{j_2 j_2 0}}\left[(x_{j_2} - a_{j_1 j_2 1})^2 - (x_{j_2 0} - a_{j_1 j_2 1})^2\right].$$

(viii$_{4a}$) For $x_{j_1}^* = a_{j_1 j_1 l_1} = a_{j_2 j_1 s_2}$ with $\bar{x}_{j_2} \neq a_{j_1 j_2 1}$,

$$\frac{dx_{j_2}}{dx_{j_1}}\bigg|_{x_{j_1}^* = a_{j_2 j_1 s_2}} = \frac{a_{j_2 j_2 0}}{a_{j_1 j_1 0}}\frac{(x_{j_1} - a_{j_2 j_1 s_1})(x_{j_1} - a_{j_2 j_1 s_2})}{(\bar{x}_{j_2} - a_{j_1 j_2 1})(x_{j_1} - a_{j_1 j_1 l_2})}\bigg|_{x_{j_1}^* = a_{j_2 j_1 s_2}} = 0.$$

If

$$\frac{d^2 x_{j_2}}{dx_{j_1}^2}\bigg|_{x_{j_1}^* = a_{j_2 j_1 s_2}} = \frac{a_{j_2 j_2 0}}{a_{j_1 j_1 0}}\frac{(a_{j_2 j_1 s_2} - a_{j_2 j_1 s_1})}{(\bar{x}_{j_2} - a_{j_1 j_2 1})(a_{j_2 j_1 s_2} - a_{j_1 j_1 l_2})} > 0,$$

there is an up-parabola flow. If

$$\frac{d^2 x_{j_2}}{dx_{j_1}^2}\bigg|_{x_{j_1}^* = a_{j_2 j_1 s_2}} = \frac{a_{j_2 j_2 0}}{a_{j_1 j_1 0}}\frac{(a_{j_2 j_1 s_2} - a_{j_2 j_1 s_1})}{(\bar{x}_{j_2} - a_{j_1 j_2 1})(a_{j_2 j_1 s_2} - a_{j_1 j_1 l_2})} < 0,$$

there is a down-parabola flow. The variational equation at $x_{j_1}^* = a_{j_1 j_1 l_1} = a_{j_2 j_1 s_2}$ with $\bar{x}_{j_2} \neq a_{j_1 j_2 1}$ is

$$\Delta\dot{x}_{j_1} = a_{j_1 j_1 0}(\bar{x}_{j_2} - a_{j_1 j_2 1})(a_{j_1 j_1 l_1} - a_{j_1 j_1 l_2})\Delta x_{j_1}.$$

The parabola flows are source and sink for $a_{j_1 j_1 0}(\bar{x}_{j_2} - a_{j_1 j_2 1})(a_{j_1 j_1 l_1} - a_{j_1 j_1 l_2}) > 0$ and $a_{j_1 j_1 0}(\bar{x}_{j_2} - a_{j_1 j_2 1})(a_{j_1 j_1 l_1} - a_{j_1 j_1 l_2}) < 0$, respectively.

Therefore, the infinite-equilibrium of $x_{j_1}^* = a_{j_1 j_1 l_1} = a_{j_2 j_1 s_2}$ with $\bar{x}_{j_2} \neq a_{j_1 j_2 1}$ has the following properties as in Eqs. (1.181)–(1.184):

- For $a_{j_1 j_1 0}(\bar{x}_{j_2} - a_{j_1 j_2 1})(a_{j_1 j_1 l_1} - a_{j_1 j_1 l_2}) > 0$ and $a_{j_2 j_2 0}(a_{j_2 j_1 s_2} - a_{j_2 j_1 s_1}) > 0$,

$$(a_{j_1 j_1 l_1}, \bar{x}_{j_2}) = \underbrace{(\text{SO}, \text{UP})}_{\text{up-parabola source}}.$$

The infinite-equilibrium of $x_{j_1}^* = a_{j_1 j_1 l_1} = a_{j_2 j_1 s_2}$ is an (SO,UP)-up-parabola source.

- For $a_{j_1j_10}(\bar{x}_{j_2} - a_{j_1j_21})(a_{j_2j_1l_1} - a_{j_1j_1l_2}) < 0$ and $a_{j_2j_20}(a_{j_2j_1s_2} - a_{j_2j_1s_1}) > 0$,

$$(a_{j_1j_1l_1}, \bar{x}_{j_2}) = \underbrace{(\text{SI}, \text{DP})}_{\text{down-parabola sink}}.$$

The infinite-equilibrium of $x_{j_1}^* = a_{j_1j_1l_1} = a_{j_2j_1s_2}$ is an (SI,DP)-down-parabola sink.

- For $a_{j_1j_10}(\bar{x}_{j_2} - a_{j_1j_21})(a_{j_2j_1l_1} - a_{j_1j_1l_2}) > 0$ and $a_{j_2j_20}(a_{j_2j_1s_2} - a_{j_2j_1s_1}) < 0$,

$$(a_{j_1j_1l_1}, \bar{x}_{j_2}) = \underbrace{(\text{SO}, \text{DP})}_{\text{down-parabola sink}}.$$

The infinite-equilibrium of $x_{j_1}^* = a_{j_1j_1l_1} = a_{j_2j_1s_2}$ is an (SO,DP)-down-parabola source.

- For $a_{j_1j_10}(\bar{x}_{j_2} - a_{j_1j_21})(a_{j_2j_1l_1} - a_{j_1j_1l_2}) < 0$ and $a_{j_2j_20}(a_{j_2j_1s_2} - a_{j_2j_1s_1}) < 0$,

$$(a_{j_1j_1l_1}, \bar{x}_{j_2}) = \underbrace{(\text{SI}, \text{UP})}_{\text{up-parabola sink}}.$$

The infinite-equilibrium of $x_{j_1}^* = a_{j_1j_1l_1} = a_{j_2j_1s_2}$ is an (SI,UP)-up-parabola sink.

(viii$_{4b}$) From case (viii$_{4a}$), for $\bar{x}_{j_2} = a_{j_1j_2l_1}$, therefore, the equilibrium of $(x_{j_2}^*, x_{j_1}^*) = (a_{j_1j_21}, a_{j_2j_1s_2})$ with $a_{j_1j_11} = a_{j_2j_1s_2}$ has the following properties as in Eqs. (1.185)–(1.188):

- For $a_{j_1j_10}(a_{j_1j_1l_1} - a_{j_1j_1l_2}) > 0$ and $a_{j_2j_20}(a_{j_2j_1s_2} - a_{j_2j_1s_1}) > 0$,

$$(a_{j_1j_21}, a_{j_2j_1s_2}) = \underbrace{(_{\text{DP}}\text{SI}:_{\text{UP}}\text{SO}, \text{DP}_+ : \text{UP}_+)}_{\text{hyperbolic sink-to-source}}.$$

The equilibrium of $(x_{j_2}^*, x_{j_1}^*) = (a_{j_1j_21}, a_{j_2j_1s_2})$ is a $(_{\text{DP}}\text{SI}:_{\text{UP}}\text{SO},\text{DP}_+:\text{UP}_+)$-hyperbolic sink-to-source.

- For $a_{j_1j_10}(a_{j_1j_1l_1} - a_{j_1j_1l_2}) < 0$ and $a_{j_2j_20}(a_{j_2j_1s_2} - a_{j_2j_1s_1}) > 0$,

$$(a_{j_1j_21}, a_{j_2j_1s_2}) = \underbrace{(_{\text{UP}}\text{SO}:_{\text{DP}}\text{SI}, \text{UP}_+ : \text{DP}_+)}_{\text{circular source-to-sink}}.$$

The equilibrium of $(x_{j_2}^*, x_{j_1}^*) = (a_{j_1j_21}, a_{j_2j_1s_2})$ is a $(_{\text{UP}}\text{SO}:_{\text{DP}}\text{SI},\text{UP}_+:\text{DP}_+)$-circular source-to-sink.

- For $a_{j_1j_10}(a_{j_1j_1l_1} - a_{j_1j_1l_2}) > 0$ and $a_{j_2j_20}(a_{j_2j_1s_2} - a_{j_2j_1s_1}) < 0$,

$$(a_{j_1j_21}, a_{j_2j_1s_2}) = \underbrace{({}_{\text{UP}}\text{SI:}_{\text{DP}}\text{SO, UP}_-\text{:DP}_-)}_{\text{circular sink-to-source}}.$$

The equilibrium of $(x_{j_2}^*, x_{j_1}^*) = (a_{j_1j_21}, a_{j_2j_1s_2})$ is a $({}_{\text{UP}}\text{SI:}_{\text{DP}}\text{SO,UP}_-\text{:DP}_-)$-circular sink-to-source.

- For $a_{j_1j_10}(a_{j_1j_1l_1} - a_{j_1j_1l_2}) < 0$ and $a_{j_2j_20}(a_{j_2j_1s_2} - a_{j_2j_1s_1}) < 0$,

$$(a_{j_1j_21}, a_{j_2j_1s_2}) = \underbrace{({}_{\text{DP}}\text{SO:}_{\text{UP}}\text{SI, DP}_-\text{:UP}_-)}_{\text{hyperbolic source-to-sink}}.$$

The equilibrium of $(x_{j_2}^*, x_{j_1}^*) = (a_{j_1j_21}, a_{j_2j_1s_2})$ is a $({}_{\text{DP}}\text{SO:}_{\text{UP}}\text{SI,DP}_-\text{:UP}_-)$-hyperbolic source-to-sink.

(viii$_5$) For $a_{j_2j_1s_1} = a_{j_2j_1s_2} = a_{j_2j_11} = a_{j_1j_11}$, in phase place,

$$\frac{dx_{j_1}}{dx_{j_2}} = \frac{a_{j_1j_10}}{a_{j_2j_20}} \frac{(x_{j_2} - a_{j_1j_21})(x_{j_1} - a_{j_1j_1l_2})}{(x_{j_1} - a_{j_2j_11})^2},$$

and the deformation of the foregoing equation is

$$[(x_{j_1} - a_{j_1j_1l_2}) + 2(a_{j_1j_1l_2} - a_{j_2j_11}) + \frac{(a_{j_1j_1l_2} - a_{j_2j_11})^2}{x_{j_1} - a_{j_1j_1l_2}}]dx_{j_1}$$
$$= \frac{a_{j_1j_10}}{a_{j_2j_20}}(x_{j_2} - a_{j_1j_21})dx_{j_2}.$$

With an initial condition of (x_{j_10}, x_{j_20}) at $t = t_0$, the integration of the above equation gives the first integral manifold for $a_{j_1j_11} \neq a_{j_2j_11}$ as

$$\frac{1}{2}\left[(x_{j_1} - a_{j_1j_1l_2})^2 - (x_{j_10} - a_{j_1j_1l_2})^2\right] + 2(a_{j_1j_1l_2} - a_{j_2j_11})(x_{j_1} - x_{j_10})$$
$$+ (a_{j_1j_1l_2} - a_{j_2j_11})^2 \ln \frac{\mid x_{j_1} - a_{j_1j_1l_2} \mid}{\mid x_{j_10} - a_{j_1j_1l_2} \mid}$$
$$= \frac{1}{2}\frac{a_{j_1j_10}}{a_{j_2j_20}}\left[(x_{j_2} - a_{j_1j_2l_1})^2 - (x_{j_20} - a_{j_1j_2l_1})^2\right].$$

(viii$_{5a}$) At $x_{j_1}^* = a_{j_2j_11} = a_{j_1j_11}$ and $\bar{x}_{j_2} \neq a_{j_1j_21}$,

$$\frac{dx_{j_2}}{dx_{j_1}}\bigg|_{x^*_{j_1}=a_{j_2j_11}} = \frac{a_{j_2j_20}}{a_{j_1j_10}}\frac{(x_{j_1}-a_{j_2j_11})^2}{(\overline{x}_{j_2}-a_{j_1j_21})(x_{j_1}-a_{j_1j_1l_2})}\bigg|_{x^*_{j_1}=a_{j_2j_11}} = 0,$$

$$\frac{d^2x_{j_2}}{dx_{j_1}^2}\bigg|_{x^*_{j_1}=a_{j_2j_11}} = 0.$$

If

$$\frac{d^3x_{j_2}}{dx_{j_1}^3}\bigg|_{x^*_{j_1}=a_{j_2j_11}} = \frac{a_{j_2j_20}}{a_{j_1j_10}}\frac{2}{(\overline{x}_{j_2}-a_{j_1j_21})(a_{j_1j_1l_1}-a_{j_1j_1l_2})} > 0,$$

there is a second-order increasing-inflection flow at $x^*_{j_1}=a_{j_2j_11}$. If

$$\frac{d^3x_{j_2}}{dx_{j_1}^3}\bigg|_{x^*_{j_1}=a_{j_2j_11}} = \frac{a_{j_2j_20}}{a_{j_1j_10}}\frac{2}{(\overline{x}_{j_2}-a_{j_1j_21})(a_{j_1j_1l_1}-a_{j_1j_1l_2})} < 0,$$

there is a second-order decreasing-inflection flow at $x^*_{j_1}=a_{j_2j_11}$. Let

$$\Delta\dot{x}_{j_1} = a_{j_1j_10}(\overline{x}_{j_2}-a_{j_1j_21})(a_{j_1j_1l_1}-a_{j_1j_1l_2})\Delta x_{j_1}.$$

The second-order inflection flows at $x^*_{j_1}=a_{j_2j_11}$ in the x_{j_1}-direction are source and sink for $a_{j_1j_10}(\overline{x}_{j_2}-a_{j_1j_21})(a_{j_1j_1l_1}-a_{j_1j_1l_2}) > 0$ and $a_{j_1j_10}(\overline{x}_{j_2}-a_{j_1j_21})(a_{j_1j_1l_1}-a_{j_1j_1l_2}) < 0$.

Therefore, the infinite-equilibrium of $x^*_{j_1}=a_{j_2j_11}=a_{j_1j_1l_1}$ with $\overline{x}_{j_2}\neq a_{j_1j_21}$ has the following properties as in Eqs. (1.190)–(1.193):

- For $a_{j_1j_10}(\overline{x}_{j_2}-a_{j_1j_21})(a_{j_1j_1l_1}-a_{j_1j_1l_2}) > 0$ and $a_{j_2j_20} > 0$,

$$(a_{j_1j_11},\overline{x}_{j_2}) = \underbrace{(\text{SO},2^{\text{nd}}\text{II})}_{\text{second-order increasing-inflection source}}.$$

The infinite-equilibrium of $x^*_{j_1}=a_{j_2j_11}=a_{j_1j_1l_1}$ is an (SO,2^{nd}II)-second-order increasing-inflection source.

- For $a_{j_1j_10}(\overline{x}_{j_2}-a_{j_1j_21})(a_{j_1j_1l_1}-a_{j_1j_1l_2}) < 0$ and $a_{j_2j_20} > 0$,

$$(a_{j_1j_1l_1},\overline{x}_{j_2}) = \underbrace{(\text{SI},2^{\text{nd}}\text{DI})}_{\text{second-order decreasing-inflection sink}}.$$

The infinite-equilibrium of $x^*_{j_1}=a_{j_2j_11}=a_{j_1j_1l_1}$ is an (SI,2^{nd}DI)-second-order decreasing-inflection sink.

- For $a_{j_1j_10}(\bar{x}_{j_2} - a_{j_1j_21})(a_{j_1j_1l_1} - a_{j_1j_1l_2}) > 0$ and $a_{j_2j_20} < 0$,

$$(a_{j_1j_1l_1}, \bar{x}_{j_2}) = \underbrace{(\text{SO}, 2^{\text{nd}}\text{DI})}_{\text{second-order decreasing-inflection source}}.$$

The infinite-equilibrium of $x_{j_1}^* = a_{j_2j_11} = a_{j_1j_1l_1}$ is an (SO,2^{nd}DI)-second-order decreasing-inflection source.

- For $a_{j_1j_10}(\bar{x}_{j_2} - a_{j_1j_21})(a_{j_1j_1l_1} - a_{j_1j_1l_2}) < 0$ and $a_{j_2j_20} < 0$,

$$(a_{j_1j_1l_1}, \bar{x}_{j_2}) = \underbrace{(\text{SI}, 2^{\text{nd}}\text{II})}_{\text{second-order increasing-inflection sink}}.$$

The infinite-equilibrium of $x_{j_1}^* = a_{j_2j_11} = a_{j_1j_1l_1}$ is an (SI,2^{nd}II)-second-order increasing-inflection sink.

(viii$_{5b}$) For $x_{j_2}^* = a_{j_1j_2l_1}$ and $\bar{x}_{j_1} \neq a_{j_2j_11}$,

$$\frac{dx_{j_1}}{dx_{j_2}}\bigg|_{x_{j_2}^* = a_{j_1j_21}} = \frac{a_{j_1j_10}}{a_{j_2j_20}} \frac{(x_{j_2} - a_{j_1j_21})(\bar{x}_{j_1} - a_{j_1j_1l_2})}{(\bar{x}_{j_1} - a_{j_2j_11})^2}\bigg|_{x_{j_1}^* = a_{j_2j_11}} = 0.$$

If

$$\frac{d^2x_{j_1}}{dx_{j_2}}\bigg|_{x_{j_2}^* = a_{j_1j_21}} = \frac{a_{j_1j_10}}{a_{j_2j_20}} \frac{\bar{x}_{j_1} - a_{j_1j_1l_2}}{(\bar{x}_{j_1} - a_{j_2j_11})^2} > 0,$$

there is an up-parabola flow at $x_{j_2}^* = a_{j_1j_21}$. If

$$\frac{d^2x_{j_1}}{dx_{j_2}}\bigg|_{x_{j_2}^* = a_{j_1j_21}} = \frac{a_{j_1j_10}}{a_{j_2j_20}} \frac{\bar{x}_{j_1} - a_{j_1j_1l_2}}{(\bar{x}_{j_1} - a_{j_2j_11})^2} < 0,$$

there is a down-parabola flow at $x_{j_2}^* = a_{j_1j_21}$. Let

$$\dot{x}_{j_2} = a_{j_2j_20}(\bar{x}_{j_1} - a_{j_2j_11})^3.$$

The parabola flows at $x_{j_1}^* = a_{j_2j_11}$ are positive and negative for $a_{j_2j_20}(\bar{x}_{j_1} - a_{j_2j_11})^3 > 0$ and $a_{j_2j_20}(\bar{x}_{j_1} - a_{j_2j_11})^3 < 0$.

From case (viii$_{5a}$) with $\bar{x}_{j_2} = a_{j_1j_21}$, the above analysis with $\bar{x}_{j_1} = a_{j_2j_11} = a_{j_1j_1l_1}$ gives the equilibrium of $(x_{j_2}^*, x_{j_1}^*) = (a_{j_1j_21}, a_{j_2j_11})$ with $a_{j_2j_11} = a_{j_1j_1l_1}$, which has the following properties as in Eqs. (1.194)–(1.197):

- For $a_{j_1 j_1 0}(a_{j_1 j_1 l_1} - a_{j_1 j_1 l_2}) > 0$ and $a_{j_2 j_2 0} > 0$,

$$(a_{j_1 j_2 1}, a_{j_2 j_1 1}) = \underbrace{(_{2^{nd}(DI:II)}UP, \ _{SI:SO}US)}_{\text{second-order up-parabola upper-saddle}}.$$

The equilibrium of $(x_{j_2}^*, x_{j_1}^*) = (a_{j_1 j_2 1}, a_{j_2 j_1 1})$ is a $(_{2^{nd}(DI:II)}UP, \ _{SI:SO}US)$-second-order up-parabola upper-saddle.

- For $a_{j_1 j_1 0}(a_{j_1 j_2 l_1} - a_{j_1 j_2 l_2}) < 0$ and $a_{j_2 j_2 0} > 0$,

$$(a_{j_1 j_2 1}, a_{j_2 j_1 1}) = \underbrace{(_{2^{nd}(II:DI)}DP, \ _{SO:SI}US)}_{\text{second-order down-parabola upper-saddle}}.$$

The equilibrium of $(x_{j_2}^*, x_{j_1}^*) = (a_{j_1 j_2 1}, a_{j_2 j_1 1})$ is a $(_{2^{nd}(II:DI)}DP, _{SI:SO}US)$-second-order down-parabola upper-saddle.

- For $a_{j_1 j_1 0}(a_{j_1 j_2 l_1} - a_{j_1 j_2 l_2}) > 0$ and $a_{j_2 j_2 0} < 0$,

$$(a_{j_1 j_2 1}, a_{j_2 j_1 1}) = \underbrace{(_{2^{nd}(II:DI)}DP, \ _{SO:SI}LS)}_{\text{second-order down-parabola lower-saddle}}.$$

The equilibrium of $(x_{j_2}^*, x_{j_1}^*) = (a_{j_1 j_2 1}, a_{j_2 j_1 1})$ is a $(_{2^{nd}(II:DI)}DP, _{SO:SI}LS)$-second-order down-parabola lower-saddler.

- For $a_{j_1 j_1 0}(a_{j_1 j_2 l_1} - a_{j_1 j_2 l_2}) < 0$ and $a_{j_2 j_2 0} < 0$,

$$(a_{j_1 j_2 1}, a_{j_2 j_1 1}) = \underbrace{(_{2^{nd}(DI:II)}UP, \ _{SO:SI}LS)}_{\text{second-order up-parabola lower-saddle}}.$$

The equilibrium of $(x_{j_2}^*, x_{j_1}^*) = (a_{j_1 j_2 1}, a_{j_2 j_1 1})$ is a $(_{2^{nd}(DI:II)}UP, _{SO:SI}LS)$-second-order up-parabola lower-saddle.

(ix) For $\Delta_{j_1 j_1} > 0$ and $\Delta_{j_2 j_1} > 0$, the standard form is

$$\dot{x}_{j_2} = a_{j_2 j_2 0}(x_{j_1} - a_{j_2 j_1 1})(x_{j_1} - a_{j_2 j_1 2})(x_{j_1} - a_{j_2 j_1 3}),$$
$$\dot{x}_{j_1} = a_{j_1 j_1 0}(x_{j_1} - a_{j_1 j_1 1})(x_{j_1} - a_{j_1 j_1 2})(x_{j_2} - a_{j_1 j_2 1})$$

where

$$b_{j_1j_11}, b_{j_1j_12} = -\frac{1}{2}(B_{j_1j_1} \pm \sqrt{\Delta_{j_1j_1}}),$$

$$\{a_{j_1j_11}, a_{j_1j_12}\} = \text{sort}\{b_{j_1j_11}, b_{j_1j_12}\}, a_{j_1j_11} < a_{j_1j_12};$$

$$b_{j_2j_12}, b_{j_2j_13} = -\frac{1}{2}(B_{j_2j_1} \pm \sqrt{\Delta_{j_2j_1}}),$$

$$\{a_{j_2j_11}, a_{j_2j_12}, a_{j_2j_13}\} = \text{sort}\{b_{j_2j_11}, b_{j_2j_12}, b_{j_2j_13}\},$$

$$a_{j_2j_1s_1} < a_{j_2j_1s_2}, s_1, s_2 \in \{1, 2, 3\}, s_1 < s_2.$$

(ix$_1$) For $a_{j_1j_1l_1} \neq a_{j_2j_1s_1}(s_1, l_1 = 1, 2)$, in phase space,

$$\frac{dx_{j_1}}{dx_{j_2}} = \frac{a_{j_1j_10}}{a_{j_2j_20}} \frac{(x_{j_1} - a_{j_1j_1l_1})(x_{j_1} - a_{j_1j_1l_2})(x_{j_2} - a_{j_1j_21})}{(x_{j_1} - a_{j_2j_11})(x_{j_1} - a_{j_2j_12})(x_{j_1} - a_{j_2j_13})},$$

and the deformation of the foregoing equation is

$$[(x_{j_1} - a_{j_2j_1s_1}) + \sum_{l_1=1, l_2 \neq l_1}^{2} \frac{\prod_{s_2=1, s_2 \neq s_1}^{3}(a_{j_1j_1l_1} - a_{j_2j_1s_1})}{a_{j_1j_1l_1} - a_{j_1j_1l_2}}$$

$$+ \sum_{l_1=1, l_2 \neq l_1}^{2} \frac{\prod_{s_1=1}^{3}(a_{j_1j_1l_1} - a_{j_2j_1s_1})}{a_{j_1j_1l_1} - a_{j_1j_1l_2}} \frac{1}{x_{j_1} - a_{j_1j_1l_1}}]dx_{j_1}$$

$$= \frac{a_{j_1j_10}}{a_{j_2j_20}}(x_{j_2} - a_{j_1j_21})dx_{j_2}.$$

With an initial condition of (x_{j_10}, x_{j_20}) at $t = t_0$, the integration of the above equation gives

$$\frac{1}{2}[(x_{j_1} - a_{j_2j_1s_1})^2 - (x_{j_10} - a_{j_2j_1s_1})^2]$$

$$+ \sum_{l_1=1, l_2 \neq l_1}^{2} \frac{\prod_{s_2=1, s_2 \neq s_1}^{3}(a_{j_1j_1l_1} - a_{j_2j_1s_2})}{a_{j_1j_1l_1} - a_{j_1j_1l_2}}(x_{j_1} - x_{j_10})$$

$$+ \sum_{l_1=1, l_2 \neq l_1}^{2} \frac{\prod_{s_1=1}^{3}(a_{j_1j_1l_1} - a_{j_2j_1s_1})}{a_{j_1j_1l_1} - a_{j_1j_1l_2}} \ln \frac{|x_{j_1} - a_{j_1j_1l_1}|}{|x_{j_10} - a_{j_1j_1l_1}|}$$

$$= \frac{1}{2}\frac{a_{j_1j_10}}{a_{j_2j_20}}[(x_{j_2} - a_{j_1j_21})^2 - (x_{j_20} - a_{j_1j_21})^2].$$

(ix$_{1a}$) For $(x_{j_1}^*, x_{j_2}^*) = (a_{j_1j_2l_1}, a_{j_2j_1s_1})$ $(s_1, s_2 \in \{1, 2, 3\}, s_1 \neq s_2, l_1, l_2 \in \{1, 2\}, l_1 \neq l_2)$, consider two cases: (I) $x_{j_1}^* = a_{j_2j_1s_1}$ and (II) $x_{j_2}^* = a_{j_1j_21}$.

Case (I): In phase space, at $x_{j_1}^* = a_{j_2j_1s_1} \neq a_{j_1j_1l_1}$ $(s_1, s_2 \in \{1, 2, 3\}, s_1 \neq s_2; l_1, l_2 \in \{1, 2\}, l_1 \neq l_2)$ with $\bar{x}_{j_2} \neq a_{j_1j_21}$,

$$\frac{dx_{j_2}}{dx_{j_1}}\bigg|_{x_{j_1}^* = a_{j_2 j_1 s_1}} = \frac{a_{j_2 j_2 0}}{a_{j_1 j_1 0}} \frac{(x_{j_1} - a_{j_2 j_1 s_1})\prod_{s_2 = 1, s_2 \neq s_1}^{3}(x_{j_1} - a_{j_2 j_1 s_2})}{(\overline{x}_{j_2} - a_{j_1 j_2 1})\prod_{l_1 = 1}^{2}(x_{j_1} - a_{j_1 j_1 l_1})}\bigg|_{x_{j_1}^* = a_{j_2 j_1 s_1}} = 0.$$

If

$$\frac{d^2 x_{j_2}}{dx_{j_1}^2}\bigg|_{x_{j_1}^* = a_{j_2 j_1 s_1}} = \frac{a_{j_2 j_2 0}}{a_{j_1 j_1 0}} \frac{\prod_{s_2 = 1, s_2 \neq s_1}^{3}(a_{j_2 j_1 s_1} - a_{j_2 j_1 s_2})}{(\overline{x}_{j_2} - a_{j_1 j_2 1})\prod_{l_1 = 1}^{2}(a_{j_2 j_1 s_1} - a_{j_1 j_1 l_1})} > 0,$$

there is an up-parabola flow. If

$$\frac{d^2 x_{j_2}}{dx_{j_1}^2}\bigg|_{x_{j_1}^* = a_{j_2 j_1 s_1}} = \frac{a_{j_2 j_2 0}}{a_{j_1 j_1 0}} \frac{\prod_{s_2 = 1, s_2 \neq s_1}^{3}(a_{j_2 j_1 s_1} - a_{j_2 j_1 s_2})}{(\overline{x}_{j_2} - a_{j_1 j_2 1})\prod_{l_1 = 1}^{2}(a_{j_2 j_1 s_1} - a_{j_1 j_1 l_1})} < 0,$$

there is a down-parabola flow. Let

$$\dot{x}_{j_1} = a_{j_1 j_1 0}(\overline{x}_{j_2} - a_{j_1 j_2 1})\prod_{l_1 = 1}^{2}(a_{j_2 j_1 s_1} - a_{j_1 j_1 l_1}).$$

For $\overline{x}_{j_2} \neq a_{j_1 j_2 1}$, the parabola flows for $\dot{x}_{j_1} > 0$ and $\dot{x}_{j_1} < 0$ at $x_{j_1}^* = a_{j_2 j_1 s_1}$ are positive and negative, respectively.

Thus, the equilibrium of $x_{j_1}^* = a_{j_2 j_1 s_1}$ has the following properties:

- For $a_{j_1 j_1 0}(\overline{x}_{j_2} - a_{j_1 j_2 1})\prod_{l_1 = 1}^{2}(a_{j_2 j_1 s_1} - a_{j_1 j_1 l_1}) > 0, a_{j_2 j_2 0}\prod_{s_2 = 1, s_2 \neq s_1}^{3}(a_{j_2 j_1 s_1} - a_{j_2 j_1 s_2}) > 0,$

$$(\dot{x}_{j_1}, a_{j_2 j_1 s_1}) = \underbrace{(\mathrm{pF, UP})}_{\text{flow}}.$$

- For $a_{j_1 j_1 0}(\overline{x}_{j_2} - a_{j_1 j_2 1})\prod_{l_1 = 1}^{2}(a_{j_2 j_1 s_1} - a_{j_1 j_1 l_1}) < 0, a_{j_2 j_2 0}\prod_{s_2 = 1, s_2 \neq s_1}^{3}(a_{j_2 j_1 s_1} - a_{j_2 j_1 s_2}) > 0,$

$$(\dot{x}_{j_1}, a_{j_2 j_1 s_1}) = \underbrace{(\mathrm{nF, DP})}_{\text{flow}}.$$

- For $a_{j_1 j_1 0}(\overline{x}_{j_2} - a_{j_1 j_2 1})\prod_{l_1 = 1}^{2}(a_{j_2 j_1 s_1} - a_{j_1 j_1 l_1}) > 0, a_{j_2 j_2 0}\prod_{s_2 = 1, s_2 \neq s_1}^{3}(a_{j_2 j_1 s_1} - a_{j_2 j_1 s_2}) < 0,$

$$(\dot{x}_{j_1}, a_{j_2 j_1 s_1}) = \underbrace{(\mathrm{pF, DP})}_{\text{flow}}.$$

- For $a_{j_1 j_1 0}(\overline{x}_{j_2} - a_{j_1 j_2 1})\prod_{l_1 = 1}^{2}(a_{j_2 j_1 s_1} - a_{j_1 j_1 l_1}) < 0, a_{j_2 j_2 0}\prod_{s_2 = 1, s_2 \neq s_1}^{3}(a_{j_2 j_1 s_1} - a_{j_2 j_1 s_2}) < 0,$

$$(\dot{x}_{j_1}, a_{j_2 j_1 s_1}) = \underbrace{(\text{nF, UP})}_{\text{flow}}.$$

Case (II): In phase space, at $x_{j_2}^* = a_{j_1 j_2 1}$ with $\bar{x}_{j_1} \neq a_{j_1 j_1 l_1}, a_{j_2 j_1 s_1}$ ($s_1 = 1, 2, 3; l_1 = 1, 2$),

$$\frac{dx_{j_1}}{dx_{j_2}} \bigg|_{x_{j_2}^* = a_{j_1 j_2 1}} = \frac{a_{j_1 j_1 0}}{a_{j_2 j_2 0}} \frac{\prod_{l_1=1}^{2}(\bar{x}_{j_1} - a_{j_1 j_1 l_1})(x_{j_2} - a_{j_1 j_2 1})}{\prod_{s_1=1}^{3}(\bar{x}_{j_1} - a_{j_2 j_1 s_1})} \bigg|_{x_{j_2}^* = a_{j_1 j_2 1}} = 0.$$

If

$$\frac{d^2 x_{j_1}}{dx_{j_2}^2} \bigg|_{x_{j_2}^* = a_{j_1 j_2 1}} = \frac{a_{j_1 j_1 0}}{a_{j_2 j_2 0}} \frac{\prod_{l_1=1}^{2}(\bar{x}_{j_1} - a_{j_1 j_1 l_1})}{\prod_{s_1=1}^{3}(\bar{x}_{j_1} - a_{j_2 j_1 s_1})} > 0,$$

there is an up-parabola flow. If

$$\frac{d^2 x_{j_1}}{dx_{j_2}^2} \bigg|_{x_{j_2}^* = a_{j_1 j_2 1}} = \frac{a_{j_1 j_1 0}}{a_{j_2 j_2 0}} \frac{\prod_{l_1=1}^{2}(\bar{x}_{j_1} - a_{j_1 j_1 l_1})}{\prod_{s_1=1}^{3}(\bar{x}_{j_1} - a_{j_2 j_1 s_1})} < 0,$$

there is a down-parabola flow. Let

$$\dot{x}_{j_2} = a_{j_2 j_2 0} \prod_{s_1=1}^{3}(\bar{x}_{j_1} - a_{j_2 j_1 s_1}).$$

For $\bar{x}_{j_1} \neq a_{11 l_1}, a_{j_2 j_1 s_1}$ ($s_1, s_2 \in \{1, 2, 3\}$, $s_1 \neq s_2$; $l_1, l_2 \in \{1, 2\}$, $l_1 \neq l_2$), the parabola flows for $\dot{x}_{j_2} > 0$ and $\dot{x}_{j_2} < 0$ at $x_{j_2}^* = a_{j_1 j_2 1}$ are positive and negative, respectively.

Therefore, the equilibrium of $x_{j_2}^* = a_{j_1 j_2 1}$ has the following properties:

- For $a_{j_1 j_1 0} \prod_{l_1=1}^{2}(\bar{x}_{j_1} - a_{j_1 j_1 l_1}) > 0$ and $a_{j_2 j_2 0} \prod_{s_1=1}^{3}(\bar{x}_{j_1} - a_{j_2 j_1 s_1}) > 0$,

$$(a_{j_1 j_2 1}, \dot{x}_{j_2}) = \underbrace{(\text{UP, pF})}_{\text{up-parabola flow } (+)} .$$

- For $a_{j_1 j_1 0} \prod_{l_1=1}^{2}(\bar{x}_{j_1} - a_{j_1 j_1 l_1}) < 0$ and $a_{j_2 j_2 0} \prod_{s_1=1}^{3}(\bar{x}_{j_1} - a_{j_2 j_1 s_1}) > 0$,

$$(a_{j_1 j_2 1}, \dot{x}_{j_2}) = \underbrace{(\text{DP, pF})}_{\text{down-parabola flow } (+)} .$$

- For $a_{j_1 j_1 0} \prod_{l_1=1}^{2}(\bar{x}_{j_1} - a_{j_1 j_1 l_1}) > 0$ and $a_{j_2 j_2 0} \prod_{s_1=1}^{3}(\bar{x}_{j_1} - a_{j_2 j_1 s_1}) < 0$,

$$(a_{j_1 j_2 1}, \dot{x}_{j_2}) = \underbrace{(\text{DP}, \text{nF})}_{\text{down-parabola flow }(-)}.$$

- For $a_{j_1 j_1 0} \prod_{l_1 = 1}^{2} (\bar{x}_{j_1} - a_{j_1 j_1 l_1}) < 0$ and $a_{j_2 j_2 0} \prod_{s_1 = 1}^{3} (\bar{x}_{j_1} - a_{j_2 j_1 s_1}) < 0$,

$$(a_{j_1 j_2 1}, \dot{x}_{j_2}) = \underbrace{(\text{UP}, \text{nF})}_{\text{up-parabola flow }(-)}.$$

From the above analysis in cases (I) and (II), the equilibrium of $(x_{j_2}^*, x_{j_1}^*) = (a_{j_1 j_2 1}, a_{j_2 j_1 s_1})$ ($s_1, s_2 \in \{1, 2, 3\}$, $s_1 \neq s_2$; $l_1, l_2 \in \{1, 2\}$, $l_1 \neq l_2$) has the following properties as in Eqs. (1.201)–(1.204):

- For $a_{j_1 j_1 0} \prod_{l_1 = 1}^{2} (a_{j_2 j_1 s_1} - a_{j_1 j_1 l_1}) > 0$ and $a_{j_2 j_2 0} \prod_{s_2 = 1, s_2 \neq s_1}^{3} (a_{j_2 j_1 s_1} - a_{j_2 j_1 s_2}) > 0$,

$$(a_{j_1 j_2 1}, a_{j_2 j_1 s_1}) = \underbrace{(\text{UP}_+, \text{UP}_+)}_{\text{positive saddle}}.$$

The equilibrium of $(x_{j_2}^*, x_{j_1}^*) = (a_{j_1 j_2 1}, a_{j_2 j_1 s_1})$ is a $(\text{UP}_+, \text{UP}_+)$-positive saddle.
- For $a_{j_1 j_1 0} \prod_{l_1 = 1}^{2} (a_{j_2 j_1 s_1} - a_{j_1 j_1 l_1}) < 0$ and $a_{j_2 j_2 0} \prod_{s_2 = 1, s_2 \neq s_1}^{3} (a_{j_2 j_1 s_1} - a_{j_2 j_1 s_2}) > 0$,

$$(a_{j_1 j_2 1}, a_{j_2 j_1 s_1}) = \underbrace{(\text{DP}_+, \text{DP}_-)}_{\text{CCW center}}.$$

The equilibrium of $(x_{j_2}^*, x_{j_1}^*) = (a_{j_1 j_2 1}, a_{j_2 j_1 s_1})$ is a $(\text{DP}_+, \text{DP}_-)$-counter-clockwise center.
- For $a_{j_1 j_1 0} \prod_{l_1 = 1}^{2} (a_{j_2 j_1 s_1} - a_{j_1 j_1 l_1}) > 0$ and $a_{j_2 j_2 0} \prod_{s_2 = 1, s_2 \neq s_1}^{3} (a_{j_2 j_1 s_1} - a_{j_2 j_1 s_2}) < 0$,

$$(a_{j_1 j_2 1}, a_{j_2 j_1 s_1}) = \underbrace{(\text{DP}_-, \text{DP}_+)}_{\text{CW center}}.$$

The equilibrium of $(x_{j_2}^*, x_{j_1}^*) = (a_{j_1 j_2 1}, a_{j_2 j_1 s_1})$ is a $(\text{DP}_-, \text{DP}_+)$-clockwise center.
- For $a_{j_1 j_1 0} \prod_{l_1 = 1}^{2} (a_{j_2 j_1 s_1} - a_{j_1 j_1 l_1}) < 0$ and $a_{j_2 j_2 0} \prod_{s_2 = 1, s_2 \neq s_1}^{3} (a_{j_2 j_1 s_1} - a_{j_2 j_1 s_2}) < 0$,

$$(a_{j_1 j_2 1}, a_{j_2 j_1 s_1}) = \underbrace{(\text{UP}_-, \text{UP}_-)}_{\text{negative saddle}}.$$

The equilibrium of $(x_{j_2}^*, x_{j_1}^*) = (a_{j_1 j_2 1}, a_{j_2 j_1 s_1})$ is a $(\text{UP}_-, \text{UP}_-)$-negative saddle.

(ix$_{1b}$) From case II of ix$_{1a}$, for $\bar{x}_{j_1} = a_{j_1 j_1 l_1}$, the equilibrium of $(x_{j_1}^*, x_{j_2}^*) = (a_{j_1 j_1 l_1}, a_{j_1 j_2 1})$ ($l_1, l_2 \in \{1, 2\}$; $l_1 \neq l_2$) with $a_{j_1 j_1 l_1} \neq a_{j_2 j_1 1}, a_{j_2 j_1 2}, a_{j_2 j_1 3}$ has the following properties as in Eqs. (1.205)–(1.208):

- For $a_{j_1j_10}(a_{j_1j_1l_1} - a_{j_1j_1l_2}) > 0$ and $a_{j_2j_20}\prod_{s_1=1}^{3}(a_{j_1j_1l_1} - a_{j_2j_1s_1}) > 0$,

$$(a_{j_1j_1l_1}, a_{j_1j_21}) = \underbrace{(\text{DP:UP, pF})}_{\text{hyperbolic flow } (+)}.$$

The equilibrium of $(x_{j_1}^*, x_{j_2}^*) = (a_{j_1j_1l_1}, a_{j_1j_21})$ is a (DP:UP,pF)-positive hyperbolic flow.

- For $a_{j_1j_10}(a_{j_1j_1l_1} - a_{j_1j_1l_2}) < 0$ and $a_{j_2j_20}\prod_{s_1=1}^{3}(a_{j_1j_1l_1} - a_{j_2j_1s_1}) > 0$,

$$(a_{j_1j_1l_1}, a_{j_1j_21}) = \underbrace{(\text{UP:DP, pF})}_{\text{hyperbolic-secant flow } (+)}.$$

The equilibrium of $(x_{j_1}^*, x_{j_2}^*) = (a_{j_1j_1l_1}, a_{j_1j_21})$ is a (UP:DP,pF)-positive hyperbolic-secant flow.

- For $a_{j_1j_10}(a_{j_1j_1l_1} - a_{j_1j_1l_2}) > 0$ and $a_{j_2j_20}\prod_{s_1=1}^{3}(a_{j_1j_1l_1} - a_{j_2j_1s_1}) < 0$,

$$(a_{j_1j_1l_1}, a_{j_1j_21}) = \underbrace{(\text{UP:DP, nF})}_{\text{hyperbolic-secant flow } (-)}.$$

The equilibrium of $(x_{j_1}^*, x_{j_2}^*) = (a_{j_1j_1l_1}, a_{j_1j_21})$ is a (UP:DP,nF)-negative hyperbolic-secant flow.

- For $a_{j_1j_10}(a_{j_1j_1l_1} - a_{j_1j_1l_2}) < 0$ and $a_{j_2j_20}\prod_{s_1=1}^{3}(a_{j_1j_1l_1} - a_{j_2j_1s_1}) < 0$,

$$(a_{j_1j_1l_1}, a_{j_1j_21}) = \underbrace{(\text{DP:UP, nF})}_{\text{hyperbolic flow } (-)}.$$

The equilibrium of $(x_{j_1}^*, x_{j_2}^*) = (a_{j_1j_1l_1}, a_{j_1j_21})$ is a (DP:UP,nF)-negative hyperbolic flow.

(ix$_2$) For $a_{j_1j_1l_1} = a_{j_2j_1s_1}$ ($s_1, s_2 \in \{1,2,3\}$, $s_1 \neq s_2$; $l_1, l_2 \in \{1,2\}$, $l_1 \neq l_2$), in phase space,

$$\frac{dx_{j_1}}{dx_{j_2}} = \frac{a_{j_1j_10}}{a_{j_2j_20}} \frac{(x_{j_2} - a_{j_1j_21})(x_{j_1} - a_{j_1j_1l_2})}{\prod_{s_2=1, s_2 \neq s_1}^{3}(x_{j_1} - a_{j_2j_1s_2})},$$

and the deformation of the foregoing equation is

$$[(x_{j_1} - a_{j_1j_1l_2}) + (2a_{j_1j_1l_2} - a_{j_2j_1s_2} - a_{j_2j_1s_3}) + \frac{\prod_{s_2=1,s_2\neq s_1}^{3}(a_{j_1j_1l_2} - a_{j_2j_1s_2})}{x_{j_1} - a_{j_1j_1l_2}}]dx_{j_1}$$

$$= \frac{a_{j_1j_10}}{a_{j_2j_20}}(x_{j_2} - a_{j_1j_2l_1})dx_{j_2}.$$

With an initial condition of (x_{j_10}, x_{j_20}) at $t = t_0$, the integration of the above equation gives

$$\frac{1}{2}[(x_{j_1} - a_{j_1j_1l_2})^2 - (x_{j_10} - a_{j_1j_1l_2})^2] + (2a_{j_1j_1l_2} - a_{j_2j_1s_2} - a_{j_2j_1s_3})(x_{j_1} - x_{j_10})$$

$$+ \prod_{s_2=1,s_2\neq s_1}^{3}(a_{j_1j_1l_2} - a_{j_2j_1s_2})\ln\frac{|x_{j_1} - a_{j_1j_1l_2}|}{|x_{j_10} - a_{j_1j_1l_2}|}$$

$$= \frac{1}{2}\frac{a_{j_1j_10}}{a_{j_2j_20}}[(x_{j_2} - a_{j_1j_2l_1})^2 - (x_{j_20} - a_{j_1j_2l_1})^2].$$

(ix$_{2a}$) For $x_{j_1}^* = a_{j_2j_1s_1} = a_{j_1j_1l_1}$ with $\bar{x}_{j_2} \neq a_{j_1j_21}(l_1, l_2 \in \{1, 2\}; l_1 \neq l_2)$,

$$\frac{dx_{j_2}}{dx_{j_1}}\Big|_{x_{j_1}^* = a_{j_2j_1s_1}} = \frac{a_{j_2j_20}}{a_{j_1j_10}}\frac{\prod_{s_2=1,s_2\neq s_1}^{3}(a_{j_2j_1s_1} - a_{j_2j_1s_2})}{(\bar{x}_{j_2} - a_{j_1j_21})(a_{j_2j_1s_1} - a_{j_1j_1l_2})} > 0,$$

there is an increasing-inflection flow, and if

$$\frac{dx_{j_2}}{dx_{j_1}}\Big|_{x_{j_1}^* = a_{j_2j_1s_1}} = \frac{a_{j_2j_20}}{a_{j_1j_10}}\frac{\prod_{s_2=1,s_2\neq s_1}^{3}(a_{j_2j_1s_1} - a_{j_2j_1s_2})}{(\bar{x}_{j_2} - a_{j_1j_21})(a_{j_2j_1s_1} - a_{j_1j_1l_2})} < 0,$$

there is a decreasing-inflection flow. The variational equation is

$$\Delta\dot{x}_{j_1} = a_{j_1j_10}(\bar{x}_{j_2} - a_{j_1j_21})(a_{j_2j_1s_1} - a_{j_1j_1l_2})\Delta x_{j_1}.$$

The inflection flows at $x_{j_1}^* = a_{j_2j_1s_1}$ in the x_{j_1} direction are source and sink for $a_{j_1j_10}(\bar{x}_{j_2} - a_{j_1j_21})(a_{j_2j_1s_1} - a_{j_1j_1l_2}) > 0$ and $a_{j_1j_10}(\bar{x}_{j_2} - a_{j_1j_21})(a_{j_2j_1s_1} - a_{j_1j_1l_2}) < 0$, respectively.

Therefore, the infinite-equilibrium of $x_{j_1}^* = a_{j_2j_1s_1} = a_{j_1j_1l_1}(l_1 = 1, 2)$ with $\bar{x}_{j_2} \neq a_{j_1j_21}$ has the following properties as in Eqs. (1.210)–(1.213):

• For $a_{j_1j_10}(\bar{x}_{j_2} - a_{j_1j_21})(a_{j_2j_1s_1} - a_{j_1j_1l_2}) > 0$ and $a_{j_2j_20}\prod_{s_2=1,s_2\neq s_1}^{3}(a_{j_2j_1s_1} - a_{j_2j_1s_2}) > 0$,

$$(a_{j_1j_1l_1}, \bar{x}_{j_2}) = \underbrace{(\text{SO, II})}_{\text{increasing-inflection source}}.$$

The infinite-equilibrium of $x_{j_1}^* = a_{j_2j_1s_1} = a_{j_1j_1l_1}$ is an (SO,II)-increasing-inflection source.

- For $a_{j_1j_10}(\bar{x}_{j_2} - a_{j_1j_21})(a_{j_2j_1s_1} - a_{j_1j_1l_2}) < 0$ and $a_{j_2j_20}\prod_{s_2=1,s_2\neq s_1}^{3}(a_{j_2j_1s_1} - a_{j_2j_1s_2}) > 0$,

$$(a_{j_1j_1l_1}, \bar{x}_{j_2}) = \underbrace{(\text{SI}, \text{DI})}_{\text{decreasing-inflection sink}}.$$

The infinite-equilibrium of $x_{j_1}^{*} = a_{j_2j_1s_1} = a_{j_1j_1l_1}$ is an (SI,DI)-decreasing-inflection sink.

- For $a_{j_1j_10}(\bar{x}_{j_2} - a_{j_1j_21})(a_{j_2j_1s_1} - a_{j_1j_1l_2}) > 0$ and $a_{j_2j_20}\prod_{s_2=1,s_2\neq s_1}^{3}(a_{j_2j_1s_1} - a_{j_2j_1s_2}) < 0$,

$$(a_{j_1j_1l_1}, \bar{x}_{j_2}) = \underbrace{(\text{SO}, \text{DI})}_{\text{decreasing-inflection source}}.$$

The infinite-equilibrium of $x_{j_1}^{*} = a_{j_2j_1s_1} = a_{j_1j_1l_1}$ is an (SO,DI)-decreasing-inflection source.

- For $a_{j_1j_10}(\bar{x}_{j_2} - a_{j_1j_21})(a_{j_2j_1s_1} - a_{j_1j_1l_2}) < 0$ and $a_{j_2j_20}\prod_{s_2=1,s_2\neq s_1}^{3}(a_{j_2j_1s_1} - a_{j_2j_1s_2}) < 0$,

$$(a_{j_1j_1l_1}, \bar{x}_{j_2}) = \underbrace{(\text{SI}, \text{II})}_{\text{increasing-inflection source}}.$$

The infinite-equilibrium of $x_{j_1}^{*} = a_{j_2j_1s_1} = a_{j_1j_1l_1}$ is an (SI,II)-increasing-inflection sink.

(ix$_{2b}$) For $x_{j_2}^{*} = a_{j_1j_21}$, $\bar{x}_{j_1} \neq a_{j_1j_1l_2}, a_{j_2j_1s_2}$ ($s_1, s_2 \in \{1, 2, 3\}$, $s_1 \neq s_2$; $l_1, l_2 \in \{1, 2\}$, $l_1 \neq l_2$),

$$\frac{dx_{j_1}}{dx_{j_2}}\bigg|_{x_{j_2}^{*}=a_{j_1j_21}} = \frac{a_{j_1j_10}}{a_{j_2j_20}} \frac{(x_{j_2} - a_{j_1j_21})(\bar{x}_{j_1} - a_{j_1j_1l_2})}{\prod_{s_2=1,s_2\neq s_1}^{3}(\bar{x}_{j_1} - a_{j_2j_1s_2})}\bigg|_{x_{j_2}^{*}=a_{j_1j_21}} = 0.$$

If

$$\frac{d^2x_{j_1}}{dx_{j_2}^2}\bigg|_{x_{j_2}^{*}=a_{j_1j_21}} = \frac{a_{j_1j_10}}{a_{j_2j_20}} \frac{(a_{j_1j_21} - a_{j_1j_21})(\bar{x}_{j_1} - a_{j_1j_1l_1})(\bar{x}_{j_1} - a_{j_1j_1l_2})}{(\bar{x}_{j_1} - a_{j_2j_1s_1})\prod_{s_2=1,s_2\neq s_1}^{3}(\bar{x}_{j_1} - a_{j_2j_1s_2})} > 0,$$

there is an up-parabola flow. If

$$\frac{d^2x_{j_1}}{dx_{j_2}^2}\bigg|_{x_{j_2}^{*}=a_{j_1j_2l_1}} = \frac{a_{j_1j_10}}{a_{j_2j_20}} \frac{(a_{j_1j_21} - a_{j_1j_21})(\bar{x}_{j_1} - a_{j_1j_1l_1})(\bar{x}_{j_1} - a_{j_1j_1l_2})}{(\bar{x}_{j_1} - a_{j_2j_1s_1})\prod_{s_2=1,s_2\neq s_1}^{3}(\bar{x}_{j_1} - a_{j_2j_1s_2})} < 0,$$

there is a down-parabola flow. Let

$$\dot{x}_{j_2} = a_{j_2 j_2 0} \prod\nolimits_{s_1 = 1}^{3} (\bar{x}_{j_1} - a_{j_2 j_1 s_1}).$$

The parabola flows in the x_{j_2}-direction are positive and negative for $\dot{x}_{j_2} > 0$ and $\dot{x}_{j_2} < 0$, respectively.

Thus, the equilibrium of $x_{j_2}^* = a_{j_1 j_2 1}$ with $a_{j_2 j_1 s_1} = a_{j_1 j_1 l_1}$ has the following properties:

- For $a_{j_1 j_1 0}(a_{j_1 j_2 1} - a_{j_1 j_2 1}) \prod_{l_1 = 1}^{2} (\bar{x}_{j_1} - a_{j_1 j_1 l_1}) > 0$ and $a_{j_2 j_2 0} \prod_{s_1 = 1}^{3} (\bar{x}_{j_1} - a_{j_2 j_1 s_1}) > 0$,

$$(a_{j_1 j_2 1}, \dot{x}_{j_2}) = \underbrace{(\text{UP}, \text{pF})}_{\text{up-parabola flow } (+)} .$$

- For $a_{j_1 j_1 0}(a_{j_1 j_2 1} - a_{j_1 j_2 1}) \prod_{l_1 = 1}^{2} (\bar{x}_{j_1} - a_{j_1 j_1 l_1}) < 0$ and $a_{j_2 j_2 0} \prod_{s_1 = 1}^{3} (\bar{x}_{j_1} - a_{j_2 j_1 s_1}) > 0$,

$$(a_{j_1 j_2 1}, \dot{x}_{j_2}) = \underbrace{(\text{DP}, \text{pF})}_{\text{down-parabola flow } (+)} .$$

- For $a_{j_1 j_1 0}(a_{j_1 j_2 1} - a_{j_1 j_2 1}) \prod_{l_1 = 1}^{2} (\bar{x}_{j_1} - a_{j_1 j_1 l_1}) > 0$ and $a_{j_2 j_2 0} \prod_{s_1 = 1}^{3} (\bar{x}_{j_1} - a_{j_2 j_1 s_1}) < 0$,

$$(a_{j_1 j_2 1}, \dot{x}_{j_2}) = \underbrace{(\text{DP}, \text{nF})}_{\text{down-parabola flow } (-)} .$$

- For $a_{j_1 j_1 0}(a_{j_1 j_2 1} - a_{j_1 j_2 1}) \prod_{l_1 = 1}^{2} (\bar{x}_{j_1} - a_{j_1 j_1 l_1}) < 0$ and $a_{j_2 j_2 0} \prod_{s_1 = 1}^{3} (\bar{x}_{j_1} - a_{j_2 j_1 s_1}) < 0$,

$$(a_{j_1 j_2 1}, \dot{x}_{j_2}) = \underbrace{(\text{UP}, \text{nF})}_{\text{up-parabola flow } (-)} .$$

From the above analysis, for $x_{j_1}^* = a_{j_2 j_1 s_1} = a_{j_1 j_1 l_1}$, therefore, the equilibrium of $(x_{j_2}^*, x_{j_1}^*) = (a_{j_1 j_2 1}, a_{j_2 j_1 s_1})$ with $a_{j_2 j_1 s_1} = a_{j_1 j_1 l_1}$ has the following properties as in Eqs. (1.214)–(1.217):

- For $a_{j_1 j_1 0}(a_{j_1 j_1 l_1} - a_{j_1 j_1 l_2}) > 0$ and $a_{j_2 j_2 0} \prod_{s_2 = 1, s_2 \neq s_1}^{3} (a_{j_2 j_1 s_1} - a_{j_2 j_1 s_2}) > 0$,

$$(a_{j_1 j_2 1}, a_{j_2 j_1 s_1}) = (\underbrace{_{\text{DI:II}}\text{UP}, \text{SI:SO} \text{US})}_{\text{up-parabola upper-saddle}} .$$

The equilibrium of $(x_{j_2}^*, x_{j_1}^*) = (a_{j_1 j_2 1}, a_{j_2 j_1 s_1})$ is a $(_{\text{DI:II}}\text{UP}, _{\text{SI:SO}}\text{US})$-up-parabola upper-saddle.

- For $a_{j_1 j_1 0}(a_{j_1 j_1 l_1} - a_{j_1 j_1 l_2}) < 0$ and $a_{j_2 j_2 0} \prod_{s_2 = 1, s_2 \neq s_1}^{3} (a_{j_2 j_1 s_1} - a_{j_2 j_1 s_2}) > 0$,

$$(a_{j_1 j_2 1}, a_{j_2 j_1 s_1}) = \underbrace{(_{\text{II:DI}}DP, _{\text{SO:SI}}LS)}_{\text{down-parabola lower-saddle}}.$$

The equilibrium of $(x^*_{j_2}, x^*_{j_1}) = (a_{j_1 j_2 1}, a_{j_2 j_1 s_1})$ is an $(_{\text{II:DI}}DP, _{\text{SO:SI}}LS)$-down-parabola lower-saddle.

- For $a_{j_1 j_1 0}(a_{j_1 j_1 l_1} - a_{j_1 j_1 l_2}) > 0$ and $a_{j_2 j_2 0}\prod_{s_2 = 1, s_2 \neq s_1}^{3}(a_{j_2 j_1 s_1} - a_{j_2 j_1 s_2}) < 0$,

$$(a_{j_1 j_2 1}, a_{j_2 j_1 s_1}) = \underbrace{(_{\text{II:DI}}DP, _{\text{SI:SO}}US)}_{\text{down-parabola upper-saddle}}.$$

The equilibrium of $(x^*_{j_2}, x^*_{j_1}) = (a_{j_1 j_2 1}, a_{j_2 j_1 s_1})$ is an $(_{\text{II:DI}}DP, _{\text{SI:SO}}US)$-down-parabola upper-saddle.

- For $a_{j_1 j_1 0}(a_{j_1 j_1 l_1} - a_{j_1 j_1 l_2}) < 0$ and $a_{j_2 j_2 0}\prod_{s_2 = 1, s_2 \neq s_1}^{3}(a_{j_2 j_1 s_1} - a_{j_2 j_1 s_2}) < 0$,

$$(a_{j_1 j_2 1}, a_{j_2 j_1 s_1}) = \underbrace{(_{\text{DI:II}}UP, _{\text{SO:SI}}LS)}_{\text{up-parabola lower-saddle}}.$$

The equilibrium of $(x^*_{j_2}, x^*_{j_1}) = (a_{j_1 j_2 1}, a_{j_2 j_1 s_1})$ is a $(_{\text{DI:II}}UP, _{\text{SO:SI}}LS)$-up-parabola upper-saddle.

In the end, this theorem is proved.   ∎

Chapter 2
Parabola-Saddles and Third-Order Centers and Saddles

In this chapter, parabola-saddles, third-order centers and saddles, and hyperbolic singular flows in the crossing and product cubic systems are presented, and the corresponding switching dynamics are discussed through infinite-equilibriums in such cubic systems. The parabola-saddles are the appearing and switching bifurcations of the saddle and center. The third-order centers are the appearing bifurcations of the center, saddle, and center. The third-order saddles are the appearing bifurcations of the saddle, center, and saddle. The hyperbolic singular flows are the appearing bifurcations of hyperbolic and hyperbolic-secant flows. The infinite-equilibriums in such a cubic system include up-down-hyperbolic upper-to-lower saddles, parabola-saddles, hyperbolic and circular upper-to-lower saddles. The up-down hyperbolic upper-to-lower saddles are the switching bifurcations of hyperbolic singular flows with a saddle and center. The parabola-saddles are the switching bifurcations of parabola-saddles with hyperbolic-to-hyperbolic-secant flows. The parabola-hyperbolic and circular upper-to-lower saddles are the switching bifurcations of a third-order saddle with a hyperbolic-to-hyperbolic-secant flow and a third-order center with a hyperbolic-secant-to hyperbolic flow.

2.1 Parabola-Saddles and Third-Order Saddles and Centers

In this section, parabola-saddles and third-order saddles and centers with hyperbolic-singular flows are discussed for the appearing bifurcations.

A. C. J. Luo, *Two-dimensional Crossing and Product Cubic Systems, Vol. II*, https://doi.org/10.1007/978-3-031-57100-8_2

2.1.1 Parabola-Saddles with Hyperbolic Singular Flows

Consider a dynamical system as

$$
\begin{aligned}
\dot{x}_1 &= a_{110}(x_1 - a_{111})^2(x_2 - a_{121}), \\
\dot{x}_2 &= a_{220}(x_1 - a_{21s_1})(x_1 - a_{21s_2})^2,
\end{aligned}
\tag{2.1}
$$

and the corresponding first integral manifold is

$$
\begin{aligned}
&\frac{1}{2}\left[(x_1 - a_{111})^2 - (x_{10} - a_{111})^2\right] \\
&+\left[2(a_{111} - a_{21s_2}) + (a_{111} - a_{21s_1})\right](x_1 - x_{10}) \\
&+\left[(a_{111} - a_{21s_2})^2 + 2(a_{111} - a_{21s_2})(a_{111} - a_{21s_1})\right]\ln\frac{|x_1 - a_{111}|}{|x_{10} - a_{111}|} \\
&- (a_{111} - a_{21s_2})^2(a_{111} - a_{21s_1})(\frac{1}{x_1 - a_{111}} - \frac{1}{x_{10} - a_{111}})\} \\
&= \frac{1}{2}\frac{a_{110}}{a_{220}}\left[(x_2 - a_{121})^2 - (x_{20} - a_{121})^2\right].
\end{aligned}
\tag{2.2}
$$

For $a_{21s_1} = a_{211}, a_{21s_2} = a_{212}$, three cases are discussed as follows.

(i) For $a_{111} < a_{211}$, the equilibriums of $x_1^* = a_{111}, a_{211}, a_{212}$ and $x_2^* = a_{121}$ are

$$
\left\{
\begin{aligned}
(a_{121}, a_{212}) \\
(a_{121}, a_{211}) \\
(a_{111}, a_{121})
\end{aligned}
\right\}
=
\left\{
\begin{aligned}
&\underbrace{(\text{UP}, \text{US})}_{\text{up-parabola upper-saddle}} \\
&\underbrace{(\text{UP}_+, \text{UP}_+)}_{\text{positive-saddle}} \\
&\underbrace{(\text{DP:DP}, \text{nF})}_{\text{HB to HS flow } (-)}
\end{aligned}
\right\}
\text{ for } a_{110} > 0 \text{ and } a_{220} > 0;
\tag{2.3}
$$

$$
\left\{
\begin{aligned}
(a_{121}, a_{212}) \\
(a_{121}, a_{211}) \\
(a_{111}, a_{121})
\end{aligned}
\right\}
=
\left\{
\begin{aligned}
&\underbrace{(\text{DP}, \text{US})}_{\text{down-parabola upper-sadddle}} \\
&\underbrace{(\text{DP}_+, \text{DP}_-)}_{\text{CCW center}} \\
&\underbrace{(\text{UP:UP}, \text{nF})}_{\text{HS to HB flow } (-)}
\end{aligned}
\right\}
\text{ for } a_{110} < 0 \text{ and } a_{220} > 0;
\tag{2.4}
$$

$$
\begin{Bmatrix} (a_{121}, a_{212}) \\ (a_{121}, a_{211}) \\ (a_{111}, a_{121}) \end{Bmatrix} = \begin{Bmatrix} \underbrace{(DP, LS)}_{\text{down-parabola lower-saddle}} \\ \underbrace{(DP_-, DP_+)}_{\text{CW center}} \\ \underbrace{(UP{:}UP, pF)}_{\text{HS to HB flow } (+)} \end{Bmatrix} \quad \text{for } a_{110} > 0 \text{ and } a_{220} < 0; \quad (2.5)
$$

$$
\begin{Bmatrix} (a_{121}, a_{212}) \\ (a_{121}, a_{211}) \\ (a_{111}, a_{121}) \end{Bmatrix} = \begin{Bmatrix} \underbrace{(UP, LS)}_{\text{up-parabola lower-saddle}} \\ \underbrace{(UP_-, UP_-)}_{\text{negtive saddle}} \\ \underbrace{(DP{:}DP, pF)}_{\text{HB to HS flow } (+)} \end{Bmatrix} \quad \text{for } a_{110} < 0 \text{ and } a_{220} < 0; \quad (2.6)
$$

as shown in Fig. 2.1. HB and HS represent hyperbolic and hyperbolic-secant, respectively. The parabola upper-saddle and lower-saddle bifurcations of the positive and negative saddles with centers are presented. The bifurcations of hyperbolic-secant and hyperbolic flows are also presented. The positive and negative saddles with centers between a parabola-saddle and the hyperbolic-to-hyperbolic-secant flow are presented.

(ii) For $a_{212} < a_{111}$, the equilibriums of $x_1^* = a_{111}, a_{211}, a_{212}$ and $x_2^* = a_{121}$ are

$$
\begin{Bmatrix} (a_{111}, a_{121}) \\ (a_{121}, a_{212}) \\ (a_{121}, a_{211}) \end{Bmatrix} = \begin{Bmatrix} \underbrace{(UP{:}UP, pF)}_{\text{HS to HB flow } (+)} \\ \underbrace{(UP, US)}_{\text{up-parabola upper-saddle}} \\ \underbrace{(UP_+, UP_+)}_{\text{positive-saddle}} \end{Bmatrix} \quad \text{for } a_{110} > 0 \text{ and } a_{220} > 0; \quad (2.7)
$$

$$
\begin{Bmatrix} (a_{111}, a_{121}) \\ (a_{121}, a_{212}) \\ (a_{121}, a_{211}) \end{Bmatrix} = \begin{Bmatrix} \underbrace{(UP{:}UP, pF)}_{\text{HB to HS flow } (+)} \\ \underbrace{(DP, US)}_{\text{down-parabola upper-sadddle}} \\ \underbrace{(DP_+, DP_-)}_{\text{CCW center}} \end{Bmatrix} \quad \text{for } a_{110} < 0 \text{ and } a_{220} > 0; \quad (2.8)
$$

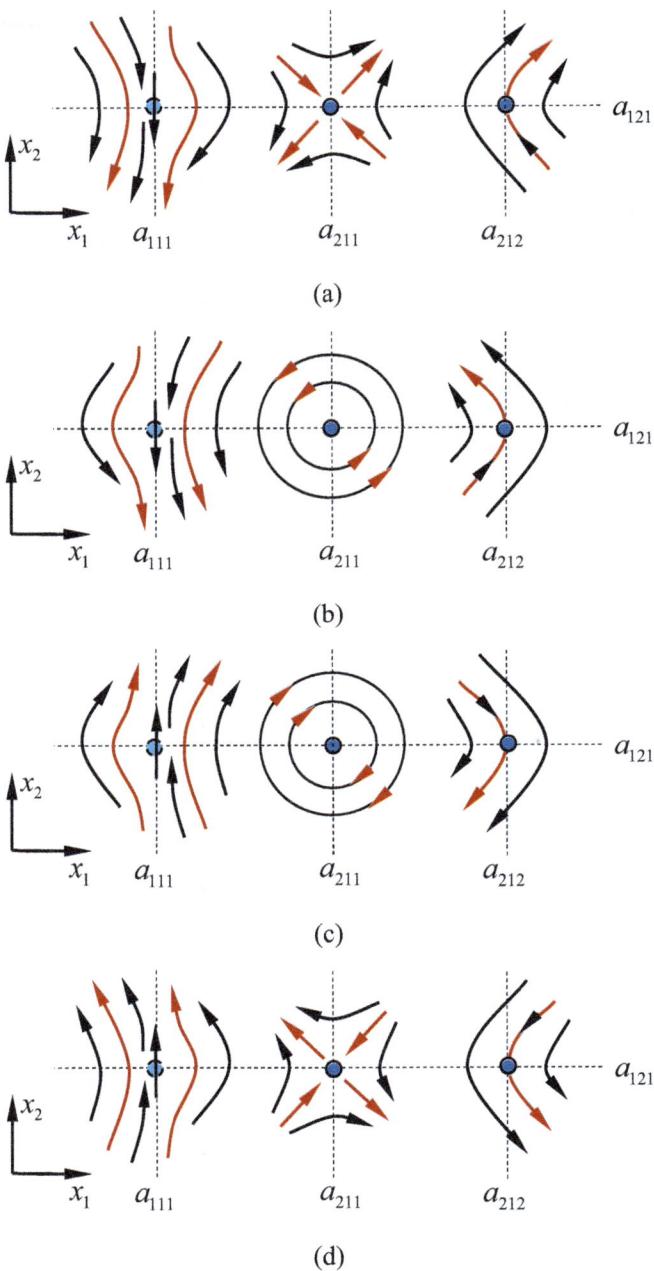

Fig. 2.1 Phase portraits ($a_{111} < a_{211}$) for two-dimensional systems on the x_1-direction with $x_1^* = a_{111}, a_{211}, a_{212}$ and on the x_2-direction with $x_2^* = a_{121}$. (**a**) ($a_{110} > 0, a_{220} > 0$), (**b**) ($a_{110} < 0$, $a_{220} > 0$), (**c**) ($a_{110} > 0, a_{220} < 0$), (**d**) ($a_{110} < 0, a_{220} < 0$)

$$\begin{Bmatrix} (a_{111}, a_{121}) \\ (a_{121}, a_{212}) \\ (a_{121}, a_{211}) \end{Bmatrix} = \begin{Bmatrix} \underbrace{(DP:DP, nF)}_{\text{HB to HS flow } (-)} \\ \underbrace{(DP, LS)}_{\text{down-parabola lower-sadddle}} \\ \underbrace{(DP_-, DP_+)}_{\text{CW center}} \end{Bmatrix} \quad \text{for } a_{110} > 0 \text{ and } a_{220} < 0; \quad (2.9)$$

$$\begin{Bmatrix} (a_{111}, a_{121}) \\ (a_{121}, a_{212}) \\ (a_{121}, a_{211}) \end{Bmatrix} = \begin{Bmatrix} \underbrace{(UP:UP, nF)}_{\text{HS to HB flow } (-)} \\ \underbrace{(UP, LS)}_{\text{up-parabola lower-saddle}} \\ \underbrace{(UP_-, UP_-)}_{\text{negtive saddle}} \end{Bmatrix} \quad \text{for } a_{110} < 0 \text{ and } a_{220} < 0; \quad (2.10)$$

as shown in Fig. 2.2. The parabola upper-saddle and lower-saddle bifurcations of the positive and negative saddles with centers are presented. The bifurcations of hyperbolic-secant and hyperbolic flows are also presented. The positive and negative saddles with centers are presented on the left side.

(iii) For $a_{111} \in (a_{211}, a_{212})$, the equilibriums of $x_1^* = a_{111}, a_{211}, a_{212}$ and $x_2^* = a_{121}$ are the same as presented in Eqs. (2.7)–(2.10). The locations of the parabola-saddle with the hyperbolic-to-hyperbolic-secant flow are switched, as presented in Fig. 2.3.

For $a_{21s_2} = a_{211}, a_{21s_1} = a_{212}$, readers can discuss three cases as an exercise.

2.1.2   Third-Order Saddles and Centers

Consider a dynamical system as

$$\begin{aligned} \dot{x}_1 &= a_{110}(x_1 - a_{111})^2(x_2 - a_{121}), \\ \dot{x}_2 &= a_{220}(x_1 - a_{211})^3, \end{aligned} \quad (2.11)$$

and the corresponding first integral manifold is

$$\begin{aligned} &\frac{1}{3}\left[(x_1 - a_{111})^3 - (x_{10} - a_{111})^3\right] \\ &+ \frac{3}{2}(a_{111} - a_{211})\left[(x_1 - a_{111})^2 - (x_{10} - a_{111})^2\right] \\ &+ 3(a_{111} - a_{211})^2\left[(x_1 - a_{111}) - (x_{10} - a_{111})\right] \\ &+ (a_{111} - a_{211})^2 \ln\frac{|x_1 - a_{111}|}{|x_{10} - a_{111}|} \\ &= \frac{1}{2}\frac{a_{110}}{a_{220}}\left[(x_2 - a_{121})^2 - (x_{20} - a_{121})^2\right] \end{aligned} \quad (2.12)$$

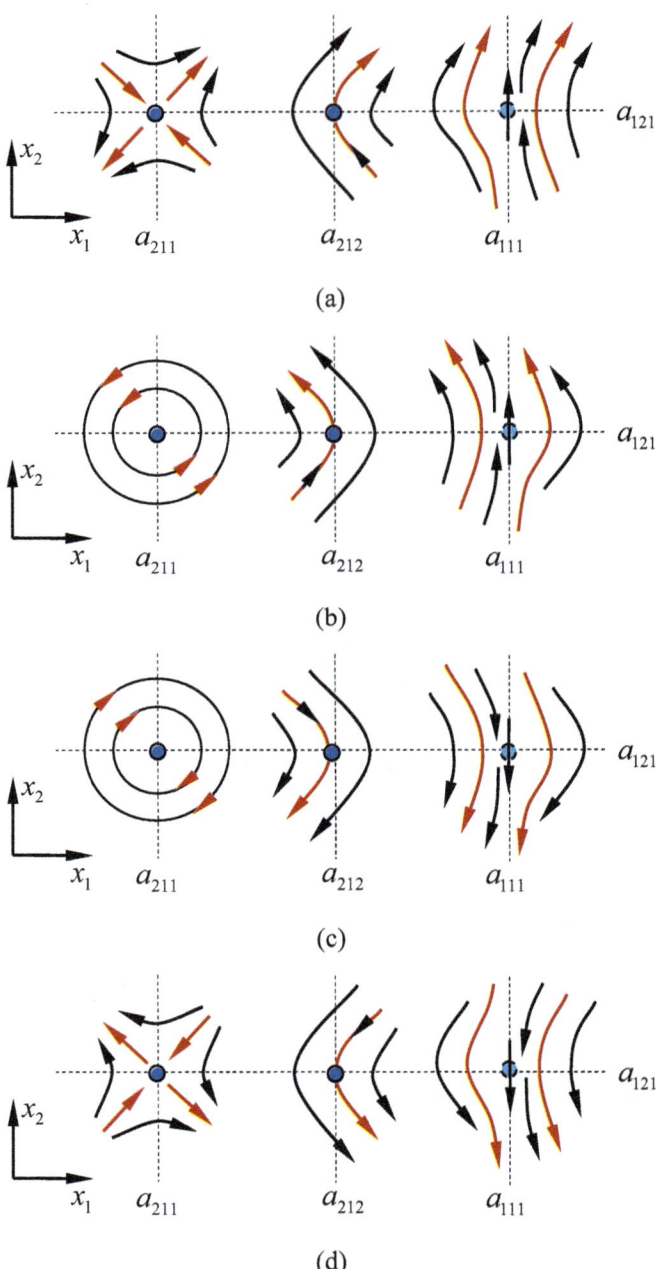

Fig. 2.2 Phase portraits $(a_{212} < a_{111})$ for two-dimensional systems on the x_1-direction with $x_1^* = a_{111}, a_{211}, a_{212}$ and on the x_2-direction with $x_2^* = a_{121}$. (**a**) $(a_{110} > 0, a_{220} > 0)$, (**b**) $(a_{110} < 0, a_{220} > 0)$, (**c**) $(a_{110} > 0, a_{220} < 0)$, (**d**) $(a_{110} < 0, a_{220} < 0)$

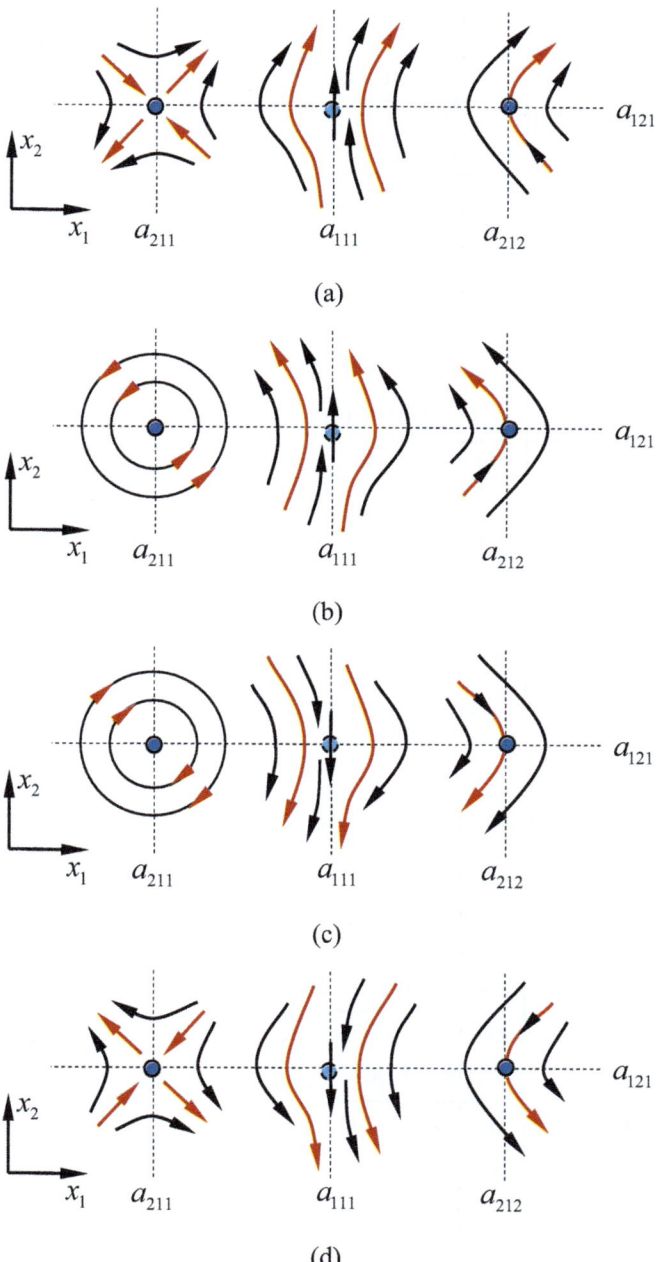

Fig. 2.3 Phase portraits ($a_{111} \in (a_{211}, a_{212})$) for two-dimensional systems on the x_1-direction with $x_1^* = a_{111}, a_{211}, a_{212}$ and on the x_2-direction with $x_2^* = a_{121}$. (**a**) ($a_{110} > 0$, $a_{220} > 0$), (**b**) ($a_{110} < 0$, $a_{220} > 0$), (**c**) ($a_{110} > 0$, $a_{220} < 0$), (**d**) ($a_{110} < 0$, $a_{220} < 0$)

(i) For $a_{111} < a_{211}$, the equilibriums of $x_1^* = a_{111}, a_{211}$ and $x_2^* = a_{121}$ are:

$$
\left\{ \begin{array}{c} (a_{121}, a_{211}) \\ (a_{111}, a_{121}) \end{array} \right\} = \left\{ \begin{array}{c} \underbrace{(\text{UP}_+, 3^{\text{rd}}\text{UP}_+)}_{\text{third-order positive saddle}} \\ \underbrace{(\text{DP}:\text{DP}, \text{nF})}_{\text{HB to HS flow }(-)} \end{array} \right\} \quad \text{for } a_{110} > 0 \text{ and } a_{220} > 0; \quad (2.13)
$$

$$
\left\{ \begin{array}{c} (a_{121}, a_{211}) \\ (a_{111}, a_{121}) \end{array} \right\} = \left\{ \begin{array}{c} \underbrace{(\text{DP}_+, 3^{\text{rd}}\text{DP}_-)}_{\text{third-order CCW center}} \\ \underbrace{(\text{UP}:\text{UP}, \text{nF})}_{\text{HB to HS flow }(-)} \end{array} \right\} \quad \text{for } a_{110} < 0 \text{ and } a_{220} > 0; \quad (2.14)
$$

$$
\left\{ \begin{array}{c} (a_{121}, a_{211}) \\ (a_{111}, a_{121}) \end{array} \right\} = \left\{ \begin{array}{c} \underbrace{(\text{DP}_-, 3^{\text{rd}}\text{DP}_+)}_{\text{third-order CW center}} \\ \underbrace{(\text{UP}:\text{UP}, \text{pF})}_{\text{HB to HS flow }(+)} \end{array} \right\} \quad \text{for } a_{110} > 0 \text{ and } a_{220} < 0; \quad (2.15)
$$

$$
\left\{ \begin{array}{c} (a_{121}, a_{211}) \\ (a_{111}, a_{121}) \end{array} \right\} = \left\{ \begin{array}{c} \underbrace{(\text{UP}_-, 3^{\text{rd}}\text{UP}_-)}_{\text{third-order negative saddle}} \\ \underbrace{(\text{DP}:\text{DP}, \text{pF})}_{\text{HB to HS flow }(+)} \end{array} \right\} \quad \text{for } a_{110} < 0 \text{ and } a_{220} < 0; \quad (2.16)
$$

as shown in Fig. 2.4.

(ii) For $a_{211} < a_{111}$, the equilibriums of $x_1^* = a_{111}, a_{211}$ and $x_2^* = a_{121}$ are

$$
\left\{ \begin{array}{c} (a_{111}, a_{121}) \\ (a_{121}, a_{211}) \end{array} \right\} = \left\{ \begin{array}{c} \underbrace{(\text{UP}:\text{UP}, \text{pF})}_{\text{HS to HB flow }(+)} \\ \underbrace{(\text{UP}_+, 3^{\text{rd}}\text{UP}_+)}_{\text{third-order positive saddle}} \end{array} \right\} \quad \text{for } a_{110} > 0 \text{ and } a_{220} > 0; \quad (2.17)
$$

$$
\left\{ \begin{array}{c} (a_{111}, a_{121}) \\ (a_{121}, a_{211}) \end{array} \right\} = \left\{ \begin{array}{c} \underbrace{(\text{DP}:\text{DP}, \text{pF})}_{\text{HS to HB flow }(+)} \\ \underbrace{(\text{DP}_+, 3^{\text{rd}}\text{DP}_-)}_{\text{third-order CCW center}} \end{array} \right\} \quad \text{for } a_{110} < 0 \text{ and } a_{220} > 0; \quad (2.18)
$$

$$
\left\{ \begin{array}{c} (a_{111}, a_{121}) \\ (a_{121}, a_{211}) \end{array} \right\} = \left\{ \begin{array}{c} \underbrace{(\text{DP}:\text{DP}, \text{nF})}_{\text{HS to HB flow }(-)} \\ \underbrace{(\text{DP}_-, 3^{\text{rd}}\text{DP}_+)}_{\text{third-order CW center}} \end{array} \right\} \quad \text{for } a_{110} > 0 \text{ and } a_{220} < 0; \quad (2.19)
$$

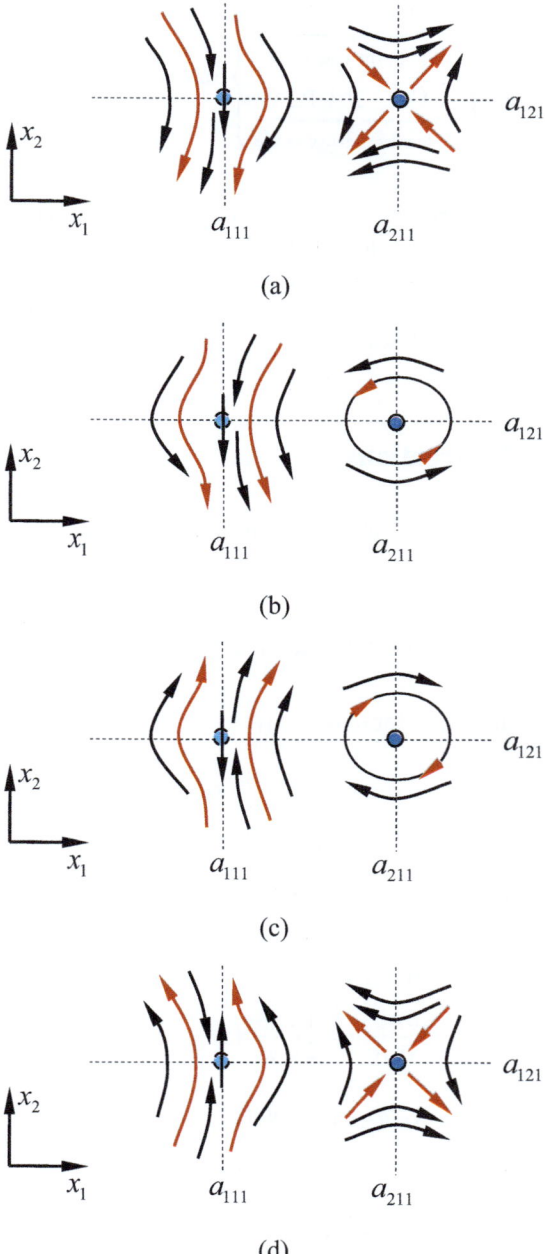

Fig. 2.4 Phase portraits ($a_{111} < a_{211}$) for two-dimensional systems on the x_1-direction with $x_1^* = a_{111}, a_{211}$ and on the x_2-direction with $x_2^* = a_{121}$. (**a**) ($a_{110} > 0$, $a_{220} > 0$), (**b**) ($a_{110} < 0$, $a_{220} > 0$), (**c**) ($a_{110} > 0$, $a_{220} < 0$), (**d**) ($a_{110} < 0$, $a_{220} < 0$)

$$\left\{ \begin{matrix} (a_{111}, a_{121}) \\ (a_{121}, a_{211}) \end{matrix} \right\} = \left\{ \begin{matrix} \overbrace{(\text{UP:UP, nF})}^{\text{HS to HB flow }(-)} \\ \underbrace{(\text{UP}_-, 3^{\text{rd}}\text{UP}_-)}_{\text{third-order negative saddle}} \end{matrix} \right\} \text{for } a_{110} < 0 \text{ and } a_{220} < 0; \qquad (2.20)$$

as shown in Fig. 2.5.

The single third-order saddles are for the appearing bifurcation from the saddle to the series of the saddle, center, and saddle. The single third-order centers are for the appearing bifurcation from the center to the series of center, saddle, and center. The hyperbolic-to-hyperbolic-secant flows are the appearing bifurcations for the hyperbolic and hyperbolic-secant flows.

2.2 Switching Bifurcations

The switching bifurcations are through the infinite equilibriums.

2.2.1 Up-Down Hyperbolic Saddle Infinite-Equilibriums

Consider a dynamical system for $a_{111} = a_{211}$ as

$$\begin{aligned} \dot{x}_1 &= a_{110}(x_1 - a_{111})^2(x_2 - a_{121}), \\ \dot{x}_2 &= a_{220}(x_1 - a_{211})(x_1 - a_{212})^2, \end{aligned} \qquad (2.21)$$

and the corresponding first integral manifold $(a_{111} = a_{211})$ is

$$\begin{aligned} &\frac{1}{2}\left[(x_1 - a_{111})^2 - (x_{10} - a_{111})^2\right] + 2(a_{111} - a_{212})(x_1 - x_{10}) \\ &+ (a_{111} - a_{212})^2 \ln \frac{|x_1 - a_{111}|}{|x_{10} - a_{111}|} = \frac{1}{2}\frac{a_{110}}{a_{220}}\left[(x_2 - a_{121})^2 - (x_{20} - a_{121})^2\right]. \end{aligned} \qquad (2.22)$$

The equilibriums of $x_1^* = a_{111}, a_{211}, a_{212}$ with $x_2^* = a_{121}$ for $a_{111} = a_{211}$ have the following properties:

$$\left\{ \begin{matrix} (a_{121}, a_{212}) \\ (a_{121}, a_{211}) \end{matrix} \right\} = \left\{ \begin{matrix} \overbrace{(\text{UP, US})}^{\text{up-parabola upper-saddle}} \\ \underbrace{(_{\text{UD}}\text{LS:}_{\text{DU}}\text{US, DP}_- : \text{UP}_+)}_{\text{hyperbolic lower-to-upper saddle}} \end{matrix} \right\} \text{for } a_{110} > 0, a_{220} > 0, \qquad (2.23)$$

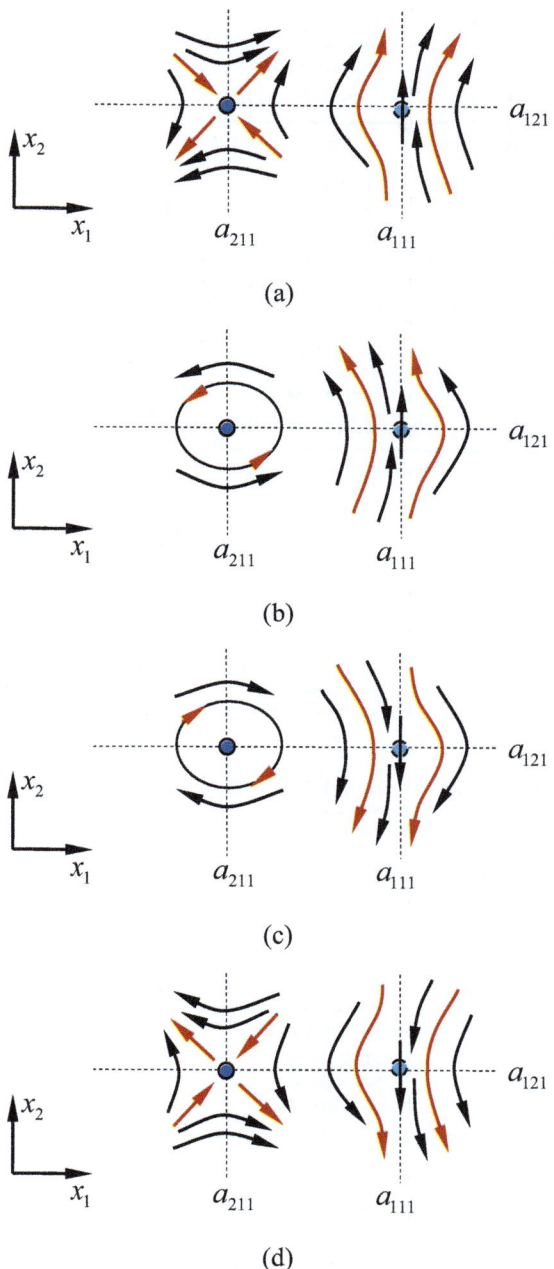

Fig. 2.5 Phase portraits ($a_{211} < a_{111}$) for two-dimensional systems on the x_1-direction with $x_1^* = a_{111}, a_{211}$ and on the x_2-direction with $x_2^* = a_{121}$. (**a**) ($a_{110} > 0$, $a_{220} > 0$), (**b**) ($a_{110} < 0$, $a_{220} > 0$), (**c**) ($a_{110} > 0$, $a_{220} < 0$), (**d**) ($a_{110} < 0$, $a_{220} < 0$)

$$\left\{ \begin{array}{l} (a_{121}, a_{212}) \\ (a_{111}, a_{121}) \end{array} \right\} = \left\{ \begin{array}{c} \underbrace{(DP, US)}_{\text{up-parabola upper-saddle}} \\ \underbrace{(_{DU}US:_{UD}LS, UP_-:DP_+)}_{\text{hyperbolic-secant upper-to-lower saddle}} \end{array} \right\} \text{ for } a_{110} < 0, a_{220} > 0, \quad (2.24)$$

$$\left\{ \begin{array}{l} (a_{121}, a_{212}) \\ (a_{111}, a_{121}) \end{array} \right\} = \left\{ \begin{array}{c} \underbrace{(DP, LS)}_{\text{up-parabola lower-saddle}} \\ \underbrace{(_{DU}LS:_{UD}US, UP_+:DP_-)}_{\text{hyperbolic-secant lower-to-upper saddle}} \end{array} \right\} \text{ for } a_{110} > 0, a_{220} < 0, \quad (2.25)$$

$$\left\{ \begin{array}{l} (a_{121}, a_{212}) \\ (a_{111}, a_{121}) \end{array} \right\} = \left\{ \begin{array}{c} \underbrace{(UP, LS)}_{\text{up-parabola lower-saddle}} \\ \underbrace{(_{UD}US:_{DU}LS, DP_+:UP_-)}_{\text{hyperbolic upper-to-lower saddle}} \end{array} \right\} \text{ for } a_{110} < 0, a_{220} < 0, \quad (2.26)$$

and the infinite equilibrium of $x_1^* = a_{112} = a_{211}$ is summarized as follows:

$$(a_{111}, \bar{x}_2) = \underbrace{(LS, UD)}_{\text{up-down lower-saddle}} \quad \text{if } \bar{x}_2 \in (-\infty, a_{121}),$$

$$(a_{111}, \bar{x}_2) = \underbrace{(US, DU)}_{\text{down-up upper-saddle}} \quad \text{if } \bar{x}_2 \in (a_{121}, \infty), \quad (2.27)$$

for $a_{110} > 0, a_{220} > 0$;

$$(a_{111}, \bar{x}_2) = \underbrace{(US, DU)}_{\text{down-up upper-saddle}} \quad \text{if } \bar{x}_2 \in (-\infty, a_{121}),$$

$$(a_{111}, \bar{x}_2) = \underbrace{(LS, UD)}_{\text{up-down lower-saddle}} \quad \text{if } \bar{x}_2 \in (a_{121}, \infty), \quad (2.28)$$

for $a_{110} < 0, a_{220} > 0$;

$$(a_{111}, \bar{x}_2) = \underbrace{(LS, DU)}_{\text{down-up lower-saddle}} \quad \text{if } \bar{x}_2 \in (-\infty, a_{121}),$$

$$(a_{111}, \bar{x}_2) = \underbrace{(US, UD)}_{\text{up-down upper-saddle}} \quad \text{if } \bar{x}_2 \in (a_{121}, \infty), \quad (2.29)$$

for $a_{110} > 0, a_{220} < 0$;

$$(a_{111}, \bar{x}_2) = \underbrace{(US, UD)}_{\text{up-down upper-saddle}} \quad \text{if } \bar{x}_2 \in (-\infty, a_{121}),$$

$$(a_{111}, \bar{x}_2) = \underbrace{(LS, DU)}_{\text{down-up lower-saddle}} \quad \text{if } \bar{x}_2 \in (a_{121}, \infty), \quad (2.30)$$

for $a_{110} < 0, a_{220} < 0$.

The infinite-equilibriums of $x_1^* = a_{111} = a_{211}$ are parabola sink and source, as presented in Fig. 2.6. The infinite equilibrium is for the switching bifurcation of

and the increasing and decreasing-inflection upper-saddle and lower-saddle infinite equilibrium of $x_1^* = a_{111} = a_{212}$ are

$$(a_{111}, \bar{x}_2) = \underbrace{(\mathrm{LS}, \mathrm{DI})}_{\text{decreasing-inflection lower-saddle}} \qquad \text{if } \bar{x}_2 \in (-\infty, a_{121}),$$

$$(a_{111}, \bar{x}_2) = \underbrace{(\mathrm{US}, \mathrm{II})}_{\text{increasing-inflection upper-saddle}} \qquad \text{if } \bar{x}_2 \in (a_{121}, \infty), \tag{2.38}$$

for $a_{110} > 0$ and $a_{220} > 0$;

$$(a_{111}, \bar{x}_2) = \underbrace{(\mathrm{US}, \mathrm{II})}_{\text{increasing-inflection upper-saddle}} \qquad \text{if } \bar{x}_2 \in (-\infty, a_{121}),$$

$$(a_{111}, \bar{x}_2) = \underbrace{(\mathrm{LS}, \mathrm{DI})}_{\text{decreasing-inflection lower-saddle}} \qquad \text{if } \bar{x}_2 \in (a_{121}, \infty), \tag{2.39}$$

for $a_{110} < 0$ and $a_{220} > 0$;

$$(a_{111}, \bar{x}_2) = \underbrace{(\mathrm{LS}, \mathrm{II})}_{\text{increasing-inflection lower-saddle}} \qquad \text{if } \bar{x}_2 \in (-\infty, a_{121}),$$

$$(a_{111}, \bar{x}_2) = \underbrace{(\mathrm{US}, \mathrm{DI})}_{\text{decreasing-inflection upper-saddle}} \qquad \text{if } \bar{x}_2 \in (a_{121}, \infty), \tag{2.40}$$

for $a_{110} > 0$ and $a_{220} < 0$;

$$(a_{111}, \bar{x}_2) = \underbrace{(\mathrm{US}, \mathrm{DI})}_{\text{decreasing-inflection upper-saddle}} \qquad \text{if } \bar{x}_2 \in (-\infty, a_{121}),$$

$$(a_{111}, \bar{x}_2) = \underbrace{(\mathrm{LS}, \mathrm{II})}_{\text{increasing-inflection lower-saddle}} \qquad \text{if } \bar{x}_2 \in (a_{121}, \infty), \tag{2.41}$$

for $a_{110} < 0$ and $a_{220} < 0$;

as shown in Fig. 2.7. The infinite equilibriums are decreasing-inflection and increasing-inflection upper-saddles and lower-saddles. The equilibriums of $(x_1^*, x_2^*) = (a_{111}, a_{121})$ are parabola-saddles.

2.2.3 Parabola Upper and Lower-Saddle Infinite-Equilibriums

Consider a dynamical system for $a_{111} = a_{211}$ as

$$\begin{aligned}
\dot{x}_1 &= a_{110}(x_1 - a_{111})^2 (x_2 - a_{121}), \\
\dot{x}_2 &= a_{220}(x_1 - a_{211})^3,
\end{aligned} \tag{2.42}$$

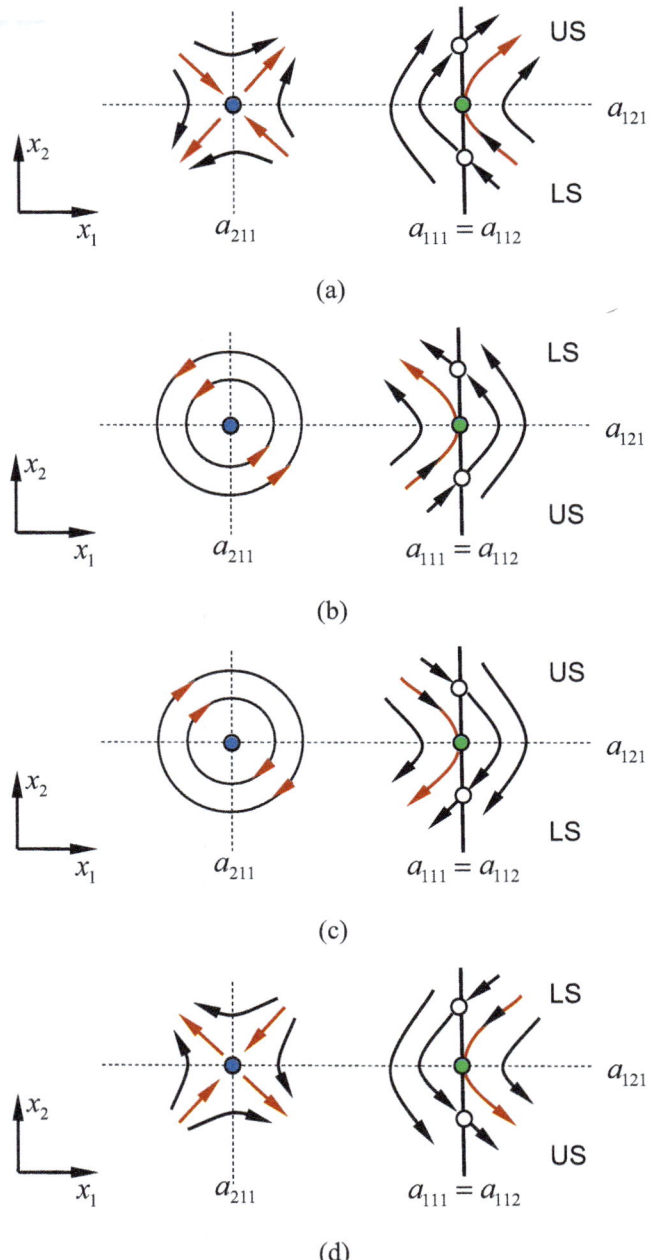

Fig. 2.7 Phase portraits ($a_{111} = a_{212}$) for two-dimensional systems on the x_1-direction with $x_1^* = a_{111}, a_{211}$ and on the x_2-direction with $x_2^* = a_{121}$. The equilibrium flows: (**a**) ($a_{110} > 0$, $a_{220} > 0$), (**b**) ($a_{110} < 0$, $a_{220} > 0$), (**c**) ($a_{110} > 0$, $a_{220} < 0$), (**d**) ($a_{110} < 0$, $a_{220} < 0$)

and the corresponding first integral manifold ($a_{111} = a_{211}$) is

$$\frac{1}{2}\left[(x_1 - a_{211})^2 - (x_{10} - a_{211})^2\right] = \frac{1}{2}\frac{a_{110}}{a_{220}}\left[(x_2 - a_{121})^2 - (x_{20} - a_{121})^2\right]. \quad (2.43)$$

The equilibriums of $x_1^* = a_{111}, a_{211}$ with $x_2^* = a_{121}$ for $a_{111} = a_{211}$ have the following properties:

$$(a_{121}, a_{211}) = \begin{cases} \underbrace{(_{DP}LS:_{UP}US, DP_-:UP_+)}_{\text{positive hyperbolic lower-to-upper saddle}} & \text{for } a_{110} > 0, a_{220} > 0, \\[2ex] \underbrace{(_{UP}US:_{DP}LS, UP_-:DP_+)}_{\text{CCW circular upper-to-lower saddle}} & \text{for } a_{110} < 0, a_{220} > 0, \\[2ex] \underbrace{(_{UP}LS:_{DP}US, UP_+:DP_-)}_{\text{CW circular lower-to-upper saddle}} & \text{for } a_{110} > 0, a_{220} < 0, \\[2ex] \underbrace{(_{DP}US:_{UP}LS, DP_+:UP_-)}_{\text{negative hyperbolic upper-to-lower saddle}} & \text{for } a_{110} < 0, a_{220} < 0 \end{cases}, \quad (2.44)$$

and the infinite-equilibrium of $x_1^* = a_{112} = a_{211}$ is summarized as follows:

$$(a_{111}, \bar{x}_2) = \underbrace{(LS, DP)}_{\text{down-parabola lower-saddle}} \quad \text{if } \bar{x}_2 \in (-\infty, a_{121}),$$

$$(a_{111}, \bar{x}_2) = \underbrace{(US, UP)}_{\text{up-parabola upper-saddle}} \quad \text{if } \bar{x}_2 \in (a_{121}, \infty), \quad (2.45)$$

for $a_{110} > 0, a_{220} > 0$;

$$(a_{111}, \bar{x}_2) = \underbrace{(US, UP)}_{\text{up-parabola upper-saddle}} \quad \text{if } \bar{x}_2 \in (-\infty, a_{121}),$$

$$(a_{111}, \bar{x}_2) = \underbrace{(LS, DP)}_{\text{down-parabola lower-saddle}} \quad \text{if } \bar{x}_2 \in (a_{121}, \infty), \quad (2.46)$$

for $a_{110} < 0, a_{220} > 0$;

$$(a_{111}, \bar{x}_2) = \underbrace{(LS, UP)}_{\text{up-parabola lower-saddle}} \quad \text{if } \bar{x}_2 \in (-\infty, a_{121}),$$

$$(a_{111}, \bar{x}_2) = \underbrace{(US, DP)}_{\text{down-parabola upper-saddle}} \quad \text{if } \bar{x}_2 \in (a_{121}, \infty), \quad (2.47)$$

for $a_{110} > 0, a_{220} < 0$;

$$(a_{111}, \bar{x}_2) = \underbrace{(US, DP)}_{\text{down-parabola upper-saddle}} \quad \text{if } \bar{x}_2 \in (-\infty, a_{121}),$$

$$(a_{111}, \bar{x}_2) = \underbrace{(LS, UP)}_{\text{up-parabola lower-saddle}} \quad \text{if } \bar{x}_2 \in (a_{121}, \infty), \quad (2.48)$$

for $a_{110} < 0, a_{220} < 0$.

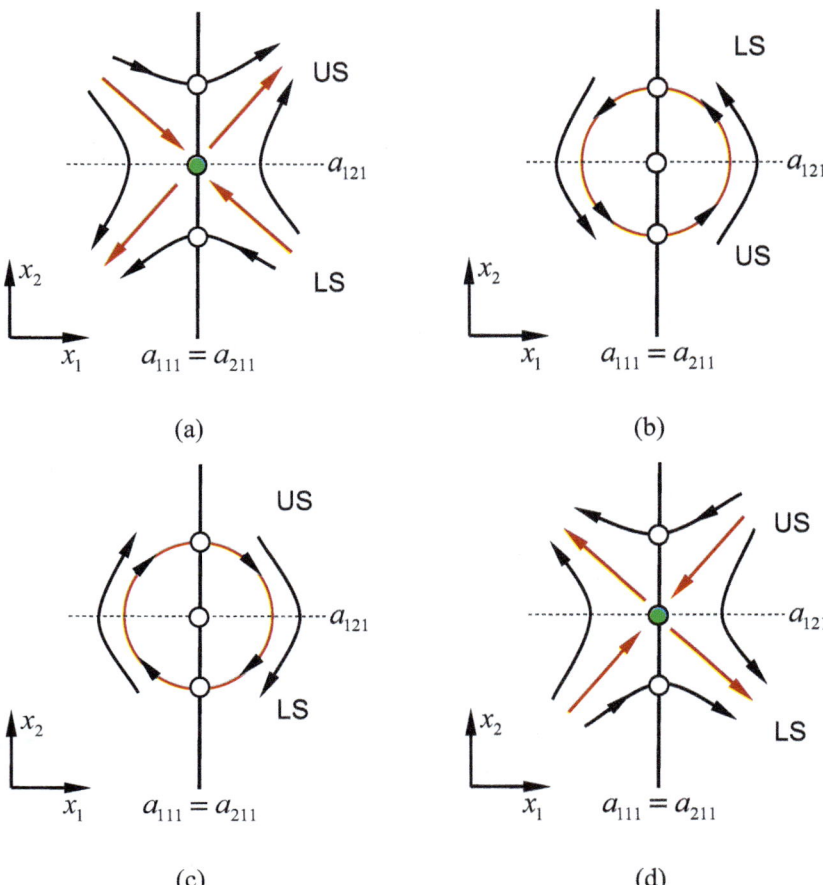

(a)   (b)

(c)   (d)

Fig. 2.8 Phase portraits of infinite-equilibrium ($a_{211} = a_{111}$) for two-dimensional systems on the x_1-direction with $x_1^* = a_{111} = a_{211}$ and on the x_2-direction with $x_2^* = a_{121}$. The infinite equilibriums: (**a**) ($a_{110} > 0$, $a_{220} > 0$), (**b**) ($a_{110} < 0$, $a_{220} > 0$), (**c**) ($a_{110} > 0$, $a_{220} < 0$), (**d**) ($a_{110} < 0$, $a_{220} < 0$)

The infinite-equilibriums of $x_1^* = a_{111} = a_{211}$ are up-parabola and down-parabola upper-saddle and lower-saddle, as shown in Fig. 2.8. On the infinite-equilibrium, there exist the positive hyperbolic lower-to-upper saddle, counter-clockwise circular upper-to-lower saddle, clockwise circular lower-to-upper saddle, and negative hyperbolic lower-to-upper saddle. Such infinite-equilibriums are for the switching bifurcations for third-order saddles and centers with hyperbolic-to-hyperbolic-secant or hyperbolic-secant-to-hyperbolic flows.

Chapter 3
Saddles, Centers, and Switching with Hyperbolic Singular Flows

In this chapter, a series of centers and saddles with hyperbolic singular flows in the crossing and product cubic systems is presented, and the corresponding switching dynamics are discussed through infinite-equilibriums. The hyperbolic singular flows are the appearing bifurcations of hyperbolic and hyperbolic-secant flows. The up-down and down-up, hyperbolic upper-to-lower saddles on the infinite-equilibriums are the switching bifurcations of parabola-saddles with saddles and centers. The same directional saddles and centers are obtained in such cubic systems.

3.1 Centers and Saddles with Hyperbolic Singular Flows

Consider a dynamical system as

$$\dot{x}_1 = a_{110}(x_1 - a_{111})^2(x_2 - a_{121}),$$
$$\dot{x}_2 = a_{220}(x_1 - a_{211})(x_1 - a_{212})(x_1 - a_{213}), \tag{3.1}$$

and the corresponding first integral manifold is

$$[(x_1 - a_{111}) - (x_{10} - a_{111})] + (2a_{111} - a_{211} - a_{212}) \ln \frac{|x_1 - a_{111}|}{|x_{10} - a_{111}|}$$
$$- (a_{111} - a_{211})(a_{111} - a_{212})(\frac{1}{x_1 - a_{111}} - \frac{1}{x_{10} - a_{111}}) \tag{3.2}$$
$$= \frac{1}{2}\frac{a_{110}}{a_{220}}\left[(x_2 - a_{121})^2 - (x_{20} - a_{121})^2\right].$$

(i) The equilibriums of $x_1^* = a_{111}, a_{211}, a_{212}, a_{213}$ and $x_2^* = a_{121}$ for $a_{111} < a_{211}$ are

© The Author(s), under exclusive license to Springer Nature Switzerland AG 2025
A. C. J. Luo, *Two-dimensional Crossing and Product Cubic Systems, Vol. II*,
https://doi.org/10.1007/978-3-031-57100-8_3

$$
\left\{
\begin{array}{ll}
(a_{212}, a_{121}) & (a_{213}, a_{121}) \\
(a_{111}, a_{121}) & (a_{211}, a_{121})
\end{array}
\right\}
=
\left\{
\begin{array}{ll}
\underbrace{(\mathrm{DP}_-, \mathrm{DP}_+)}_{\text{CW center}} & \underbrace{(\mathrm{UP}_+, \mathrm{UP}_+)}_{\text{positive saddle}} \\
\underbrace{(\mathrm{DP}:\mathrm{DP}, \mathrm{nF})}_{\text{HB to HS flow } (-)} & \underbrace{(\mathrm{UP}_+, \mathrm{UP}_+)}_{\text{positive saddle}}
\end{array}
\right\}
\tag{3.3}
$$

for $a_{110} > 0$ and $a_{220} > 0$;

$$
\left\{
\begin{array}{ll}
(a_{212}, a_{121}) & (a_{213}, a_{121}) \\
(a_{111}, a_{121}) & (a_{211}, a_{121})
\end{array}
\right\}
=
\left\{
\begin{array}{ll}
\underbrace{(\mathrm{UP}_-, \mathrm{UP}_-)}_{\text{negative saddle}} & \underbrace{(\mathrm{DP}_+, \mathrm{DP}_-)}_{\text{CCW center}} \\
\underbrace{(\mathrm{UP}:\mathrm{UP}, \mathrm{nF})}_{\text{HS to HB flow } (-)} & \underbrace{(\mathrm{DP}_+, \mathrm{DP}_-)}_{\text{CCW center}}
\end{array}
\right\}
\tag{3.4}
$$

for $a_{110} < 0$ and $a_{220} > 0$;

$$
\left\{
\begin{array}{ll}
(a_{212}, a_{121}) & (a_{213}, a_{121}) \\
(a_{111}, a_{121}) & (a_{211}, a_{121})
\end{array}
\right\}
=
\left\{
\begin{array}{ll}
\underbrace{(\mathrm{UP}_+, \mathrm{UP}_+)}_{\text{positive saddle}} & \underbrace{(\mathrm{DP}_-, \mathrm{DP}_+)}_{\text{CW center}} \\
\underbrace{(\mathrm{UP}:\mathrm{UP}, \mathrm{pF})}_{\text{HS to HB flow } (+)} & \underbrace{(\mathrm{DP}_-, \mathrm{DP}_+)}_{\text{CW center}}
\end{array}
\right\}
\tag{3.5}
$$

for $a_{110} > 0$ and $a_{220} < 0$;

$$
\left\{
\begin{array}{ll}
(a_{212}, a_{121}) & (a_{213}, a_{121}) \\
(a_{111}, a_{121}) & (a_{211}, a_{121})
\end{array}
\right\}
=
\left\{
\begin{array}{ll}
\underbrace{(\mathrm{DP}_+, \mathrm{DP}_-)}_{\text{CCW center}} & \underbrace{(\mathrm{UP}_-, \mathrm{UP}_-)}_{\text{negative saddle}} \\
\underbrace{(\mathrm{DP}:\mathrm{DP}, \mathrm{pF})}_{\text{HB to HS flow } (+)} & \underbrace{(\mathrm{UP}_-, \mathrm{UP}_-)}_{\text{negative saddle}}
\end{array}
\right\}
\tag{3.6}
$$

for $a_{110} < 0$ and $a_{220} < 0$;

as shown in Fig. 3.1. The hyperbolic-secant-to-hyperbolic or hyperbolic-secant-to-hyperbolic flow bifurcations for hyperbolic and hyperbolic-secant flows are also presented. The center, saddle, and center with hyperbolic-to-hyperbolic-secant flows exist in the equilibrium series. The saddle, center, and saddle with hyperbolic-secant-to-hyperbolic flows exist in the equilibrium series.

(ii) For $a_{111} \in (a_{211}, a_{212})$, the equilibriums of $x_1^* = a_{111}, a_{211}, a_{212}, a_{213}$ and $x_2^* = a_{121}$ are

$$
\left\{
\begin{array}{ll}
(a_{212}, a_{121}) & (a_{213}, a_{121}) \\
(a_{211}, a_{121}) & (a_{111}, a_{121})
\end{array}
\right\}
=
\left\{
\begin{array}{ll}
\underbrace{(\mathrm{DP}_-, \mathrm{DP}_+)}_{\text{CW center}} & \underbrace{(\mathrm{UP}_+, \mathrm{UP}_+)}_{\text{positive saddle}} \\
\underbrace{(\mathrm{UP}_+, \mathrm{UP}_+)}_{\text{positive saddle}} & \underbrace{(\mathrm{UP}:\mathrm{UP}, \mathrm{pF})}_{\text{HS to HB flow } (+)}
\end{array}
\right\}
\tag{3.7}
$$

for $a_{110} > 0$ and $a_{220} > 0$;

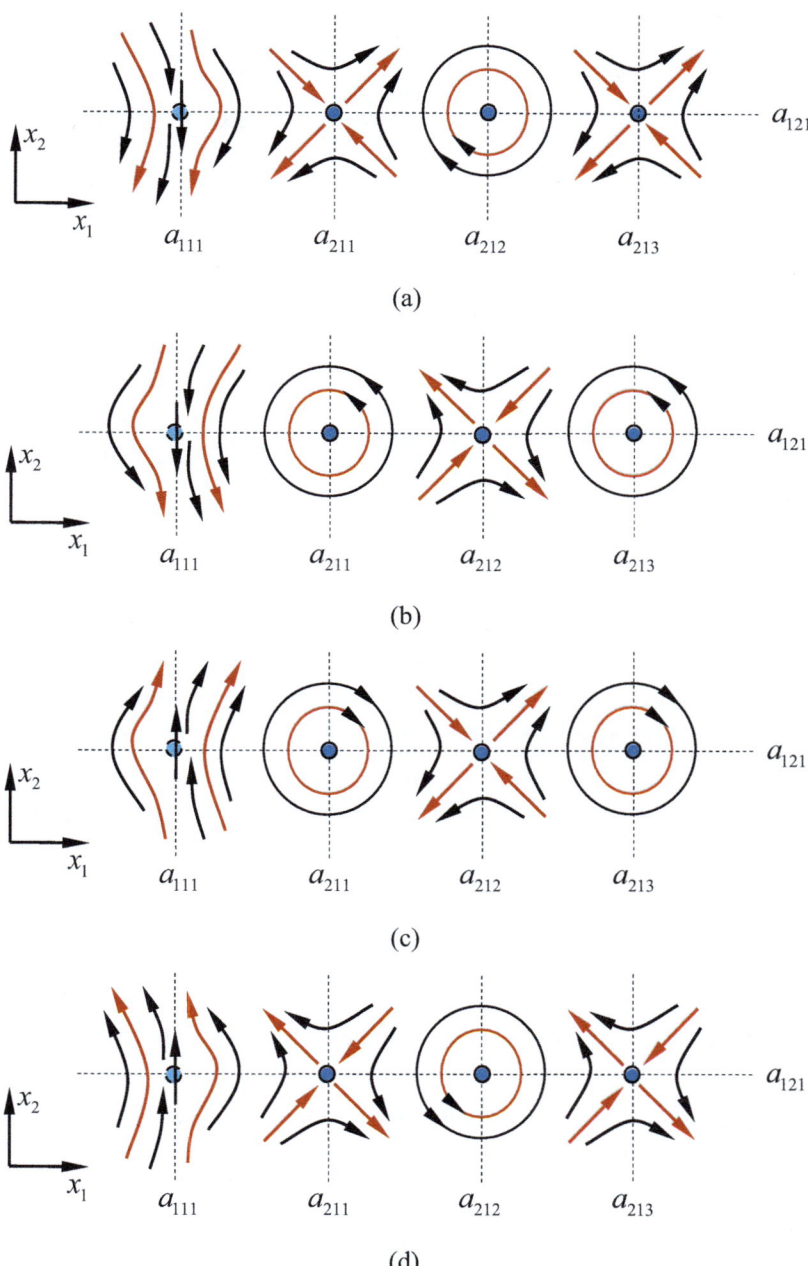

Fig. 3.1 Phase portraits ($a_{111} < a_{211}$) for two-dimensional systems on the x_1-direction with $x_1^* = a_{111}, a_{211}, a_{212}, a_{213}$ and on the x_2-direction with $x_2^* = a_{121}$. (**a**) ($a_{110} > 0$, $a_{220} > 0$), (**b**) ($a_{110} < 0$, $a_{220} > 0$), (**c**) ($a_{110} > 0$, $a_{220} < 0$), (**d**) ($a_{110} < 0$, $a_{220} < 0$)

$$
\left\{
\begin{array}{ll}
(a_{212}, a_{121}) & (a_{213}, a_{121}) \\
(a_{211}, a_{121}) & (a_{111}, a_{121})
\end{array}
\right\}
=
\left\{
\begin{array}{cc}
\underbrace{(\mathrm{UP}_-, \mathrm{UP}_-)}_{\text{negative saddle}} & \underbrace{(\mathrm{DP}_+, \mathrm{DP}_-)}_{\text{CCW center}} \\
\underbrace{(\mathrm{DP}_+, \mathrm{DP}_-)}_{\text{CCW center}} & \underbrace{(\mathrm{DP:DP}, \mathrm{pF})}_{\text{HB to HS flow }(+)}
\end{array}
\right\}
\tag{3.8}
$$

for $a_{110} < 0$ and $a_{220} > 0$;

$$
\left\{
\begin{array}{ll}
(a_{212}, a_{121}) & (a_{213}, a_{121}) \\
(a_{211}, a_{121}) & (a_{111}, a_{121})
\end{array}
\right\}
=
\left\{
\begin{array}{cc}
\underbrace{(\mathrm{UP}_+, \mathrm{UP}_+)}_{\text{positive saddle}} & \underbrace{(\mathrm{DP}_-, \mathrm{DP}_+)}_{\text{CW center}} \\
\underbrace{(\mathrm{DP}_-, \mathrm{DP}_+)}_{\text{CW center}} & \underbrace{(\mathrm{DP:DP}, \mathrm{nF})}_{\text{HB to HS flow }(-)}
\end{array}
\right\}
\tag{3.9}
$$

for $a_{110} > 0$ and $a_{220} < 0$;

$$
\left\{
\begin{array}{ll}
(a_{212}, a_{121}) & (a_{213}, a_{121}) \\
(a_{211}, a_{121}) & (a_{111}, a_{121})
\end{array}
\right\}
=
\left\{
\begin{array}{cc}
\underbrace{(\mathrm{DP}_+, \mathrm{DP}_-)}_{\text{CCW center}} & \underbrace{(\mathrm{UP}_-, \mathrm{UP}_-)}_{\text{negative saddle}} \\
\underbrace{(\mathrm{UP}_-, \mathrm{UP}_-)}_{\text{negative saddle}} & \underbrace{(\mathrm{UP:UP}, \mathrm{nF})}_{\text{HS to HB flow }(-)}
\end{array}
\right\}
\tag{3.10}
$$

for $a_{110} < 0$ and $a_{220} < 0$;

as shown in Fig. 3.2. The center, saddle, and center with hyperbolic-secant-to-hyperbolic flows exist in the equilibrium series. The saddle, center, and saddle with hyperbolic-to-hyperbolic-secant flows exist in the equilibrium series.

(iii) The equilibrium series of $x_1^* = a_{111}, a_{211}, a_{212}, a_{213}$ and $x_2^* = a_{121}$ for $a_{111} \in (a_{212}, a_{213})$ has the same as in Eqs. (3.3)–(3.6) for $a_{111} < a_{211}$, and only the locations of the hyperbolic-to-hyperbolic-secant flows are switched. The saddle, center, and saddle and the center, saddle, and center are the same. Such equilibrium series are presented in Fig. 3.3.

(iv) The equilibrium series of $x_1^* = a_{111}, a_{211}, a_{212}, a_{213}$ and $x_2^* = a_{121}$ for $a_{111} > a_{213}$ has the same as in Eqs. (3.7)–(3.10) for $a_{111} \in (a_{211}, a_{212})$, and only the locations of the hyperbolic-to-hyperbolic-secant flows are switched. The saddle, center, and saddle and the center, saddle, and center are the same. Such equilibrium series are presented in Fig. 3.4.

The hyperbolic-secant-to-hyperbolic or hyperbolic-to-hyperbolic-secant flow bifurcations are the appearing bifurcations for hyperbolic and hyperbolic-secant flows, which are also presented. The saddle, center, and saddle and the center, saddle, and center are invariant. The center, saddle, and center with hyperbolic-to-hyperbolic-secant flows exist in the equilibrium series. The saddle, center, and saddle with hyperbolic-secant-to-hyperbolic flows exist in the equilibrium series. For the four cases, only the locations of the hyperbolic-to-hyperbolic-secant flows are switched.

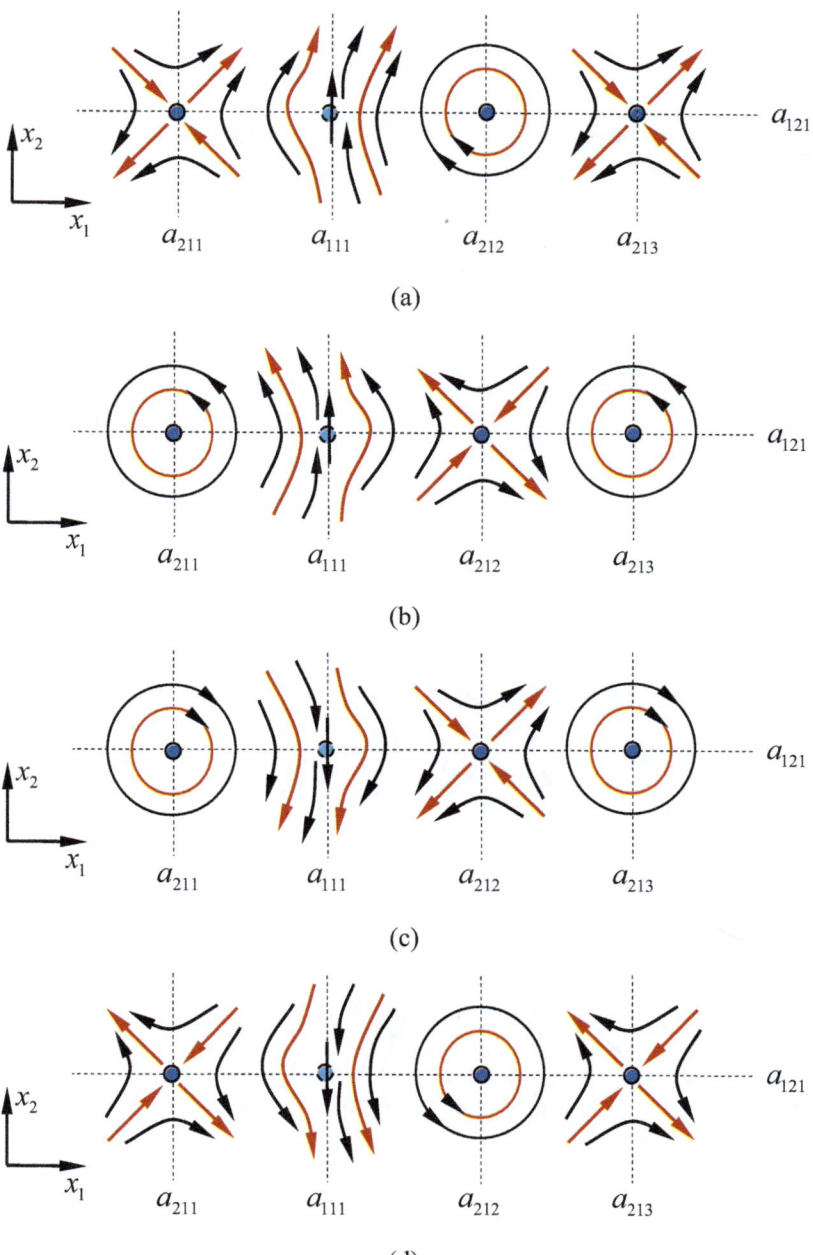

Fig. 3.2 Phase portraits ($a_{111} \in (a_{211}, a_{212})$) for two-dimensional systems on the x_1-direction with $x_1^* = a_{111}, a_{211}, a_{212}, a_{213}$ and on the x_2-direction with $x_2^* = a_{121}$. (**a**) ($a_{110} > 0, a_{220} > 0$), (**b**) ($a_{110} < 0, a_{220} > 0$), (**c**) ($a_{110} > 0, a_{220} < 0$), (**d**) ($a_{110} < 0, a_{220} < 0$)

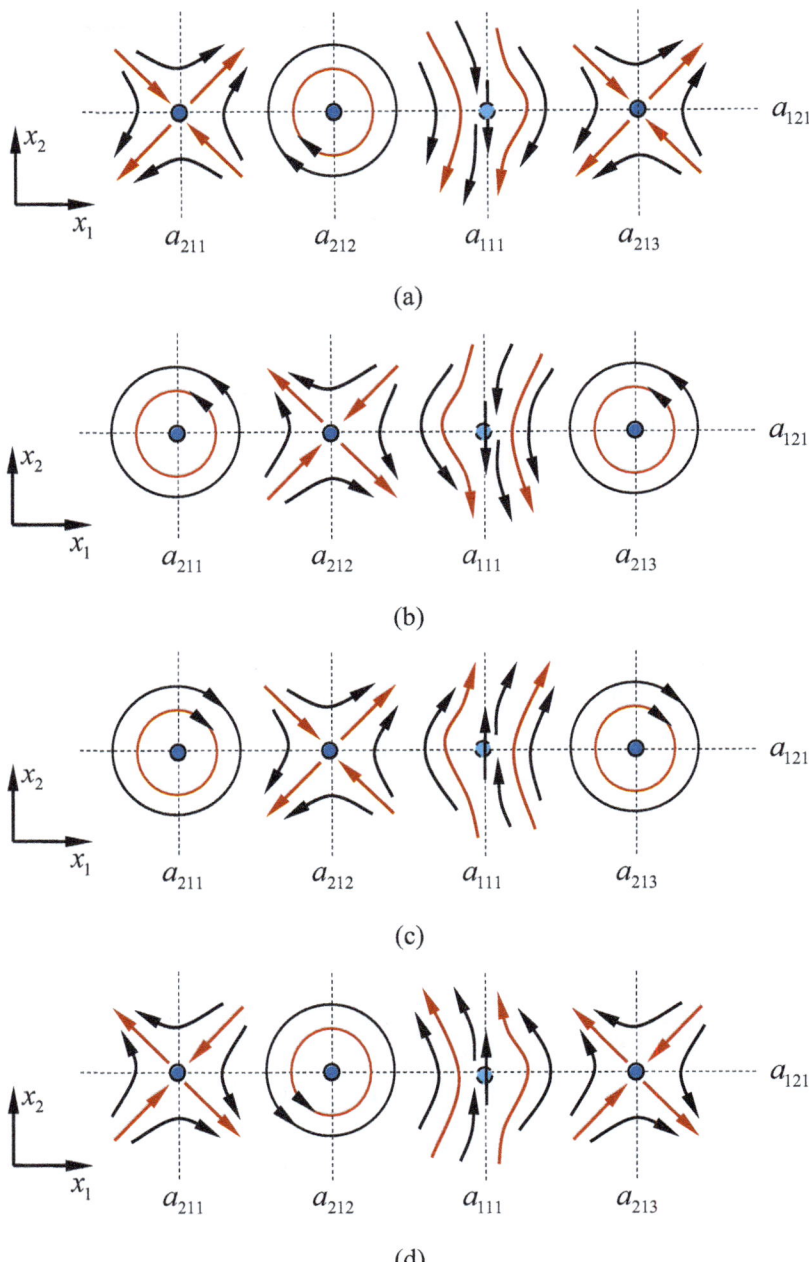

Fig. 3.3 Phase portraits ($a_{111} \in (a_{212}, a_{213})$) for two-dimensional systems on the x_1-direction with $x_1^* = a_{111}, a_{211}, a_{212}, a_{213}$ and on the x_2-direction with $x_2^* = a_{121}$. (**a**) ($a_{110} > 0$, $a_{220} > 0$), (**b**) ($a_{110} < 0$, $a_{220} > 0$), (**c**) ($a_{110} > 0$, $a_{220} < 0$), (**d**) ($a_{110} < 0$, $a_{220} < 0$)

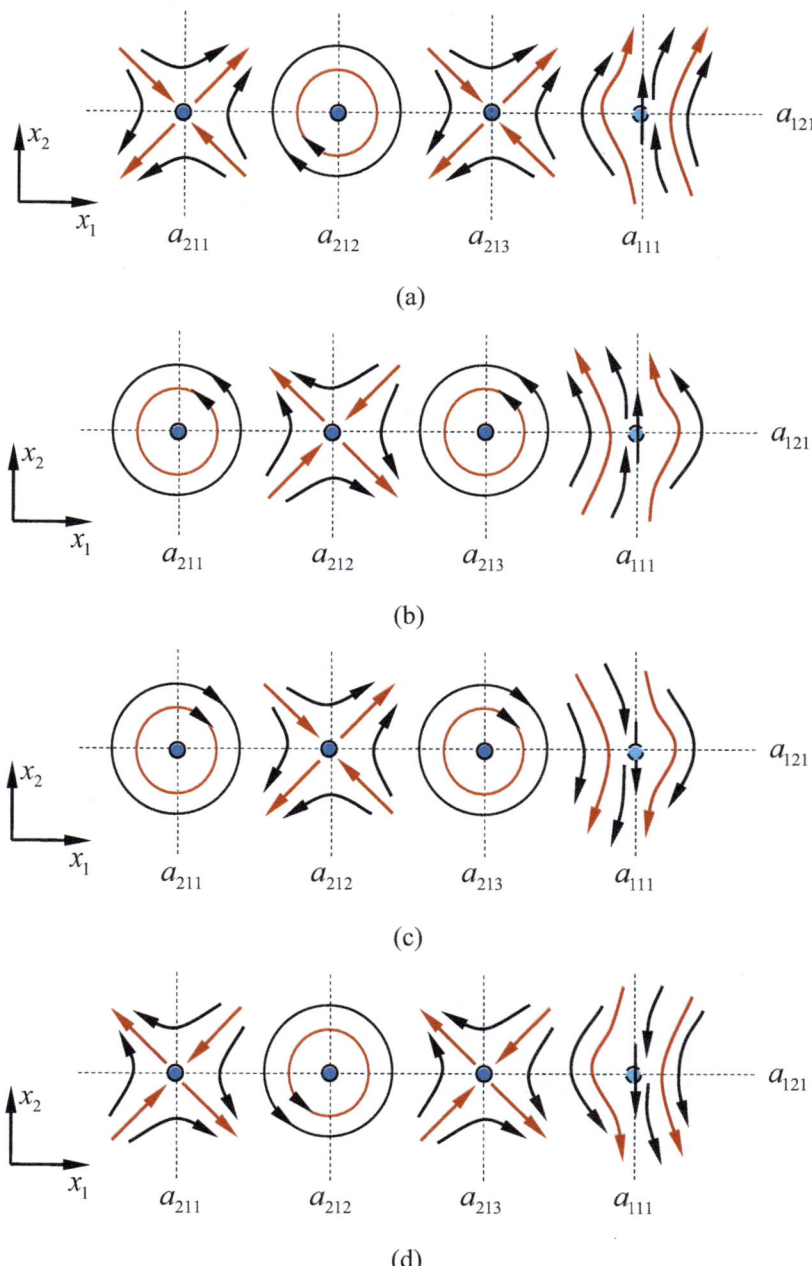

Fig. 3.4 Phase portraits ($a_{213} < a_{111}$) for two-dimensional systems on the x_1-direction with $x_1^* = a_{111}, a_{211}, a_{212}, a_{213}$ and on the x_2-direction with $x_2^* = a_{121}$. (**a**) ($a_{110} > 0$, $a_{220} > 0$), (**b**) ($a_{110} < 0$, $a_{220} > 0$), (**c**) ($a_{110} > 0$, $a_{220} < 0$), (**d**) ($a_{110} < 0$, $a_{220} < 0$)

3.2 Up-Down and Down-Up Saddle Infinite-Equilibriums

For the above four cases, the three switching bifurcations will be discussed through the parabola upper-saddle and lower-saddle infinite-equilibriums.

Consider a dynamical system as

$$\dot{x}_1 = a_{110}(x_1 - a_{111})^2(x_2 - a_{121}),$$
$$\dot{x}_2 = a_{220}(x_1 - a_{211})(x_1 - a_{212})(x_1 - a_{213}), \tag{3.11}$$

and the corresponding first integral manifold for $a_{21s_1} = a_{111}$ is

$$\frac{1}{2}\left[(x_1 - a_{111})^2 - (x_{10} - a_{111})^2\right] + (2a_{111} - a_{21s_2} - a_{21s_3})(x_1 - x_{10})$$
$$+ (a_{111} - a_{21s_2})(a_{111} - a_{21s_3}) \ln \frac{|x_1 - a_{111}|}{|x_{10} - a_{111}|} \tag{3.12}$$
$$= \frac{1}{2}\frac{a_{110}}{a_{220}}\left[(x_2 - a_{121})^2 - (x_{20} - a_{121})^2\right].$$

(i) For $a_{211} = a_{111}$, the equilibriums of $x_1^* = a_{111}, a_{211}, a_{212}, a_{213}$ with $x_2^* = a_{121}$ are

$$\left\{\begin{array}{l}(a_{121}, a_{213})\\(a_{121}, a_{212})\\(a_{121}, a_{211})\end{array}\right\} = \left\{\begin{array}{c}\underbrace{(UP_+, UP_+)}_{\text{positive saddle}}\\\underbrace{(DP_-, DP_+)}_{\text{CW center}}\\\underbrace{(_{UD}LS:_{DU}US, DP_-:UP_+)}_{\text{hyperbolic lower-to-upper saddle}}\end{array}\right\} \text{ for } a_{110} > 0, a_{220} > 0, \tag{3.13}$$

$$\left\{\begin{array}{l}(a_{121}, a_{213})\\(a_{121}, a_{212})\\(a_{121}, a_{211})\end{array}\right\} = \left\{\begin{array}{c}\underbrace{(DP_+, DP_-)}_{\text{CCW center}}\\\underbrace{(UP_-, UP_-)}_{\text{negative saddle}}\\\underbrace{(_{DU}US:_{UD}LS, UP_-:DP_+)}_{\text{hyperbolic-secant upper-to-lower saddle}}\end{array}\right\} \text{ for } a_{110} < 0, a_{220} > 0, \tag{3.14}$$

$$\left\{\begin{array}{l}(a_{121}, a_{213})\\(a_{121}, a_{212})\\(a_{121}, a_{211})\end{array}\right\} = \left\{\begin{array}{c}\underbrace{(DP_-, DP_+)}_{\text{CW center}}\\\underbrace{(UP_+, UP_+)}_{\text{positive saddle}}\\\underbrace{(_{DU}LS:_{UD}US, UP_+:DP_-)}_{\text{hyperbolic-secant lower-to-upper saddle}}\end{array}\right\} \text{ for } a_{110} > 0, a_{220} < 0, \tag{3.15}$$

$$\left.\begin{array}{c}(a_{121},a_{213})\\(a_{121},a_{212})\\(a_{121},a_{211})\end{array}\right\}=\left\{\begin{array}{c}\underbrace{(\text{UP}_-,\text{UP}_-)}_{\text{negative saddle}}\\[2pt]\underbrace{(\text{DP}_+,\text{DP}_-)}_{\text{CCW center}}\\[2pt]\underbrace{(_{\text{UD}}\text{US}:_{\text{DU}}\text{LS},\text{DP}_+:\text{UP}_-)}_{\text{hyperbolic upper-to-lower saddle}}\end{array}\right\}\quad\text{for }a_{110}<0,a_{220}<0.\quad(3.16)$$

The infinite-equilibrium of $x_1^*=a_{111}=a_{211}$ is summarized as follows:

$$(a_{111},\bar{x}_2)=\underbrace{(\text{LS},\text{UD})}_{\text{up-down lower-saddle}}\quad\text{if }\bar{x}_2\in(-\infty,a_{121}),$$

$$(a_{111},\bar{x}_2)=\underbrace{(\text{US},\text{DU})}_{\text{down-up upper-saddle}}\quad\text{if }\bar{x}_2\in(a_{121},\infty),\quad(3.17)$$

for $a_{110}>0,a_{220}>0$;

$$(a_{111},\bar{x}_2)=\underbrace{(\text{US},\text{DU})}_{\text{down-up upper-saddle}}\quad\text{if }\bar{x}_2\in(-\infty,a_{121}),$$

$$(a_{111},\bar{x}_2)=\underbrace{(\text{LS},\text{UD})}_{\text{up-down lower-saddle}}\quad\text{if }\bar{x}_2\in(a_{121},\infty),\quad(3.18)$$

for $a_{110}<0,a_{220}>0$;

$$(a_{111},\bar{x}_2)=\underbrace{(\text{LS},\text{DU})}_{\text{down-up lower-saddle}}\quad\text{if }\bar{x}_2\in(-\infty,a_{121}),$$

$$(a_{111},\bar{x}_2)=\underbrace{(\text{US},\text{UD})}_{\text{up-down upper-saddle}}\quad\text{if }\bar{x}_2\in(a_{121},\infty),\quad(3.19)$$

for $a_{110}>0,a_{220}<0$;

$$(a_{111},\bar{x}_2)=\underbrace{(\text{US},\text{UD})}_{\text{up-down upper-saddle}}\quad\text{if }\bar{x}_2\in(-\infty,a_{121}),$$

$$(a_{111},\bar{x}_2)=\underbrace{(\text{LS},\text{DU})}_{\text{down-up lower-saddle}}\quad\text{if }\bar{x}_2\in(a_{121},\infty),\quad(3.20)$$

for $a_{110}<0,a_{220}<0$.

The up-down and down-up, upper-saddle and lower-saddle infinite-equilibriums of $x_1^*=a_{111}=a_{211}$ are presented in Fig. 3.5. The infinite-equilibriums are for the switching bifurcation for the equilibriums for $a_{111}<a_{211}$ and $a_{111}\in(a_{211},a_{212})$. The hyperbolic and hyperbolic-secant lower-to-upper saddle and upper-to-lower-saddles are also presented at $(x_2^*,x_1^*)=(a_{121},a_{211})$.

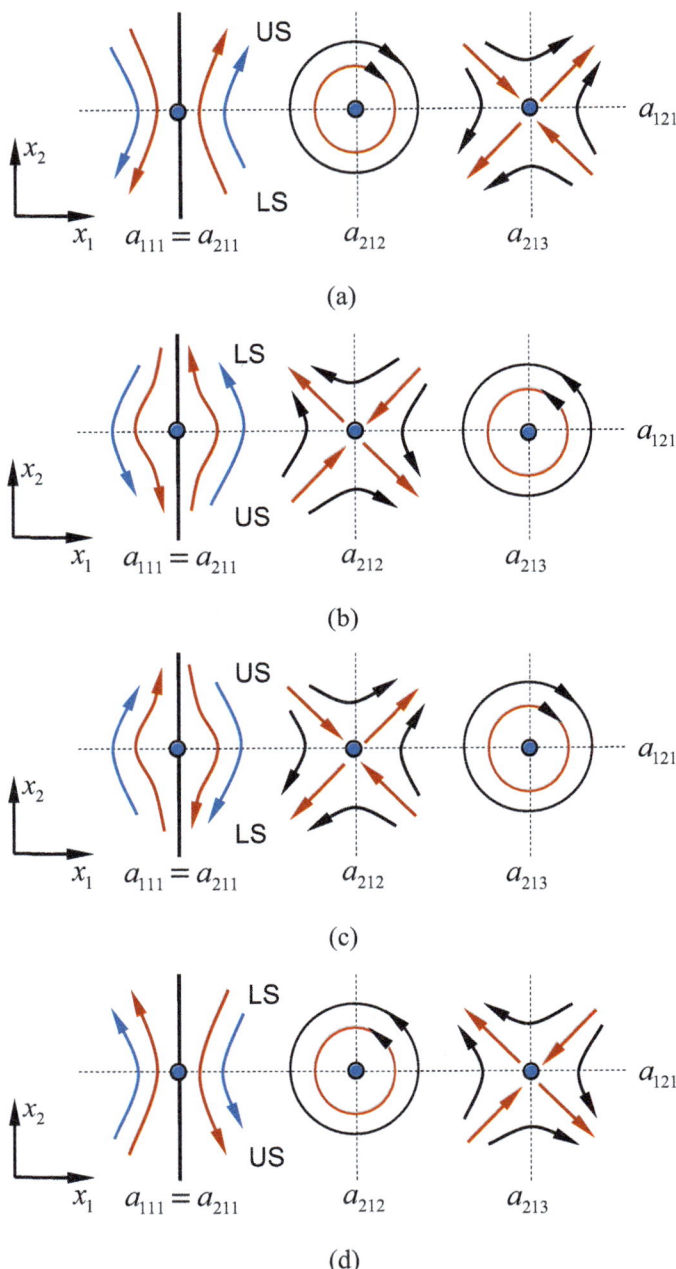

Fig. 3.5 Phase portraits ($a_{111} = a_{211}$) for two-dimensional systems on the x_1-direction with $x_1^* = a_{111}, a_{211}, a_{212}, a_{213}$ and on the x_2-direction with $x_2^* = a_{121}$. The equilibrium flows: (**a**) ($a_{110} > 0$, $a_{220} > 0$), (**b**) ($a_{110} < 0$, $a_{220} > 0$), (**c**) ($a_{110} > 0$, $a_{220} < 0$), (**d**) ($a_{110} < 0$, $a_{220} < 0$)

(ii) The equilibriums of $x_1^* = a_{111}, \ a_{211}, \ a_{212}$ with $x_2^* = a_{121}$ for $a_{212} = a_{111}$ are

$$\left. \begin{array}{c} (a_{121}, a_{211}) \\ (a_{121}, a_{212}) \\ (a_{121}, a_{211}) \end{array} \right\} = \left\{ \begin{array}{c} \underbrace{(\mathrm{UP_+}, \mathrm{UP_+})}_{\text{positive saddle}} \\ \underbrace{(_{\mathrm{DU}}\mathrm{LS}{:}_{\mathrm{UD}}\mathrm{US}, \ \mathrm{UP_+}{:}\mathrm{DP_-})}_{\text{hyperbolic-secant lower-to-upper saddle}} \\ \underbrace{(\mathrm{UP_+}, \mathrm{UP_+})}_{\text{positive saddle}} \end{array} \right\} \quad \text{for } a_{110} > 0, a_{220} > 0, \quad (3.21)$$

$$\left. \begin{array}{c} (a_{121}, a_{211}) \\ (a_{121}, a_{212}) \\ (a_{121}, a_{211}) \end{array} \right\} = \left\{ \begin{array}{c} \underbrace{(\mathrm{DP_+}, \mathrm{DP_-})}_{\text{CCW center}} \\ \underbrace{(_{\mathrm{UD}}\mathrm{US}{:}_{\mathrm{DU}}\mathrm{LS}, \ \mathrm{DP_+}{:}\mathrm{UP_-})}_{\text{hyperbolic upper-to-lower saddle}} \\ \underbrace{(\mathrm{DP_+}, \mathrm{DP_-})}_{\text{CCW center}} \end{array} \right\} \quad \text{for } a_{110} < 0, a_{220} > 0, \quad (3.22)$$

$$\left. \begin{array}{c} (a_{121}, a_{211}) \\ (a_{121}, a_{212}) \\ (a_{121}, a_{211}) \end{array} \right\} = \left\{ \begin{array}{c} \underbrace{(\mathrm{DP_-}, \mathrm{DP_+})}_{\text{CW center}} \\ \underbrace{(_{\mathrm{UD}}\mathrm{LS}{:}_{\mathrm{DU}}\mathrm{US}, \ \mathrm{DP_-}{:}\mathrm{UP_+})}_{\text{hyperbolic lower-to-upper saddle}} \\ \underbrace{(\mathrm{DP_-}, \mathrm{DP_+})}_{\text{CW center}} \end{array} \right\} \quad \text{for } a_{110} > 0, a_{220} < 0, \quad (3.23)$$

$$\left. \begin{array}{c} (a_{121}, a_{211}) \\ (a_{121}, a_{212}) \\ (a_{121}, a_{211}) \end{array} \right\} = \left\{ \begin{array}{c} \underbrace{(\mathrm{UP_-}, \mathrm{UP_-})}_{\text{negative saddle}} \\ \underbrace{(_{\mathrm{DU}}\mathrm{US}{:}_{\mathrm{UD}}\mathrm{LS}, \ \mathrm{UP_-}{:}\mathrm{DP_+})}_{\text{hyperbolic-secant lower-to-upper saddle}} \\ \underbrace{(\mathrm{UP_-}, \mathrm{UP_-})}_{\text{negative saddle}} \end{array} \right\} \quad \text{for } a_{110} < 0, a_{220} < 0; \quad (3.24)$$

and the infinite-equilibrium of $x_1^* = a_{111} = a_{212}$ is summarized as follows:

$$(a_{111}, \bar{x}_2) = \underbrace{(\mathrm{LS}, \mathrm{DU})}_{\text{down-up lower-saddle}} \quad \text{if } \bar{x}_2 \in (-\infty, a_{121}),$$

$$(a_{111}, \bar{x}_2) = \underbrace{(\mathrm{US}, \mathrm{UD})}_{\text{up-down upper-saddle}} \quad \text{if } \bar{x}_2 \in (a_{121}, \infty), \quad (3.25)$$

$$\text{for } a_{110} > 0, a_{220} > 0;$$

$$(a_{111}, \bar{x}_2) = \underbrace{(\mathrm{US}, \mathrm{UD})}_{\text{up-down upper-saddle}} \quad \text{if } \bar{x}_2 \in (-\infty, a_{121}),$$

$$(a_{111}, \bar{x}_2) = \underbrace{(\mathrm{LS}, \mathrm{DU})}_{\text{down-up lower-saddle}} \quad \text{if } \bar{x}_2 \in (a_{121}, \infty), \quad (3.26)$$

$$\text{for } a_{110} < 0, a_{220} > 0;$$

$$(a_{212}, \bar{x}_2) = \underbrace{(\text{LS}, \text{UD})}_{\text{up-down lower-saddle}} \quad \text{if } \bar{x}_2 \in (-\infty, a_{121}),$$

$$(a_{212}, \bar{x}_2) = \underbrace{(\text{US}, \text{DU})}_{\text{down-up upper-saddle}} \quad \text{if } \bar{x}_2 \in (a_{121}, \infty), \tag{3.27}$$

for $a_{110} > 0, a_{220} < 0$;

$$(a_{111}, \bar{x}_2) = \underbrace{(\text{US}, \text{DU})}_{\text{down-up upper-saddle}} \quad \text{if } \bar{x}_2 \in (-\infty, a_{121}),$$

$$(a_{111}, \bar{x}_2) = \underbrace{(\text{LS}, \text{UD})}_{\text{up-down lower-saddle}} \quad \text{if } \bar{x}_2 \in (a_{121}, \infty), \tag{3.28}$$

for $a_{110} < 0, a_{220} < 0$.

The up-down and down-up, lower-saddle and upper-saddle infinite-equilibriums of $x_1^* = a_{111} = a_{212}$ are for the switching bifurcation for the equilibriums for $a_{111} \in (a_{211}, a_{212})$ and $a_{111} \in (a_{212}, a_{213})$, as presented in Fig. 3.6. The hyperbolic and circular lower-to-upper saddle and upper-to-lower-saddles are also presented at $(x_2^*, x_1^*) = (a_{121}, a_{212})$.

(iii) For $a_{213} = a_{111}$, the equilibriums of $x_1^* = a_{111}, a_{211}, a_{212}, a_{213}$ with $x_2^* = a_{121}$ are

$$\left.\begin{matrix} (a_{121}, a_{213}) \\ (a_{121}, a_{212}) \\ (a_{121}, a_{211}) \end{matrix}\right\} = \left\{\begin{matrix} \underbrace{(_{\text{UD}}\text{LS}:_{\text{DU}}\text{US}, \text{DP}_- : \text{UP}_+)}_{\text{hyperbolic lower-to-upper saddle}} \\ \underbrace{(\text{DP}_-, \text{DP}_+)}_{\text{CW center}} \\ \underbrace{(\text{UP}_+, \text{UP}_+)}_{\text{positive saddle}} \end{matrix}\right\} \quad \text{for } a_{110} > 0, a_{220} > 0, \tag{3.29}$$

$$\left.\begin{matrix} (a_{121}, a_{213}) \\ (a_{121}, a_{212}) \\ (a_{121}, a_{211}) \end{matrix}\right\} = \left\{\begin{matrix} \underbrace{(_{\text{DU}}\text{US}:_{\text{UD}}\text{LS}, \text{UP}_- : \text{DP}_+)}_{\text{hyperbolic-secant upper-to-lower saddle}} \\ \underbrace{(\text{UP}_-, \text{UP}_-)}_{\text{negative saddle}} \\ \underbrace{(\text{DP}_+, \text{DP}_-)}_{\text{CCW center}} \end{matrix}\right\} \quad \text{for } a_{110} < 0, a_{220} > 0, \tag{3.30}$$

$$\left.\begin{matrix} (a_{121}, a_{213}) \\ (a_{121}, a_{212}) \\ (a_{121}, a_{211}) \end{matrix}\right\} = \left\{\begin{matrix} \underbrace{(_{\text{DU}}\text{LS}:_{\text{UD}}\text{US}, \text{UP}_+ : \text{DP}_-)}_{\text{hyperbolic-secant lower-to-upper saddle}} \\ \underbrace{(\text{UP}_+, \text{UP}_+)}_{\text{positive saddle}} \\ \underbrace{(\text{DP}_-, \text{DP}_+)}_{\text{CW center}} \end{matrix}\right\} \quad \text{for } a_{110} > 0, a_{220} < 0, \tag{3.31}$$

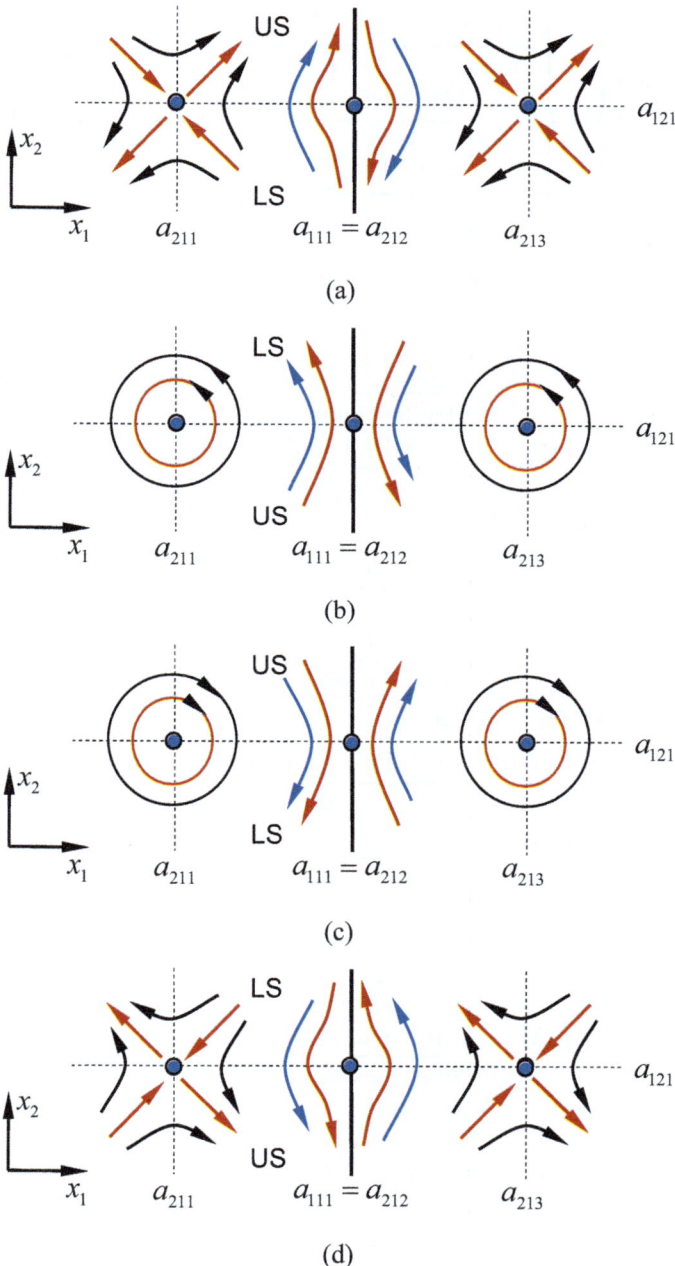

Fig. 3.6 Phase portraits ($a_{111} = a_{212}$) for two-dimensional systems on the x_1-direction with $x_1^* = a_{111}, a_{211}, a_{212}, a_{213}$ and on the x_2-direction with $x_2^* = a_{121}$. The equilibrium flows: (**a**) ($a_{110} > 0, a_{220} > 0$), (**b**) ($a_{110} < 0, a_{220} > 0$), (**c**) ($a_{110} > 0, a_{220} < 0$), (**d**) ($a_{110} < 0, a_{220} < 0$)

$$\left\{\begin{array}{c} (a_{121}, a_{213}) \\ (a_{121}, a_{212}) \\ (a_{121}, a_{211}) \end{array}\right\} = \left\{\begin{array}{c} \underbrace{(_{UD}US : _{DU}LS, DP_{+} : UP_{-})}_{\text{hyperbolic upper-to-lower saddle}} \\ \underbrace{(DP_{+}, DP_{-})}_{\text{CCW center}} \\ \underbrace{(UP_{-}, UP_{-})}_{\text{negative saddle}} \end{array}\right\} \quad \text{for } a_{110} < 0, a_{220} < 0. \quad (3.32)$$

and the infinite-equilibrium of $x_1^* = a_{111} = a_{213}$ is the same as for $x_1^* = a_{111} = a_{211}$, as follows:

$$(a_{111}, \bar{x}_2) = \underbrace{(LS, UD)}_{\text{up-down lower-saddle}} \quad \text{if } \bar{x}_2 \in (-\infty, a_{121}),$$

$$(a_{111}, \bar{x}_2) = \underbrace{(US, DU)}_{\text{down-up upper-saddle}} \quad \text{if } \bar{x}_2 \in (a_{121}, \infty), \quad (3.33)$$

for $a_{110} > 0, a_{220} > 0$;

$$(a_{111}, \bar{x}_2) = \underbrace{(US, DU)}_{\text{down-up upper-saddle}} \quad \text{if } \bar{x}_2 \in (-\infty, a_{121}),$$

$$(a_{111}, \bar{x}_2) = \underbrace{(LS, UD)}_{\text{up-down lower-saddle}} \quad \text{if } \bar{x}_2 \in (a_{121}, \infty), \quad (3.34)$$

for $a_{110} < 0, a_{220} > 0$;

$$(a_{111}, \bar{x}_2) = \underbrace{(LS, DU)}_{\text{down-up lower-saddle}} \quad \text{if } \bar{x}_2 \in (-\infty, a_{121}),$$

$$(a_{111}, \bar{x}_2) = \underbrace{(US, UD)}_{\text{up-down upper-saddle}} \quad \text{if } \bar{x}_2 \in (a_{121}, \infty), \quad (3.35)$$

for $a_{110} > 0, a_{220} < 0$;

$$(a_{111}, \bar{x}_2) = \underbrace{(US, UD)}_{\text{up-down upper-saddle}} \quad \text{if } \bar{x}_2 \in (-\infty, a_{121}),$$

$$(a_{111}, \bar{x}_2) = \underbrace{(LS, DU)}_{\text{down-up lower-saddle}} \quad \text{if } \bar{x}_2 \in (a_{121}, \infty), \quad (3.36)$$

for $a_{110} < 0, a_{220} < 0$.

The up-down and down-up, upper-saddle and lower-saddle infinite-equilibriums of $x_1^* = a_{111} = a_{213}$ are presented in Fig. 3.7. The infinite-equilibrium is for the switching bifurcation for the equilibriums for $a_{111} \in (a_{212}, a_{213})$ and $a_{111} > a_{213}$. The hyperbolic and hyperbolic-secant lower-to-upper saddle and upper-to-lower-saddles are also presented at $(x_2^*, x_1^*) = (a_{121}, a_{213})$.

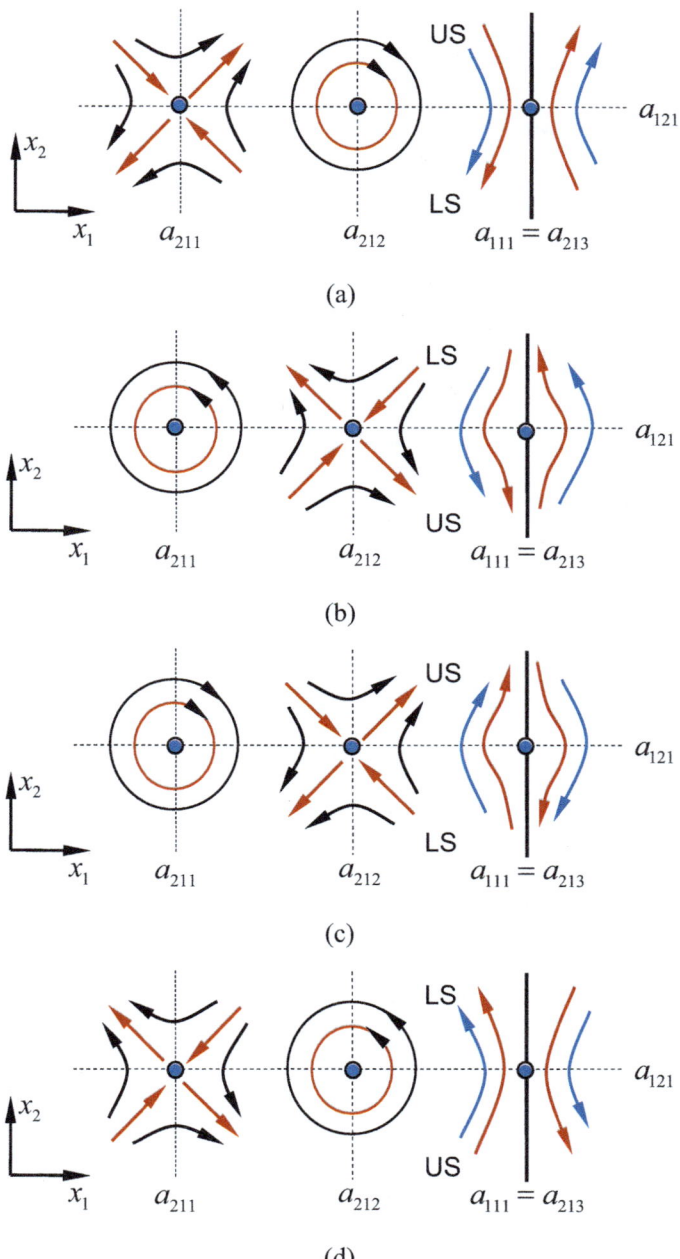

Fig. 3.7 Phase portraits ($a_{111} = a_{213}$) for two-dimensional systems on the x_1-direction with $x_1^* = a_{111}, a_{211}, a_{212}, a_{213}$ and on the x_2-direction with $x_2^* = a_{121}$. The equilibrium flows: (**a**) ($a_{110} > 0, a_{220} > 0$), (**b**) ($a_{110} < 0, a_{220} > 0$), (**c**) ($a_{110} > 0, a_{220} < 0$), (**d**) ($a_{110} < 0, a_{220} < 0$)

Chapter 4
Parabola-Saddles, Third-Order Saddles and Centers, and Switching with Hyperbolic Flows

In this chapter, parabola-saddles and hyperbolic-singular flows in the crossing and product cubic systems are presented, and the corresponding switching dynamics are discussed through infinite-equilibriums. The parabola-saddles are the appearing and switching bifurcations of saddle and center. The third-order centers are the appearing bifurcations of center, saddle, and center. The third-order saddles are the appearing bifurcations of saddle, center, and saddle. The parabola-saddles on the inflection-source and sink infinite-equilibriums are the switching bifurcations of a saddle and a hyperbolic-secant flow, a center and hyperbolic flow, respectively. The parabola-hyperbolic sink-to-source on the parabola sink and source infinite-equilibriums are the switching bifurcations of a parabola-saddle and a hyperbolic flow, another parabola-saddle and a hyperbolic-secant flow. The third-order parabola-saddles are the switching bifurcations of a third-order saddle and a hyperbolic-secant flow, a third-order center and a hyperbolic flow.

4.1 Parabola-Saddles, Third-Order Saddles and Centers, and Hyperbolic Flows

In this section, equilibriums with parabola-saddles and third-order saddles and centers will be discussed.

4.1.1 Parabola-Saddle with Hyperbolic Flows

Consider a dynamical system as

$$\dot{x}_1 = a_{110}(x_1 - a_{111})(x_1 - a_{112})(x_2 - a_{121}),$$

$$\dot{x}_2 = a_{220}(x_1 - a_{211})(x_1 - a_{212})^2,$$

(4.1)

and the corresponding first integral manifold is

$$\frac{1}{2}\left[(x_1 - a_{211})^2 - (x_{10} - a_{211})^2\right] + (a_{111} + a_{112} - 2a_{212})(x_1 - x_{10})$$

$$+ \frac{(a_{111} - a_{211})(a_{111} - a_{212})^2}{a_{111} - a_{112}} \ln \frac{|x_1 - a_{111}|}{|x_{10} - a_{111}|}$$

$$+ \frac{(a_{112} - a_{211})(a_{112} - a_{211})^2}{a_{112} - a_{111}} \ln \frac{|x_1 - a_{112}|}{|x_{10} - a_{112}|}$$

(4.2)

$$= \frac{1}{2}\frac{a_{110}}{a_{220}}\left[(x_2 - a_{121})^2 - (x_{20} - a_{121})^2\right].$$

(i) For the case of $a_{112} < a_{211}$, the parabola upper-saddles and lower-saddles with hyperbolic and hyperbolic-secant flows of $x_1^* = a_{111}, a_{112}, a_{211}, a_{212}$ with $x_2^* = a_{121}$ have the following properties:

$$\left\{\begin{matrix}(a_{121}, a_{211}) & (a_{121}, a_{212})\\(a_{111}, a_{121}) & (a_{112}, a_{121})\end{matrix}\right\} = \left\{\begin{matrix}\underbrace{(UP_+ : UP_+)}_{\text{positive saddle}} & \underbrace{(UP, US)}_{\text{up-parabola upper-saddle}}\\\underbrace{(DP : UP, nF)}_{\text{hyperbolic flow } (-)} & \underbrace{(UP : DP, nF)}_{\text{hyperbolic-secant flow } (-)}\end{matrix}\right\}$$

(4.3)

for $a_{110} > 0$ and $a_{220} > 0$;

$$\left\{\begin{matrix}(a_{121}, a_{211}) & (a_{121}, a_{212})\\(a_{111}, a_{121}) & (a_{112}, a_{121})\end{matrix}\right\} = \left\{\begin{matrix}\underbrace{(DP_+ : DP_-)}_{\text{CCW center}} & \underbrace{(DP, US)}_{\text{down-parabola upper-saddle}}\\\underbrace{(UP : DP, nF)}_{\text{hyperbolic-secant flow } (-)} & \underbrace{(DP : UP, nF)}_{\text{hyperbolic flow } (-)}\end{matrix}\right\}$$

for $a_{110} < 0$ and $a_{220} > 0$;

(4.4)

$$\left\{\begin{matrix}(a_{121}, a_{211}) & (a_{121}, a_{212})\\(a_{111}, a_{121}) & (a_{112}, a_{121})\end{matrix}\right\} = \left\{\begin{matrix}\underbrace{(DP_- : DP_+)}_{\text{CW center}} & \underbrace{(DP, LS)}_{\text{down-parabola lower-saddle}}\\\underbrace{(UP : DP, pF)}_{\text{hyperbolic-secant flow } (+)} & \underbrace{(DP : UP, pF)}_{\text{hyperbolic flow } (+)}\end{matrix}\right\}$$

for $a_{110} > 0$ and $a_{220} < 0$;

(4.5)

$$\left\{\begin{matrix}(a_{121}, a_{211}) & (a_{121}, a_{212})\\(a_{111}, a_{121}) & (a_{112}, a_{121})\end{matrix}\right\} = \left\{\begin{matrix}\underbrace{(UP_- : UP_-)}_{\text{negative saddle}} & \underbrace{(UP, LS)}_{\text{up-parabola lower-saddle}}\\\underbrace{(DP : UP, pF)}_{\text{hyperbolic flow } (+)} & \underbrace{(UP : DP, pF)}_{\text{hyperbolic-secant flow } (+)}\end{matrix}\right\}$$

(4.6)

for $a_{110} < 0$ and $a_{220} < 0$.

In Fig. 4.1, from the first integration manifold, the phase portraits of the equilibriums are sketched, and the corresponding singularity characteristics of equilibriums are presented.

(ii) For the case of $a_{211} \in (a_{111}, a_{112})$, the parabola upper-saddles and lower-saddles with hyperbolic and hyperbolic-secant flows of $x_1^* = a_{111}, a_{112}, a_{211}, a_{212}$ with $x_2^* = a_{121}$ have the following properties:

$$
\left\{
\begin{array}{ll}
(a_{112}, a_{121}) & (a_{121}, a_{212}) \\
(a_{111}, a_{121}) & (a_{121}, a_{211})
\end{array}
\right\}
=
\left\{
\begin{array}{ll}
\underbrace{(\mathrm{DP:UP, pF})}_{\text{hyperbolic flow } (+)} & \underbrace{(\mathrm{UP, US})}_{\text{up-parabola upper-saddle}} \\
\underbrace{(\mathrm{DP:UP, nF})}_{\text{hyperbolic flow } (-)} & \underbrace{(\mathrm{DP_+ : DP_-})}_{\text{CCW center}}
\end{array}
\right\}
\tag{4.7}
$$

for $a_{110} > 0$ and $a_{220} > 0$;

$$
\left\{
\begin{array}{ll}
(a_{112}, a_{121}) & (a_{121}, a_{212}) \\
(a_{111}, a_{121}) & (a_{121}, a_{211})
\end{array}
\right\}
=
\left\{
\begin{array}{ll}
\underbrace{(\mathrm{UP:DP, pF})}_{\text{hyperbolic-secant flow } (+)} & \underbrace{(\mathrm{DP, US})}_{\text{down-parabola upper-saddle}} \\
\underbrace{(\mathrm{UP:DP, nF})}_{\text{hyperbolic-secant flow } (-)} & \underbrace{(\mathrm{UP_+, UP_+})}_{\text{positive saddle}}
\end{array}
\right\}
$$

for $a_{110} < 0$ and $a_{220} > 0$;

$$\tag{4.8}$$

$$
\left\{
\begin{array}{ll}
(a_{112}, a_{121}) & (a_{121}, a_{212}) \\
(a_{111}, a_{121}) & (a_{121}, a_{211})
\end{array}
\right\}
=
\left\{
\begin{array}{ll}
\underbrace{(\mathrm{UP:DP, nF})}_{\text{hyperbolic-secant flow } (-)} & \underbrace{(\mathrm{DP, LS})}_{\text{down-parabola lower-saddle}} \\
\underbrace{(\mathrm{UP:DP, pF})}_{\text{hyperbolic-secant flow } (+)} & \underbrace{(\mathrm{UP_-, UP_-})}_{\text{neagitve saddle}}
\end{array}
\right\}
$$

for $a_{110} > 0$ and $a_{220} < 0$;

$$\tag{4.9}$$

$$
\left\{
\begin{array}{ll}
(a_{112}, a_{121}) & (a_{121}, a_{212}) \\
(a_{111}, a_{121}) & (a_{121}, a_{211})
\end{array}
\right\}
=
\left\{
\begin{array}{ll}
\underbrace{(\mathrm{DP:UP, nF})}_{\text{hyperbolic flow } (-)} & \underbrace{(\mathrm{UP, LS})}_{\text{up-parabola lower-saddle}} \\
\underbrace{(\mathrm{DP:UP, pF})}_{\text{hyperbolic flow } (+)} & \underbrace{(\mathrm{DP_- : DP_+})}_{\text{CW center}}
\end{array}
\right\}
\tag{4.10}
$$

for $a_{110} < 0$ and $a_{220} < 0$.

In Fig. 4.2, from the first integration manifold, the phase portraits of the equilibriums are sketched, and the corresponding singularity characteristics of equilibriums are presented.

(iii) For the case of $a_{212} \in (a_{211}, a_{112})$ with $a_{111} < a_{211}$, the parabola upper-saddles and lower-saddles with hyperbolic and hyperbolic-secant flows of $x_1^* = a_{111}, a_{112}$, a_{211}, a_{212} with $x_2^* = a_{121}$ have the following properties:

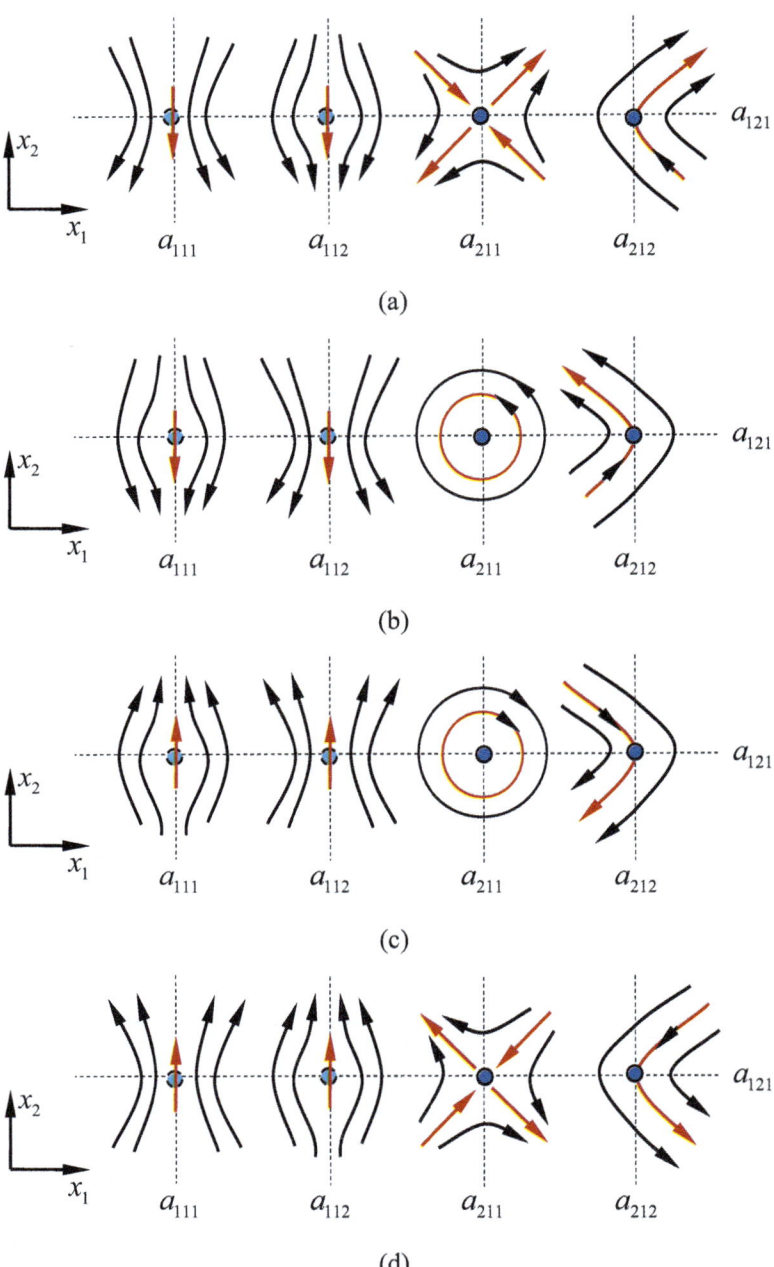

Fig. 4.1 Phase portraits ($a_{112} < a_{211}$) for two-dimensional systems on the x_1-direction with $x_1^* = a_{111}, a_{112}, a_{211}, a_{212}$ and on the x_2-direction with $x_2^* = a_{121}$. (**a**) ($a_{110} > 0$, $a_{220} > 0$), (**b**) ($a_{110} < 0$, $a_{220} > 0$), (**c**) ($a_{110} > 0$, $a_{220} < 0$), (**d**) ($a_{110} < 0$, $a_{220} < 0$)

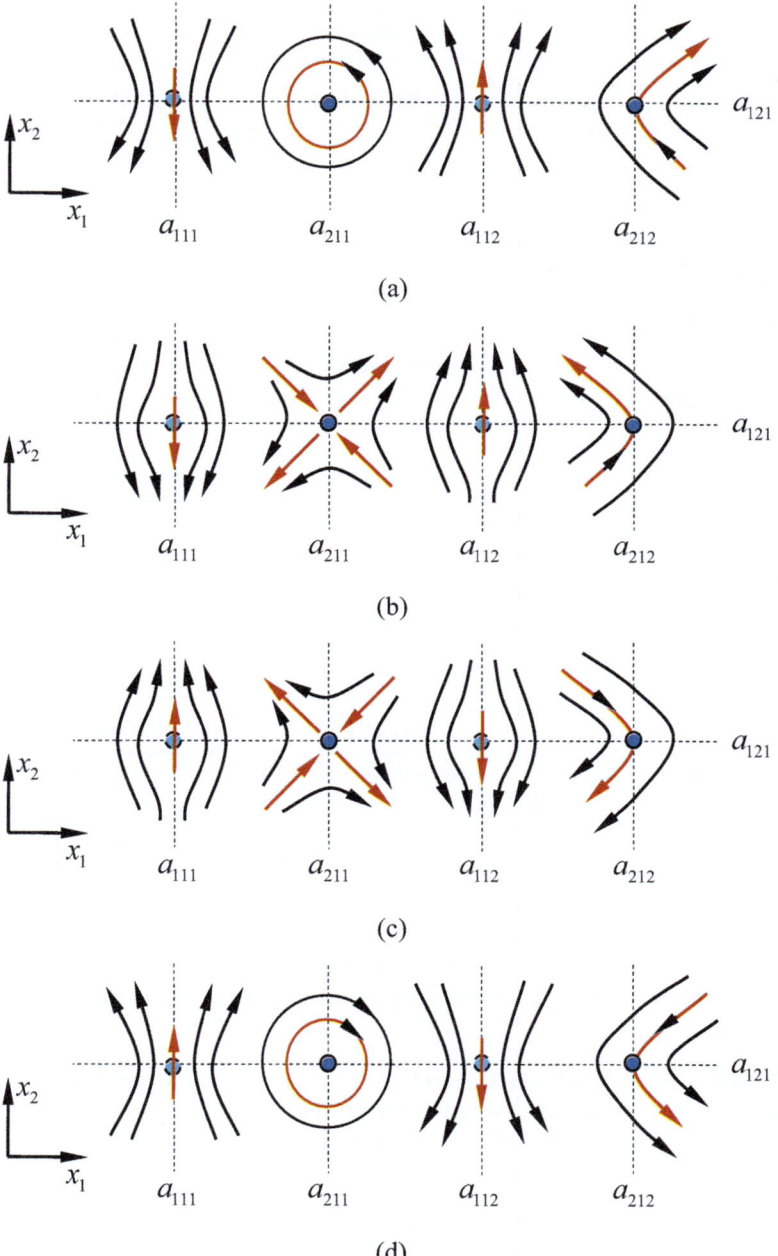

Fig. 4.2 Phase portraits $(a_{211} \in (a_{111}, a_{112})$ with $a_{112} < a_{212})$ for two-dimensional systems on the x_1-direction with $x_1^* = a_{111}, a_{112}, a_{211}, a_{212}$ and on the x_2-direction with $x_2^* = a_{121}$. (**a**) $(a_{110} > 0, a_{220} > 0)$, (**b**) $(a_{110} < 0, a_{220} > 0)$, (**c**) $(a_{110} > 0, a_{220} < 0)$, (**d**) $(a_{110} < 0, a_{220} < 0)$

$$\left\{\begin{matrix}(a_{121},a_{212}) & (a_{112},a_{121})\\ (a_{111},a_{121}) & (a_{121},a_{211})\end{matrix}\right\} = \left\{\begin{matrix}\underbrace{(\text{DP, US})}_{\text{down-parabola upper-saddle}} & \underbrace{(\text{DP:UP, pF})}_{\text{hyperbolic flow }(+)}\\ \underbrace{(\text{DP:UP, nF})}_{\text{hyperbolic flow }(-)} & \underbrace{(\text{DP}_+:\text{DP}_-)}_{\text{CCW center}}\end{matrix}\right\} \quad (4.11)$$

for $a_{110} > 0$ and $a_{220} > 0$;

$$\left\{\begin{matrix}(a_{121},a_{212}) & (a_{112},a_{121})\\ (a_{111},a_{121}) & (a_{121},a_{211})\end{matrix}\right\} = \left\{\begin{matrix}\underbrace{(\text{UP, US})}_{\text{up-parabola upper-saddle}} & \underbrace{(\text{UP:DP, pF})}_{\text{hyperbolic-secant flow }(+)}\\ \underbrace{(\text{UP:DP, nF})}_{\text{hyperbolic-secant flow }(-)} & \underbrace{(\text{UP}_+,\text{UP}_+)}_{\text{positive saddle}}\end{matrix}\right\}$$

$$(4.12)$$

for $a_{110} < 0$ and $a_{220} > 0$;

$$\left\{\begin{matrix}(a_{121},a_{212}) & (a_{112},a_{121})\\ (a_{111},a_{121}) & (a_{121},a_{211})\end{matrix}\right\} = \left\{\begin{matrix}\underbrace{(\text{UP, LS})}_{\text{up-parabola lower-saddle}} & \underbrace{(\text{UP:DP, nF})}_{\text{hyperbolic-secant flow }(-)}\\ \underbrace{(\text{UP:DP, pF})}_{\text{hyperbolic-secant flow }(+)} & \underbrace{(\text{UP}_-,\text{UP}_-)}_{\text{neagitve saddle}}\end{matrix}\right\}$$

$$(4.13)$$

for $a_{110} > 0$ and $a_{220} < 0$;

$$\left\{\begin{matrix}(a_{121},a_{212}) & (a_{112},a_{121})\\ (a_{111},a_{121}) & (a_{121},a_{211})\end{matrix}\right\} = \left\{\begin{matrix}\underbrace{(\text{DP, LS})}_{\text{down-parabola lower-saddle}} & \underbrace{(\text{DP:UP, nF})}_{\text{hyperbolic flow }(-)}\\ \underbrace{(\text{DP:UP, pF})}_{\text{hyperbolic flow }(+)} & \underbrace{(\text{DP}_-:\text{DP}_+)}_{\text{CW center}}\end{matrix}\right\} \quad (4.14)$$

for $a_{110} < 0$ and $a_{220} < 0$.

In Fig. 4.3, from the first integration manifold, the phase portraits of the equilibriums are sketched, and the corresponding singularity characteristics of equilibriums are presented.

(iv) For the case of $a_{112} \in (a_{111}, a_{212})$ with $a_{211} < a_{111}$, the parabola upper-saddles and lower-saddles with hyperbolic and hyperbolic-secant flows of $x_1^* = a_{111}, a_{112}$, a_{211}, a_{212} with $x_2^* = a_{121}$ have the following properties:

$$\left\{\begin{matrix}(a_{112},a_{121}) & (a_{121},a_{212})\\ (a_{121},a_{211}) & (a_{111},a_{121})\end{matrix}\right\} = \left\{\begin{matrix}\underbrace{(\text{DP:UP, pF})}_{\text{hyperbolic flow }(+)} & \underbrace{(\text{UP, US})}_{\text{up-parabola upper-saddle}}\\ \underbrace{(\text{UP}_+,\text{UP}_+)}_{\text{positive saddle}} & \underbrace{(\text{UP:DP, pF})}_{\text{hyperbolic-secant flow }(+)}\end{matrix}\right\} \quad (4.15)$$

for $a_{110} > 0$ and $a_{220} > 0$;

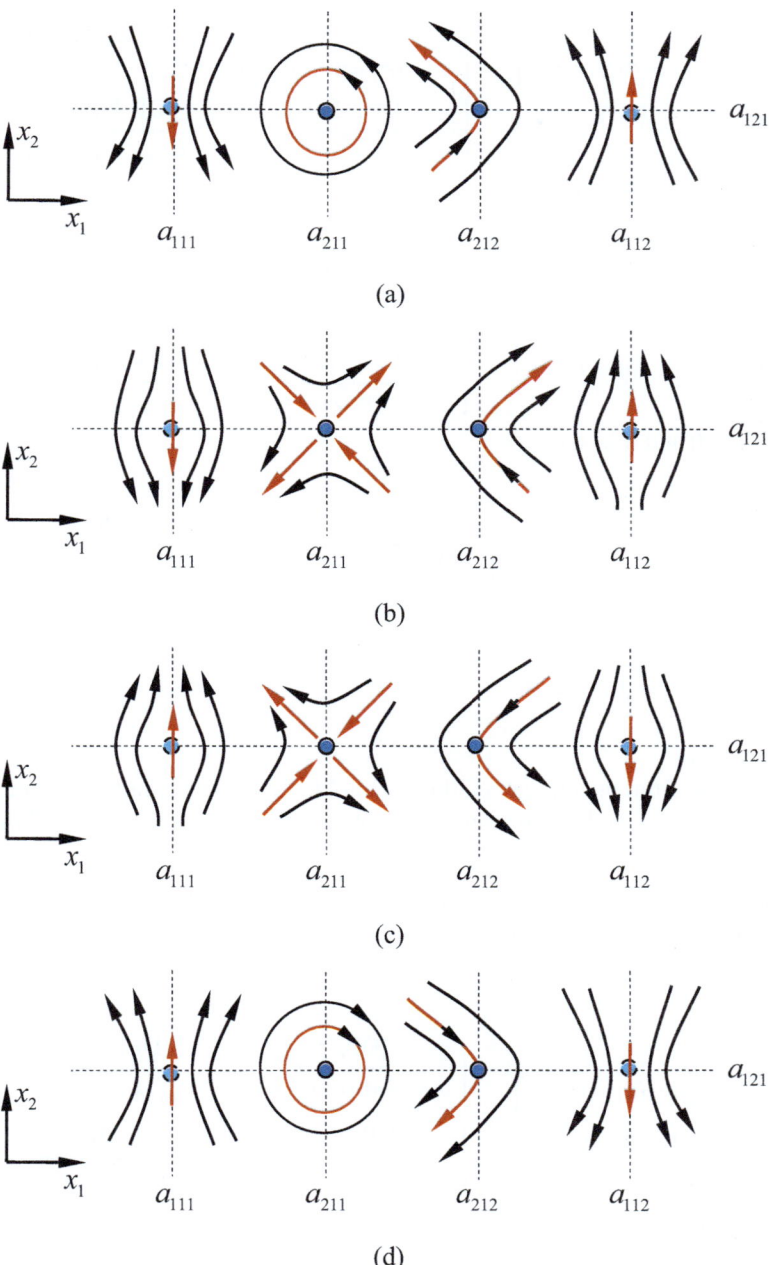

Fig. 4.3 Phase portraits ($a_{212} \in (a_{211}, a_{112})$ with $a_{111} < a_{211}$) for two-dimensional systems on the x_1-direction with $x_1^* = a_{111}, a_{112}, a_{211}, a_{212}$ and on the x_2-direction with $x_2^* = a_{121}$. (**a**) ($a_{110} > 0$, $a_{220} > 0$), (**b**) ($a_{110} < 0, a_{220} > 0$), (**c**) ($a_{110} > 0, a_{220} < 0$), (**d**) ($a_{110} < 0, a_{220} < 0$)

$$
\left\{\begin{array}{ll}(a_{112}, a_{121}) & (a_{121}, a_{212}) \\ (a_{121}, a_{211}) & (a_{111}, a_{121})\end{array}\right\} = \left\{\begin{array}{ll}\underbrace{(\text{UP}:\text{DP}, \text{pF})}_{\text{hyperbolic-secant flow } (+)} & \underbrace{(\text{DP}, \text{US})}_{\text{down-parabola upper-saddle}} \\ \underbrace{(\text{DP}_+, \text{DP}_-)}_{\text{CCW center}} & \underbrace{(\text{DP}:\text{UP}, \text{pF})}_{\text{hyperbolic flow } (+)}\end{array}\right\}
$$

for $a_{110} < 0$ and $a_{220} > 0$;

$$(4.16)$$

$$
\left\{\begin{array}{ll}(a_{112}, a_{121}) & (a_{121}, a_{212}) \\ (a_{121}, a_{211}) & (a_{111}, a_{121})\end{array}\right\} = \left\{\begin{array}{ll}\underbrace{(\text{UP}:\text{DP}, \text{nF})}_{\text{hyperbolic-secant flow } (-)} & \underbrace{(\text{DP}, \text{LS})}_{\text{down-parabola lower-saddle}} \\ \underbrace{(\text{DP}_-, \text{DP}_+)}_{\text{CW center}} & \underbrace{(\text{DP}:\text{UP}, \text{nF})}_{\text{hyperbolic flow } (-)}\end{array}\right\}
$$

for $a_{110} > 0$ and $a_{220} < 0$;

$$(4.17)$$

$$
\left\{\begin{array}{ll}(a_{112}, a_{121}) & (a_{121}, a_{212}) \\ (a_{121}, a_{211}) & (a_{111}, a_{121})\end{array}\right\} = \left\{\begin{array}{ll}\underbrace{(\text{DP}:\text{UP}, \text{nF})}_{\text{hyperbolic flow } (-)} & \underbrace{(\text{UP}, \text{LS})}_{\text{up-parabola lower-saddle}} \\ \underbrace{(\text{UP}_-, \text{UP}_-)}_{\text{negative saddle}} & \underbrace{(\text{UP}:\text{DP}, \text{nF})}_{\text{hyperbolic-secant flow } (-)}\end{array}\right\}
$$

for $a_{110} < 0$ and $a_{220} < 0$.

$$(4.18)$$

In Fig. 4.4, from the first integration manifold, the phase portraits of the equilibriums are sketched, and the corresponding singularity characteristics of equilibriums are presented.

(v) For the case of $a_{212} \in (a_{111}, a_{112})$, the parabola upper-saddles and lower-saddles with hyperbolic and hyperbolic-secant flows of $x_1^* = a_{111}, a_{112}, a_{211}, a_{212}$ with $x_2^* = a_{121}$ have the following properties:

$$
\left\{\begin{array}{ll}(a_{121}, a_{212}) & (a_{112}, a_{121}) \\ (a_{121}, a_{211}) & (a_{111}, a_{121})\end{array}\right\} = \left\{\begin{array}{ll}\underbrace{(\text{DP}, \text{US})}_{\text{down-parabola upper-saddle}} & \underbrace{(\text{DP}:\text{UP}, \text{pF})}_{\text{hyperbolic flow } (+)} \\ \underbrace{(\text{UP}_+, \text{UP}_+)}_{\text{positive saddle}} & \underbrace{(\text{UP}:\text{DP}, \text{pF})}_{\text{hyperbolic-secant flow } (+)}\end{array}\right\}
$$

for $a_{110} > 0$ and $a_{220} > 0$;

$$(4.19)$$

$$
\left\{\begin{array}{ll}(a_{121}, a_{212}) & (a_{112}, a_{121}) \\ (a_{121}, a_{211}) & (a_{111}, a_{121})\end{array}\right\} = \left\{\begin{array}{ll}\underbrace{(\text{UP}, \text{US})}_{\text{up-parabola upper-saddle}} & \underbrace{(\text{UP}:\text{DP}, \text{pF})}_{\text{hyperbolic-secant flow } (+)} \\ \underbrace{(\text{DP}_+, \text{DP}_-)}_{\text{CCW center}} & \underbrace{(\text{DP}:\text{UP}, \text{pF})}_{\text{hyperbolic flow } (+)}\end{array}\right\}
$$

for $a_{110} < 0$ and $a_{220} > 0$;

$$(4.20)$$

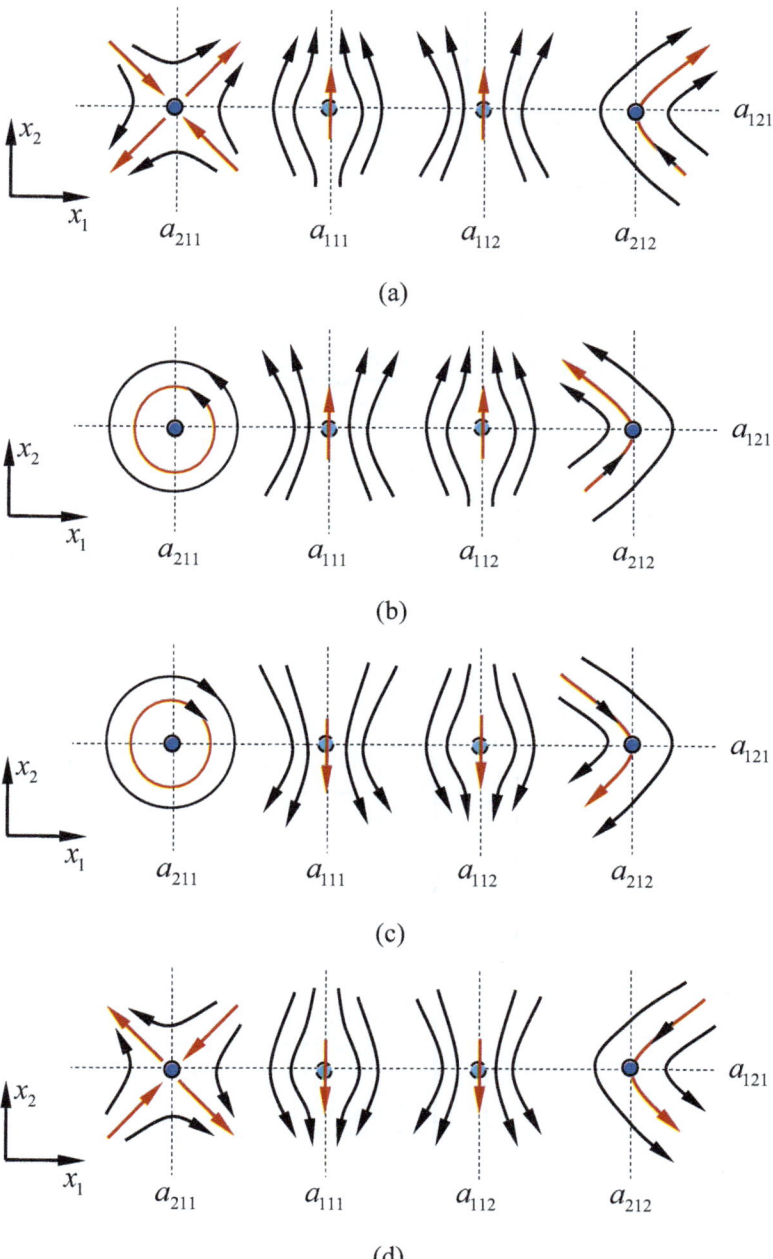

Fig. 4.4 Phase portraits ($a_{112} \in (a_{111}, a_{212})$) with $a_{211} < a_{111}$ for two-dimensional systems on the x_1-direction with $x_1^* = a_{111}, a_{112}, a_{211}, a_{212}$ and on the x_2-direction with $x_2^* = a_{121}$. (**a**) ($a_{110} > 0$, $a_{220} > 0$), (**b**) ($a_{110} < 0, a_{220} > 0$), (**c**) ($a_{110} > 0, a_{220} < 0$), (**d**) ($a_{110} < 0, a_{220} < 0$)

$$
\left\{ \begin{array}{ll} (a_{121}, a_{212}) & (a_{112}, a_{121}) \\ (a_{121}, a_{211}) & (a_{111}, a_{121}) \end{array} \right\} = \left\{ \begin{array}{cc} \underbrace{(\mathrm{UP, LS})}_{\text{up-parabola lower-saddle}} & \underbrace{(\mathrm{UP:DP, nF})}_{\text{hyperbolic-secant flow } (-)} \\ \underbrace{(\mathrm{DP_-, DP_+})}_{\text{CW center}} & \underbrace{(\mathrm{DP:UP, nF})}_{\text{hyperbolic flow } (-)} \end{array} \right\}
$$

for $a_{110} > 0$ and $a_{220} < 0$;

$$(4.21)$$

$$
\left\{ \begin{array}{ll} (a_{121}, a_{212}) & (a_{112}, a_{121}) \\ (a_{121}, a_{211}) & (a_{111}, a_{121}) \end{array} \right\} = \left\{ \begin{array}{cc} \underbrace{(\mathrm{DP, LS})}_{\text{down-parabola lower-saddle}} & \underbrace{(\mathrm{DP:UP, nF})}_{\text{hyperbolic flow } (-)} \\ \underbrace{(\mathrm{UP_-, UP_-})}_{\text{negative saddle}} & \underbrace{(\mathrm{UP:DP, nF})}_{\text{hyperbolic-secant flow } (-)} \end{array} \right\}
$$

for $a_{110} < 0$ and $a_{220} < 0$.

$$(4.22)$$

In Fig. 4.5, from the first integration manifold, the phase portraits of the equilibriums are sketched, and the corresponding singularity characteristics of equilibriums are presented.

(vi) For the case of $a_{212} < a_{111}$, the parabola upper-saddles and lower-saddles with hyperbolic and hyperbolic-secant flows of $x_1^* = a_{111}, a_{112}, a_{211}, a_{212}$ with $x_2^* = a_{121}$ have the following properties:

$$
\left\{ \begin{array}{ll} (a_{111}, a_{121}) & (a_{112}, a_{121}) \\ (a_{121}, a_{211}) & (a_{121}, a_{212}) \end{array} \right\} = \left\{ \begin{array}{cc} \underbrace{(\mathrm{UP:DP, pF})}_{\text{hyperbolic-secant flow } (+)} & \underbrace{(\mathrm{DP:UP, pF})}_{\text{hyperbolic flow } (+)} \\ \underbrace{(\mathrm{UP_+, UP_+})}_{\text{positive saddle}} & \underbrace{(\mathrm{UP, US})}_{\text{up-parabola upper-saddle}} \end{array} \right\}
$$

for $a_{110} > 0$ and $a_{220} > 0$;

$$(4.23)$$

$$
\left\{ \begin{array}{ll} (a_{111}, a_{121}) & (a_{112}, a_{121}) \\ (a_{121}, a_{211}) & (a_{121}, a_{212}) \end{array} \right\} = \left\{ \begin{array}{cc} \underbrace{(\mathrm{DP:UP, pF})}_{\text{hyperbolic flow } (+)} & \underbrace{(\mathrm{UP:DP, pF})}_{\text{hyperbolic-secant flow } (+)} \\ \underbrace{(\mathrm{DP_+, DP_-})}_{\text{CCW center}} & \underbrace{(\mathrm{DP, US})}_{\text{down-parabola upper-saddle}} \end{array} \right\}
$$

for $a_{110} < 0$ and $a_{220} > 0$;

$$(4.24)$$

$$
\left\{ \begin{array}{ll} (a_{111}, a_{121}) & (a_{112}, a_{121}) \\ (a_{121}, a_{211}) & (a_{121}, a_{212}) \end{array} \right\} = \left\{ \begin{array}{cc} \underbrace{(\mathrm{DP:UP, nF})}_{\text{hyperbolic flow } (-)} & \underbrace{(\mathrm{UP:DP, nF})}_{\text{hyperbolic-secant flow } (-)} \\ \underbrace{(\mathrm{DP_-, DP_+})}_{\text{CW center}} & \underbrace{(\mathrm{DP, LS})}_{\text{down-parabola lower-saddle}} \end{array} \right\}
$$

for $a_{110} > 0$ and $a_{220} < 0$;

$$(4.25)$$

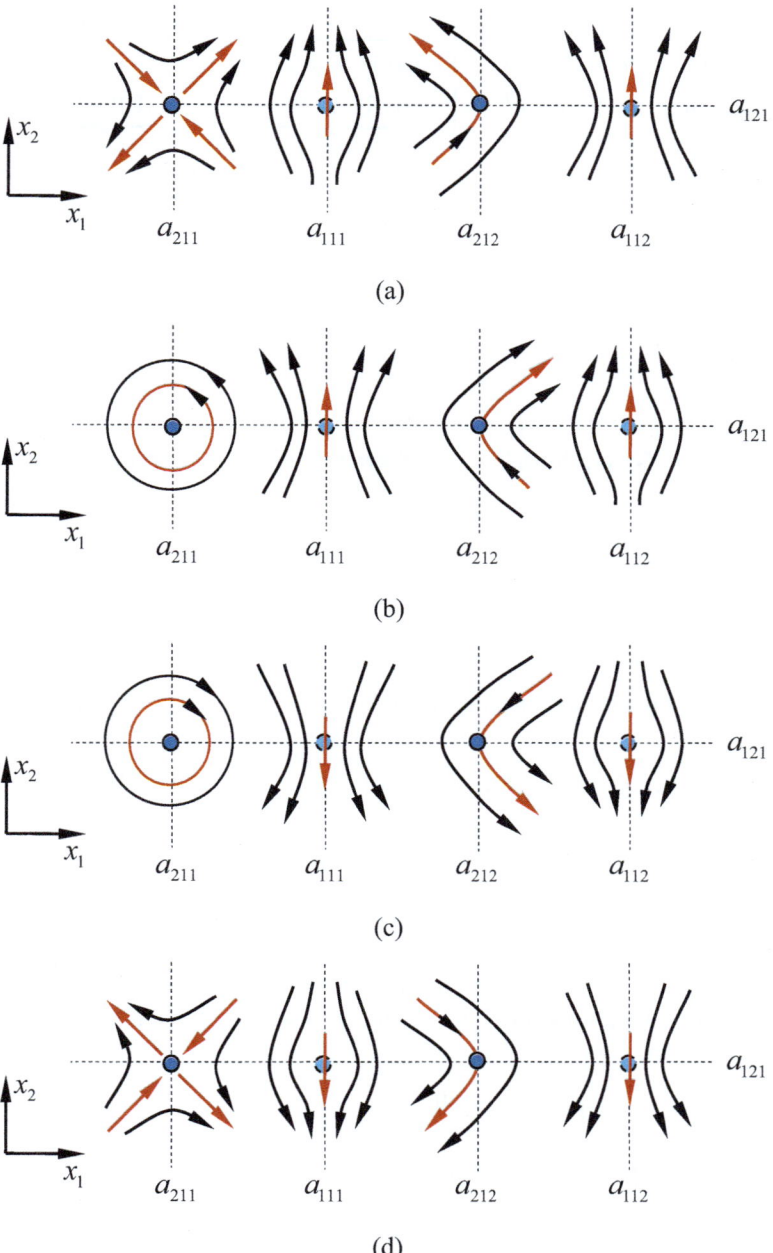

Fig. 4.5 Phase portraits ($a_{212} \in (a_{111}, a_{112})$) for two-dimensional systems on the x_1-direction with $x_1^* = a_{111}, a_{112}, a_{211}, a_{212}$ and on the x_2-direction with $x_2^* = a_{121}$. (**a**) ($a_{110} > 0$, $a_{220} > 0$), (**b**) ($a_{110} < 0$, $a_{220} > 0$), (**c**) ($a_{110} > 0$, $a_{220} < 0$), (**d**) ($a_{110} < 0$, $a_{220} < 0$)

$$
\left\{
\begin{array}{ll}
(a_{111}, a_{121}) & (a_{112}, a_{121}) \\
(a_{121}, a_{211}) & (a_{121}, a_{212})
\end{array}
\right\}
=
\left\{
\begin{array}{cc}
\overbrace{(\text{UP}:\text{DP}, \text{nF})} & \overbrace{(\text{DP}:\text{UP}, \text{nF})} \\
\text{hyperbolic-secant flow } (-) & \text{hyperbolic flow } (-) \\
\underbrace{(\text{UP}_-, \text{UP}_-)} & \underbrace{(\text{UP}, \text{LS})} \\
\text{negative saddle} & \text{up-parabola lower-saddle}
\end{array}
\right\}
$$

for $a_{110} < 0$ and $a_{220} < 0$.

$$(4.26)$$

In Fig. 4.6, from the first integration manifold, the phase portraits of the equilibriums are sketched, and the corresponding singularity characteristics of equilibriums are presented.

4.1.2 Third-Order Saddles and Centers with Hyperbolic Flows

Consider a dynamical system as

$$
\begin{aligned}
\dot{x}_1 &= a_{110}(x_1 - a_{111})(x_1 - a_{112})(x_2 - a_{121}), \\
\dot{x}_2 &= a_{220}(x_1 - a_{211})^3,
\end{aligned}
\tag{4.27}
$$

and the corresponding first integral manifold is

$$
\begin{aligned}
&\frac{1}{2}\left[(x_1 - a_{211})^2 - (x_{10} - a_{211})^2\right] + (a_{111} + a_{112} - 2a_{211})(x_1 - x_{10}) \\
&+ \frac{(a_{111} - a_{211})^3}{a_{111} - a_{112}} \ln \frac{|x_1 - a_{111}|}{|x_{10} - a_{111}|} + \frac{(a_{112} - a_{211})^3}{a_{112} - a_{111}} \ln \frac{|x_1 - a_{112}|}{|x_{10} - a_{112}|} \\
&= \frac{1}{2}\frac{a_{110}}{a_{220}}\left[(x_2 - a_{121})^2 - (x_{20} - a_{121})^2\right].
\end{aligned}
\tag{4.28}
$$

(i) For the case of $a_{112} < a_{211}$, the third-order saddles and centers with hyperbolic and hyperbolic-secant flows of $x_1^* = a_{111}, a_{112}, a_{211}$ with $x_2^* = a_{121}$ have the following properties:

$$
\left\{
\begin{array}{l}
(a_{121}, a_{211}) \\
(a_{112}, a_{121}) \\
(a_{111}, a_{121})
\end{array}
\right\}
=
\left\{
\begin{array}{c}
\overbrace{(\text{UP}_+, 3^{\text{rd}}\text{UP}_+)} \\
\text{third-order positive saddle} \\
\overbrace{(\text{UP}:\text{DP}, \text{nF})} \\
\text{hyperbolic-secant flow } (-) \\
\overbrace{(\text{DP}:\text{UP}, \text{nF})} \\
\text{hyperbolic flow } (-)
\end{array}
\right\}
\quad \text{for } a_{110} > 0 \text{ and } a_{220} > 0;
\tag{4.29}
$$

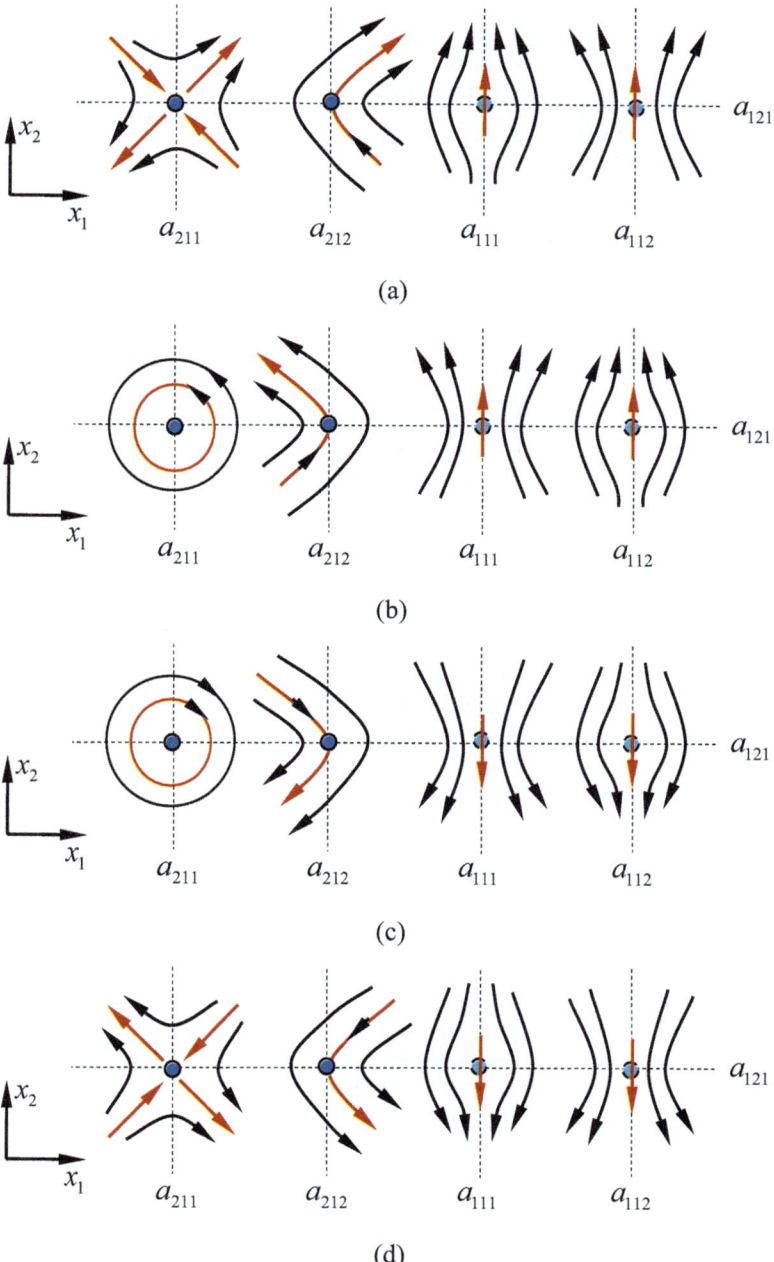

Fig. 4.6 Phase portraits ($a_{112} < a_{111}$) for two-dimensional systems on the x_1-direction with $x_1^* = a_{111}, a_{112}, a_{211}, a_{212}$ and on the x_2-direction with $x_2^* = a_{121}$. (**a**) ($a_{110} > 0$, $a_{220} > 0$), (**b**) ($a_{110} < 0$, $a_{220} > 0$), (**c**) ($a_{110} > 0$, $a_{220} < 0$), (**d**) ($a_{110} < 0$, $a_{220} < 0$)

$$
\begin{Bmatrix} (a_{121}, a_{211}) \\ (a_{112}, a_{121}) \\ (a_{111}, a_{121}) \end{Bmatrix} = \left\{ \begin{array}{c} \underbrace{(DP_+, 3^{rd}DP_-)}_{\text{third-order CCW center}} \\ \underbrace{(DP:UP, nF)}_{\text{hyperbolic flow }(-)} \\ \underbrace{(UP:DP, nF)}_{\text{hyperbolic-secant flow }(-)} \end{array} \right\} \text{ for } a_{110} < 0 \text{ and } a_{220} > 0; \qquad (4.30)
$$

$$
\begin{Bmatrix} (a_{121}, a_{211}) \\ (a_{112}, a_{121}) \\ (a_{111}, a_{121}) \end{Bmatrix} = \left\{ \begin{array}{c} \underbrace{(DP_-, 3^{rd}DP_+)}_{\text{third-order CW center}} \\ \underbrace{(DP:UP, pF)}_{\text{hyperbolic flow }(+)} \\ \underbrace{(UP:DP, pF)}_{\text{hyperbolic-secant flow }(+)} \end{array} \right\} \text{ for } a_{110} > 0 \text{ and } a_{220} < 0; \qquad (4.31)
$$

$$
\begin{Bmatrix} (a_{121}, a_{211}) \\ (a_{112}, a_{121}) \\ (a_{111}, a_{121}) \end{Bmatrix} = \left\{ \begin{array}{c} \underbrace{(UP_-, 3^{rd}UP_-)}_{\text{third-order negative saddle}} \\ \underbrace{(UP:DP, pF)}_{\text{hyperbolic-secant flow }(+)} \\ \underbrace{(DP:UP, pF)}_{\text{hyperbolic flow }(+)} \end{array} \right\} \text{ for } a_{110} < 0 \text{ and } a_{220} < 0. \qquad (4.32)
$$

In Fig. 4.7, the phase portraits of the equilibriums are sketched, and the corresponding singularity characteristics of equilibriums are presented.

(ii) For $a_{211} < a_{111}$, the equilibriums of $x_1^* = a_{111}, a_{112}, a_{211}$ with $x_2^* = a_{121}$ have the following properties:

$$
\begin{Bmatrix} (a_{112}, a_{121}) \\ (a_{111}, a_{121}) \\ (a_{121}, a_{211}) \end{Bmatrix} = \left\{ \begin{array}{c} \underbrace{(DP:UP, pF)}_{\text{hyperbolic flow }(+)} \\ \underbrace{(UP:DP, pF)}_{\text{hyperbolic-secant flow }(+)} \\ \underbrace{(UP_+, 3^{rd}UP_+)}_{\text{third-order positive saddle}} \end{array} \right\} \text{ for } a_{110} > 0 \text{ and } a_{220} > 0; \qquad (4.33)
$$

$$
\begin{Bmatrix} (a_{112}, a_{121}) \\ (a_{111}, a_{121}) \\ (a_{121}, a_{211}) \end{Bmatrix} = \left\{ \begin{array}{c} \underbrace{(UP:DP, pF)}_{\text{hyperbolic-secant flow }(+)} \\ \underbrace{(DP:UP, pF)}_{\text{hyperbolic flow }(+)} \\ \underbrace{(DP_+, 3^{rd}DP_-)}_{\text{Third-order CCW center}} \end{array} \right\} \text{ for } a_{110} < 0 \text{ and } a_{220} > 0; \qquad (4.34)
$$

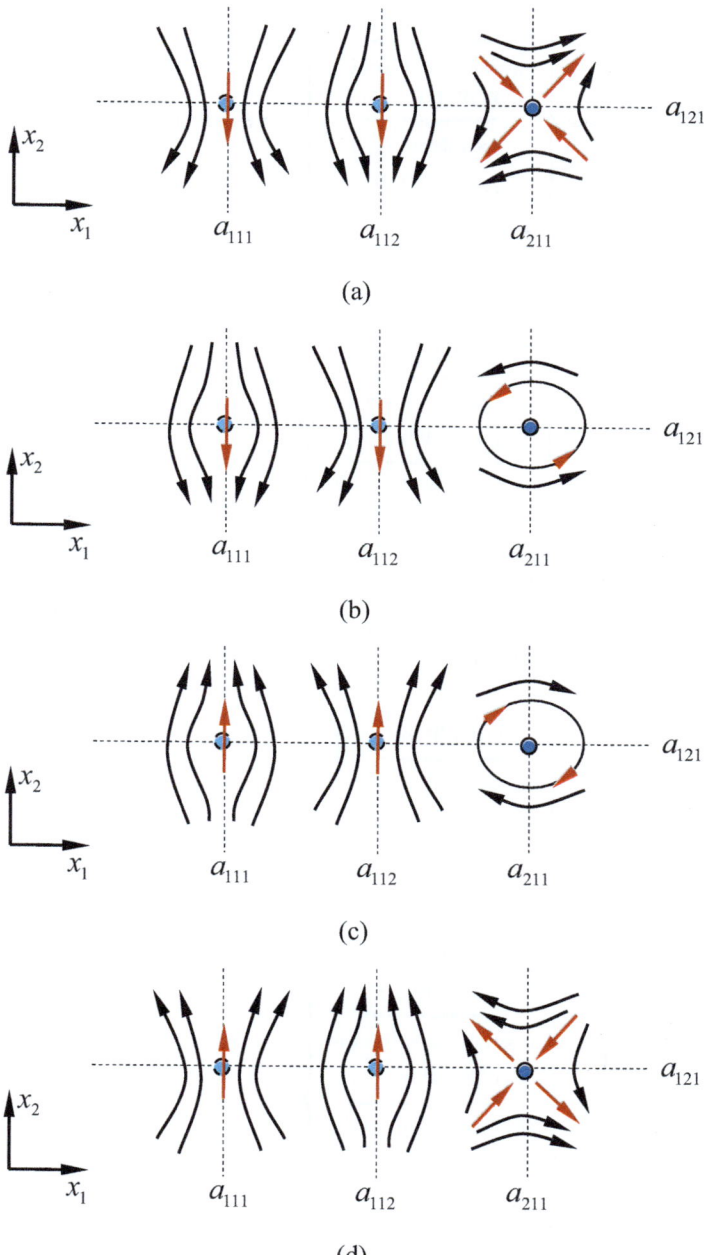

Fig. 4.7 Phase portraits $(a_{112} < a_{211})$ for two-dimensional systems on the x_1-direction with $x_1^* = a_{111}, a_{112}, a_{211}$ and on the x_2-direction with $x_2^* = a_{121}$. (**a**) $(a_{110} > 0, a_{220} > 0)$, (**b**) $(a_{110} < 0, a_{220} > 0)$, (**c**) $(a_{110} > 0, a_{220} < 0)$, (**d**) $(a_{110} < 0, a_{220} < 0)$

$$\left\{\begin{array}{c} (a_{112}, a_{121}) \\ (a_{111}, a_{121}) \\ (a_{121}, a_{211}) \end{array}\right\} = \left\{\begin{array}{c} \underbrace{(\mathrm{UP:DP, nF})}_{\text{hyperbolic-secant flow }(-)} \\ \underbrace{(\mathrm{DP:UP, nF})}_{\text{hyperbolic flow }(-)} \\ \underbrace{(\mathrm{DP}_-, 3^{\mathrm{rd}}\mathrm{DP}_+)}_{\text{third-order CW center}} \end{array}\right\} \text{ for } a_{110} > 0 \text{ and } a_{220} < 0; \quad (4.35)$$

$$\left\{\begin{array}{c} (a_{112}, a_{121}) \\ (a_{111}, a_{121}) \\ (a_{121}, a_{211}) \end{array}\right\} = \left\{\begin{array}{c} \underbrace{(\mathrm{DP:UP, nF})}_{\text{hyperbolic flow }(-)} \\ \underbrace{(\mathrm{UP:DP, nF})}_{\text{hyperbolic-secant flow }(-)} \\ \underbrace{(\mathrm{UP}_-, 3^{\mathrm{rd}}\mathrm{UP}_-)}_{\text{third-order negative saddle}} \end{array}\right\} \text{ for } a_{110} < 0 \text{ and } a_{220} < 0. \quad (4.36)$$

In Fig. 4.8, the phase portraits of the equilibriums are sketched, and the corresponding singularity characteristics of equilibriums are presented.

(iii) For the case of $a_{211} \in (a_{111}, a_{112})$, the third-order saddles and centers with hyperbolic and hyperbolic-secant flows of $x_1^* = a_{111}, a_{112}, a_{211}$ with $x_2^* = a_{121}$ have the following properties:

$$\left\{\begin{array}{c} (a_{112}, a_{121}) \\ (a_{121}, a_{211}) \\ (a_{111}, a_{121}) \end{array}\right\} = \left\{\begin{array}{c} \underbrace{(\mathrm{DP:UP, pF})}_{\text{hyperbolic flow }(+)} \\ \underbrace{(\mathrm{DP}_+, 3^{\mathrm{rd}}\mathrm{DP}_-)}_{\text{third-order CCW center}} \\ \underbrace{(\mathrm{DP:UP, nF})}_{\text{hyperbolic flow }(-)} \end{array}\right\} \text{ for } a_{110} > 0 \text{ and } a_{220} > 0; \quad (4.37)$$

$$\left\{\begin{array}{c} (a_{112}, a_{121}) \\ (a_{121}, a_{211}) \\ (a_{111}, a_{121}) \end{array}\right\} = \left\{\begin{array}{c} \underbrace{(\mathrm{UP:DP, pF})}_{\text{hyperbolic-secant flow }(+)} \\ \underbrace{(\mathrm{UP}_+, 3^{\mathrm{rd}}\mathrm{UP}_+)}_{\text{third-order positive saddle}} \\ \underbrace{(\mathrm{UP:DP, nF})}_{\text{hyperbolic-secant flow }(-)} \end{array}\right\} \text{ for } a_{110} < 0 \text{ and } a_{220} > 0; \quad (4.38)$$

$$\left\{\begin{array}{c} (a_{112}, a_{121}) \\ (a_{121}, a_{211}) \\ (a_{111}, a_{121}) \end{array}\right\} = \left\{\begin{array}{c} \underbrace{(\mathrm{UP:DP, nF})}_{\text{hyperbolic-secant flow }(-)} \\ \underbrace{(\mathrm{UP}_-, 3^{\mathrm{rd}}\mathrm{UP}_-)}_{\text{third-order negative saddle}} \\ \underbrace{(\mathrm{DP:UP, pF})}_{\text{hyperbolic flow }(+)} \end{array}\right\} \text{ for } a_{110} > 0 \text{ and } a_{220} < 0; \quad (4.39)$$

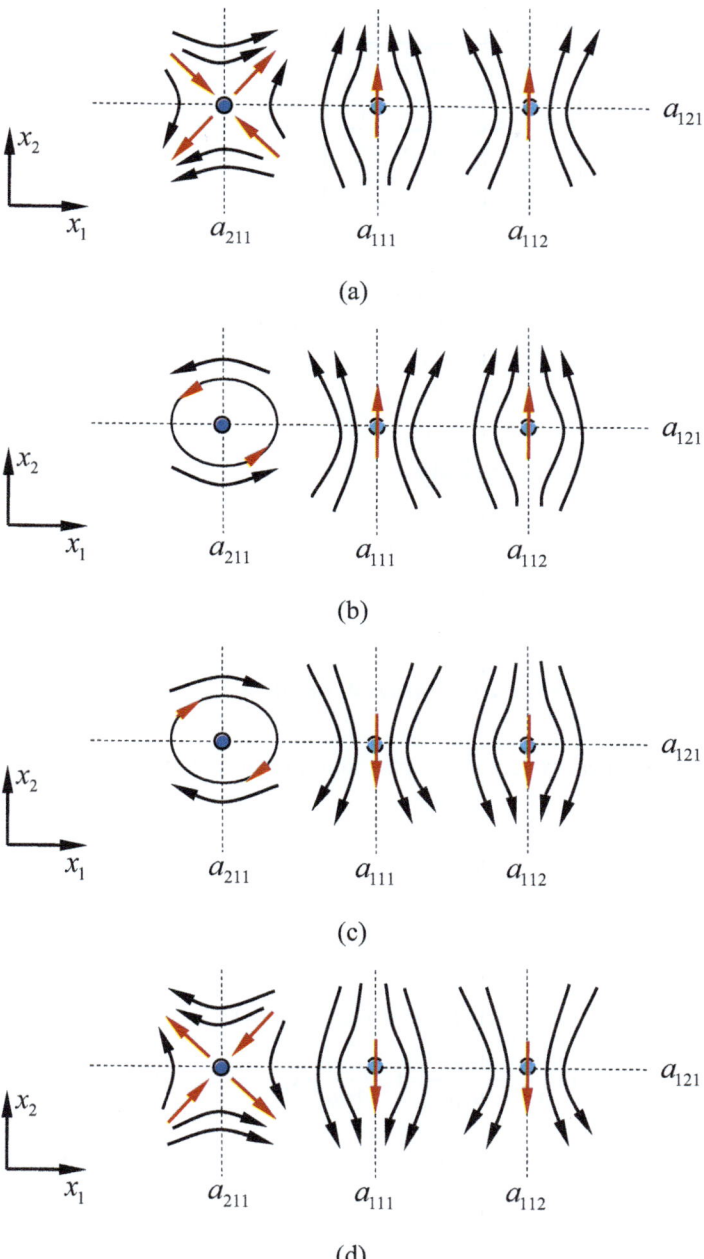

Fig. 4.8 Phase portraits ($a_{211} < a_{111}$) for two-dimensional systems on the x_1-direction with $x_1^* = a_{111}, a_{112}, a_{211}$ and on the x_2-direction with $x_2^* = a_{121}$. (**a**) ($a_{110} > 0, a_{220} > 0$), (**b**) ($a_{110} < 0, a_{220} > 0$), (**c**) ($a_{110} > 0, a_{220} < 0$), (**d**) ($a_{110} < 0, a_{220} < 0$)

$$\left.\begin{cases} (a_{112}, a_{121}) \\ (a_{121}, a_{211}) \\ (a_{111}, a_{121}) \end{cases}\right\} = \begin{cases} \underbrace{(DP:UP, nF)}_{\text{hyperbolic flow } (-)} \\ \underbrace{(DP_-, 3^{rd}DP_+)}_{\text{third-order CW center}} \\ \underbrace{(DP:UP, pF)}_{\text{hyperbolic flow } (+)} \end{cases} \text{ for } a_{110} < 0 \text{ and } a_{220} < 0. \quad (4.40)$$

In Fig. 4.9, from the first integration manifold, the phase portraits of the equilibriums are sketched, and the corresponding singularity characteristics of equilibriums are presented.

4.2 Switching Bifurcations

In this section, switching bifurcations are discussed through the inflection and parabola sinks and sources.

4.2.1 Inflection-Sink and Source Infinite-Equilibriums

Consider a dynamical system with a parabola sink and source infinite-equilibrium as

$$\begin{aligned} \dot{x}_1 &= a_{110}(x_1 - a_{111})(x_1 - a_{112})(x_2 - a_{121}), \\ \dot{x}_2 &= a_{220}(x_1 - a_{211})(x_1 - a_{212})^2. \end{aligned} \quad (4.41)$$

For $a_{211} = a_{11l_1}$ ($l_1, l_2 \in \{1, 2\}$, $l_1 \neq l_2$), the corresponding first integral manifold is

$$\begin{aligned} &\frac{1}{2}\left[(x_1 - a_{11l_2})^2 - (x_{10} - a_{11l_2})^2\right] + 2(a_{11l_2} - a_{212})(x_1 - x_{10}) \\ &+ (a_{11l_2} - a_{212})^2 \ln \frac{|x_1 - a_{11l_2}|}{|x_{10} - a_{11l_2}|} = \frac{1}{2}\frac{a_{110}}{a_{220}}\left[(x_2 - a_{121})^2 - (x_{20} - a_{121})^2\right]. \end{aligned} \quad (4.42)$$

(i) For $a_{211} = a_{111} = a_{11l_1}$ with $a_{112} < a_{212}$, the equilibriums of $x_1^* = a_{111}, a_{112}$, a_{211}, a_{212} with $x_2^* = a_{121}$ have the following properties:

$$\left.\begin{cases} (a_{121}, a_{212}) \\ (a_{112}, a_{121}) \\ (a_{121}, a_{211}) \end{cases}\right\} = \begin{cases} \underbrace{(UP, US)}_{\text{up-parabola upper-saddle}} \\ \underbrace{(DP:UP, pF)}_{\text{hyperbolic flow } (+)} \\ \underbrace{(_{\text{II:DI}}DP, _{\text{SO:SI}}LS)}_{\text{down-parabola lower-saddle}} \end{cases} \text{ for } a_{110} > 0, a_{220} > 0, \quad (4.43)$$

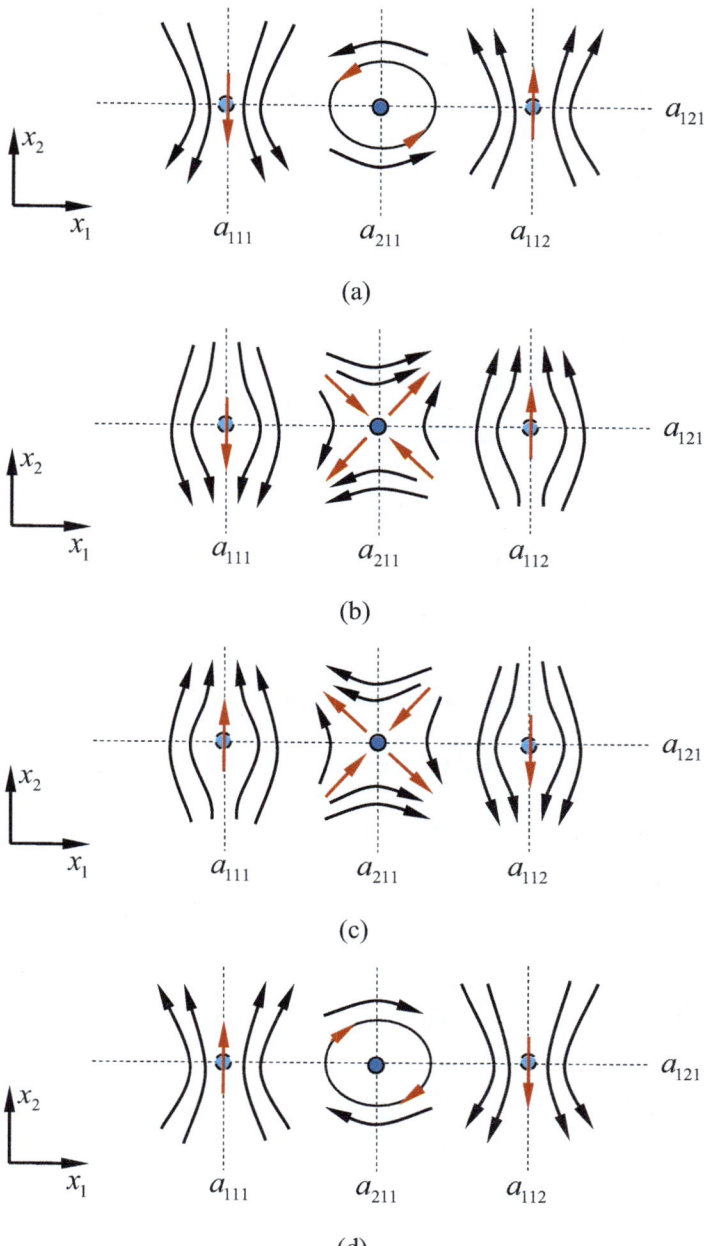

Fig. 4.9 Phase portraits ($a_{211} \in (a_{111}, a_{112})$) for two-dimensional systems on the x_1-direction with $x_1^* = a_{111}, a_{112}, a_{211}$ and on the x_2-direction with $x_2^* = a_{121}$. (**a**) ($a_{110} > 0, a_{220} > 0$), (**b**) ($a_{110} < 0, a_{220} > 0$), (**c**) ($a_{110} > 0, a_{220} < 0$), (**d**) ($a_{110} < 0, a_{220} < 0$)

$$
\left.\begin{array}{l} (a_{121}, a_{212}) \\ (a_{112}, a_{121}) \\ (a_{121}, a_{211}) \end{array}\right\} = \left\{\begin{array}{c} \underbrace{(DP, US)}_{\text{down-parabola upper-saddle}} \\ \underbrace{(UP:DP, pF)}_{\text{hyperbolic-secant flow (+)}} \\ \underbrace{(_{DI:II}UP, {}_{SI:SO}US)}_{\text{up-parabola upper-saddle}} \end{array}\right\} \text{ for } a_{110} < 0, a_{220} > 0, \qquad (4.44)
$$

$$
\left.\begin{array}{l} (a_{121}, a_{212}) \\ (a_{112}, a_{121}) \\ (a_{121}, a_{211}) \end{array}\right\} = \left\{\begin{array}{c} \underbrace{(DP, LS)}_{\text{down-parabola lower-saddle}} \\ \underbrace{(UP:DP, nF)}_{\text{hyperbolic-secant flow (-)}} \\ \underbrace{(_{DI:II}UP, {}_{SO:SI}LS)}_{\text{up-parabola lower-saddle}} \end{array}\right\} \text{ for } a_{110} > 0, a_{220} < 0, \qquad (4.45)
$$

$$
\left.\begin{array}{l} (a_{121}, a_{212}) \\ (a_{112}, a_{121}) \\ (a_{121}, a_{211}) \end{array}\right\} = \left\{\begin{array}{c} \underbrace{(UP, LS)}_{\text{up-parabola lower-saddle}} \\ \underbrace{(DP:UP, nF)}_{\text{hyperbolic flow (-)}} \\ \underbrace{(_{II:DI}DP, {}_{SI:SO}US)}_{\text{down-parabola upper-saddle}} \end{array}\right\} \text{ for } a_{110} < 0, a_{220} < 0. \qquad (4.46)
$$

and the infinite-equilibrium of $x_1^* = a_{111} = a_{211}$ is summarized as follows:

$$
(a_{111}, \bar{x}_2) = \underbrace{(SO, II)}_{\text{increasing-inflection source}} \quad \text{if } \bar{x}_2 \in (-\infty, a_{121}),
$$

$$
(a_{111}, \bar{x}_2) = \underbrace{(SI, DI)}_{\text{decreasing-infleciton sink}} \quad \text{if } \bar{x}_2 \in (a_{121}, \infty), \qquad (4.47)
$$

for $a_{110} > 0, a_{220} > 0$;

$$
(a_{111}, \bar{x}_2) = \underbrace{(SI, DI)}_{\text{decreasing-infleciton sink}} \quad \text{if } \bar{x}_2 \in (-\infty, a_{121}),
$$

$$
(a_{111}, \bar{x}_2) = \underbrace{(SO, II)}_{\text{increasing-inflection source}} \quad \text{if } \bar{x}_2 \in (a_{121}, \infty), \qquad (4.48)
$$

for $a_{110} < 0, a_{220} > 0$;

$$
(a_{111}, \bar{x}_2) = \underbrace{(SO, DI)}_{\text{decreasing-infleciton source}} \quad \text{if } \bar{x}_2 \in (-\infty, a_{121}),
$$

$$
(a_{111}, \bar{x}_2) = \underbrace{(SI, II)}_{\text{increasing-inflection sink}} \quad \text{if } \bar{x}_2 \in (a_{121}, \infty), \qquad (4.49)
$$

for $a_{110} > 0, a_{220} < 0$;

$$(a_{111}, \bar{x}_2) = \underbrace{(\text{SI, II})}_{\text{increasing-inflection sink}} \quad \text{if } \bar{x}_2 \in (-\infty, a_{121}),$$

$$(a_{111}, \bar{x}_2) = \underbrace{(\text{SO, DI})}_{\text{decreasing-infleciton source}} \quad \text{if } \bar{x}_2 \in (a_{121}, \infty), \tag{4.50}$$

for $a_{110} < 0, a_{220} < 0$.

The inflection-source and sink infinite-equilibriums of $x_1^* = a_{111} = a_{211}$ are presented in Fig. 4.10. The infinite-equilibriums are for the switching bifurcation for the equilibriums for $a_{111} < a_{211}$ and $a_{111} \in (a_{211}, a_{212})$ with $a_{112} < a_{212}$. The parabola-saddles on the finite-equilibriums are for the switching bifurcations of a saddle and centers with hyperbolic flows at $(x_2^*, x_1^*) = (a_{121}, a_{211})$. The parabola-saddles not on the infinite-equilibriums are for the appearing bifurcations of saddle and center at $(x_2^*, x_1^*) = (a_{121}, a_{212})$.

(ii) For $a_{211} = a_{111} = a_{11l_1}$ with $a_{112} < a_{212}$, the equilibriums of $x_1^* = a_{111}, a_{112}, a_{211}, a_{212}$ with $x_2^* = a_{121}$ have the following properties:

$$\left\{ \begin{array}{l} (a_{112}, a_{121}) \\ (a_{121}, a_{212}) \\ (a_{121}, a_{211}) \end{array} \right\} = \left\{ \begin{array}{l} \underbrace{(\text{DP:UP, pF})}_{\text{hyperbolic flow } (+)} \\ \underbrace{(\text{DP, US})}_{\text{down-parabola upper-saddle}} \\ \underbrace{(_{\text{II:DI}}\text{DP, }_{\text{SO:SI}}\text{LS})}_{\text{down-parabola lower-saddle}} \end{array} \right\} \quad \text{for } a_{110} > 0, a_{220} > 0, \tag{4.51}$$

$$\left\{ \begin{array}{l} (a_{112}, a_{121}) \\ (a_{121}, a_{212}) \\ (a_{121}, a_{211}) \end{array} \right\} = \left\{ \begin{array}{l} \underbrace{(\text{UP:DP, pF})}_{\text{hyperbolic-secant flow } (+)} \\ \underbrace{(\text{UP, US})}_{\text{up-parabola upper-saddle}} \\ \underbrace{(_{\text{DI:II}}\text{UP, }_{\text{SI:SO}}\text{US})}_{\text{up-parabola upper-saddle}} \end{array} \right\} \quad \text{for } a_{110} < 0, a_{220} > 0, \tag{4.52}$$

$$\left\{ \begin{array}{l} (a_{112}, a_{121}) \\ (a_{121}, a_{212}) \\ (a_{121}, a_{211}) \end{array} \right\} = \left\{ \begin{array}{l} \underbrace{(\text{UP:DP, nF})}_{\text{hyperbolic-secant flow } (-)} \\ \underbrace{(\text{UP, LS})}_{\text{up-parabola lower-saddle}} \\ \underbrace{(_{\text{DI:II}}\text{UP, }_{\text{SO:SI}}\text{LS})}_{\text{up-parabola lower-saddle}} \end{array} \right\} \quad \text{for } a_{110} > 0, a_{220} < 0, \tag{4.53}$$

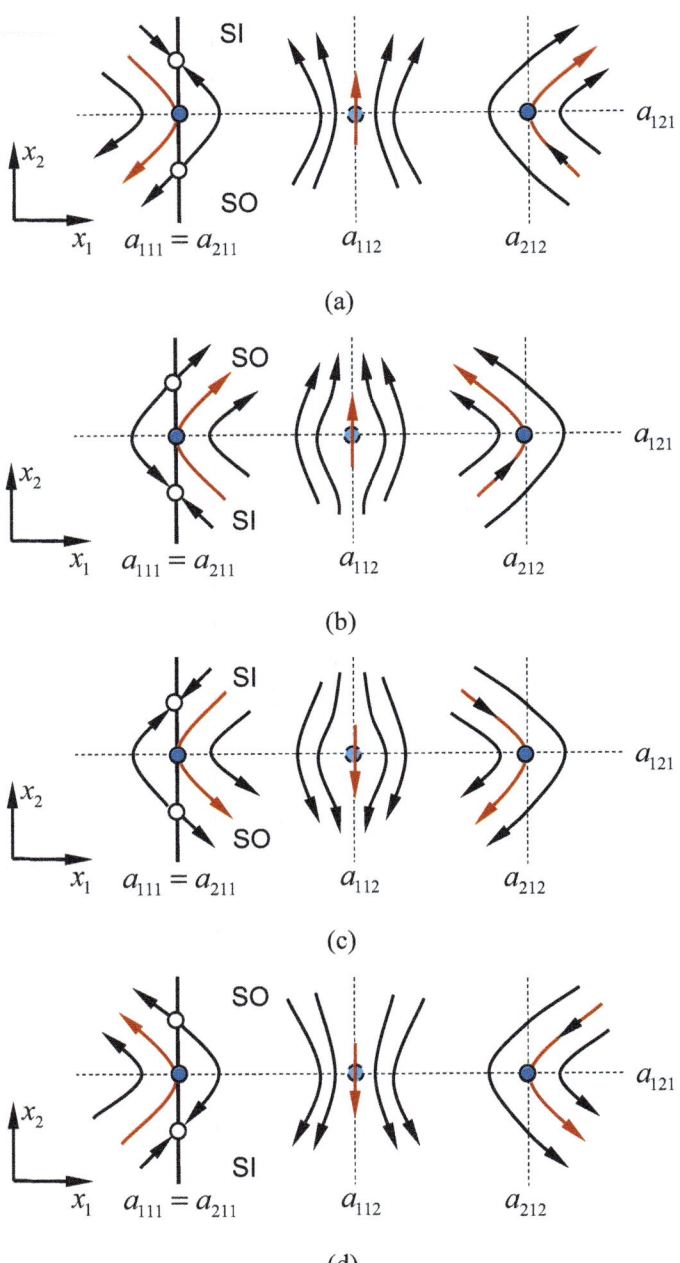

Fig. 4.10 Phase portraits ($a_{111} = a_{211}$ and $a_{112} < a_{212}$) for two-dimensional systems on the x_1-direction with $x_1^* = a_{111}, a_{112}, a_{211}, a_{212}$ and on the x_2-direction with $x_2^* = a_{121}$. (**a**) ($a_{110} > 0$, $a_{220} > 0$), (**b**) ($a_{110} < 0$, $a_{220} > 0$), (**c**) ($a_{110} > 0$, $a_{220} < 0$), (**d**) ($a_{110} < 0$, $a_{220} < 0$)

$$\left\{\begin{array}{l}(a_{112},a_{121})\\(a_{121},a_{212})\\(a_{121},a_{211})\end{array}\right\}=\left\{\begin{array}{c}\underbrace{(\mathrm{DP:UP,nF})}_{\text{hyperbolic flow }(-)}\\\underbrace{(\mathrm{DP,LS})}_{\text{down-parabola lower-saddle}}\\\underbrace{(_{\mathrm{II:DI}}\mathrm{DP},_{\mathrm{SI:SO}}\mathrm{US})}_{\text{down-parabola upper-saddle}}\end{array}\right\}\ \text{for } a_{110}<0, a_{220}<0. \qquad (4.54)$$

The infinite-equilibrium of $x_1^*=a_{111}=a_{211}$ has the same properties as in Eqs. (4.47)–(4.50). The parabola-saddles are for the appearing bifurcations of saddles and centers, and such parabola-saddles with hyperbolic flows are changed. The corresponding phase portraits are presented in Fig. 4.11. The infinite-equilibriums for two cases are invariant. The infinite-equilibriums are for the switching bifurcation for the equilibriums for $a_{111} < a_{211}$ and $a_{111} \in (a_{211},a_{212})$ with $a_{212} < a_{112}$.

(iii) For $a_{211}=a_{112}=a_{11l_1}$, the equilibriums of $x_1^*=a_{111},a_{112},a_{211},a_{212}$ with $x_2^*=a_{121}$ have the following properties:

$$\left\{\begin{array}{l}(a_{121},a_{212})\\(a_{121},a_{112})\\(a_{111},a_{121})\end{array}\right\}=\left\{\begin{array}{c}\underbrace{(\mathrm{UP,US})}_{\text{up-parabola upper-saddle}}\\\underbrace{(_{\mathrm{DI:II}}\mathrm{UP},_{\mathrm{SI:SO}}\mathrm{US})}_{\text{up-parabola lower-saddle}}\\\underbrace{(\mathrm{DP:UP,nF})}_{\text{hyperbolic flow }(-)}\end{array}\right\}\ \text{for } a_{110}>0, a_{220}>0, \qquad (4.55)$$

$$\left\{\begin{array}{l}(a_{121},a_{212})\\(a_{121},a_{112})\\(a_{111},a_{121})\end{array}\right\}=\left\{\begin{array}{c}\underbrace{(\mathrm{DP,US})}_{\text{doun-parabola upper-saddle}}\\\underbrace{(_{\mathrm{II:DI}}\mathrm{DP},_{\mathrm{SO:SI}}\mathrm{LS})}_{\text{up-parabola lower-saddle}}\\\underbrace{(\mathrm{UP:DP,nF})}_{\text{hyperbolic-secant flow }(-)}\end{array}\right\}\ \text{for } a_{110}<0, a_{220}>0, \qquad (4.56)$$

$$\left\{\begin{array}{l}(a_{121},a_{212})\\(a_{121},a_{112})\\(a_{111},a_{121})\end{array}\right\}=\left\{\begin{array}{c}\underbrace{(\mathrm{DP,LS})}_{\text{doun-parabola lower-saddle}}\\\underbrace{(_{\mathrm{II:DI}}\mathrm{DP},_{\mathrm{SI:SO}}\mathrm{US})}_{\text{down-parabola upper-saddle}}\\\underbrace{(\mathrm{UP:DP,pF})}_{\text{hyperbolic-secant flow }(+)}\end{array}\right\}\ \text{for } a_{110}>0, a_{220}<0, \qquad (4.57)$$

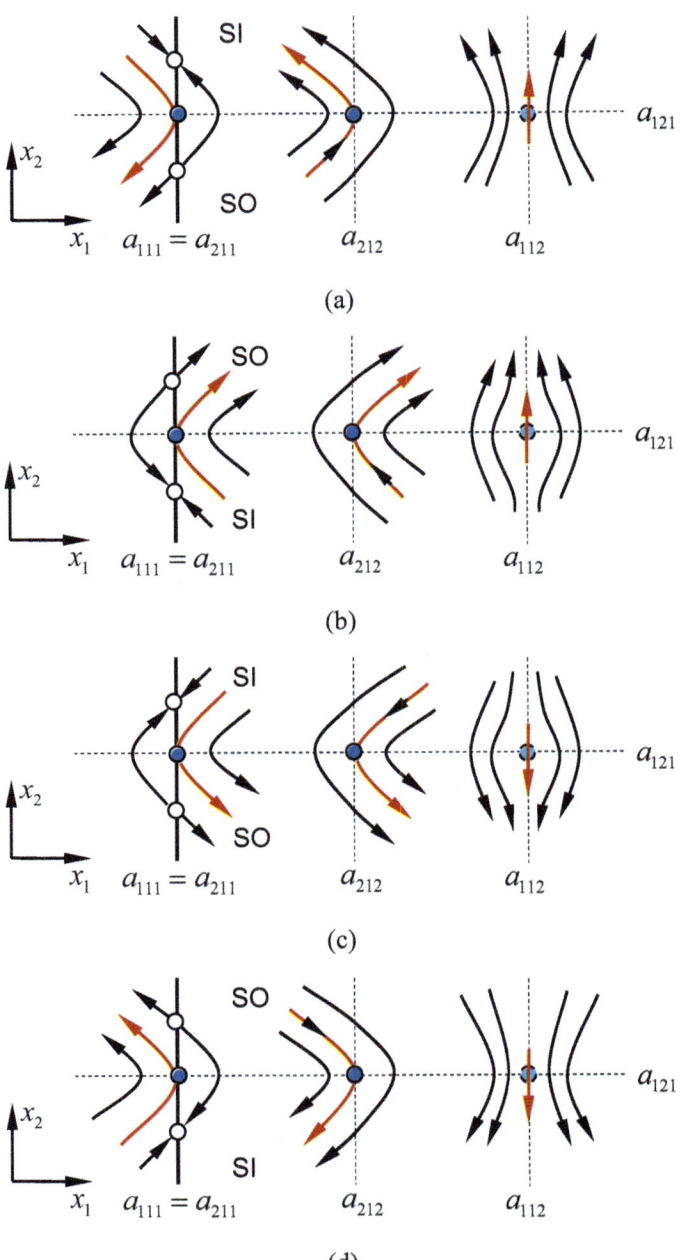

Fig. 4.11 Phase portraits ($a_{111} = a_{211}$ and $a_{212} < a_{112}$) for two-dimensional systems on the x_1-direction with $x_1^* = a_{111}, a_{112}, a_{211}, a_{212}$ and on the x_2-direction with $x_2^* = a_{121}$. (**a**) ($a_{110} > 0$, $a_{220} > 0$), (**b**) ($a_{110} < 0$, $a_{220} > 0$), (**c**) ($a_{110} > 0$, $a_{220} < 0$), (**d**) ($a_{110} < 0$, $a_{220} < 0$)

$$
\left\{
\begin{array}{c}
(a_{121}, a_{212}) \\
(a_{121}, a_{112}) \\
(a_{111}, a_{121})
\end{array}
\right\}
=
\left\{
\begin{array}{c}
\underset{\text{up-parabola lower-saddle}}{\underbrace{(\text{UP}, \text{LS})}} \\
\underset{\text{up-parabola lower-saddle}}{\underbrace{(_{\text{DI:II}}\text{UP}, _{\text{SO:SI}}\text{LS})}} \\
\underset{\text{hyperbolic flow } (+)}{\underbrace{(\text{DP:UP}, \text{pF})}}
\end{array}
\right\}
\quad \text{for } a_{110} < 0, a_{220} < 0.
\qquad (4.58)
$$

And the infinite-equilibrium of $x_1^* = a_{111} = a_{211}$ is summarized as follows:

$$
(a_{112}, \bar{x}_2) = \underset{\text{decreasing-inflection sink}}{\underbrace{(\text{SI}, \text{DI})}} \quad \text{if } \bar{x}_2 \in (-\infty, a_{121}),
$$

$$
(a_{112}, \bar{x}_2) = \underset{\text{increasing-inflection source}}{\underbrace{(\text{SO}, \text{II})}} \quad \text{if } \bar{x}_2 \in (a_{121}, \infty),
\qquad (4.59)
$$

for $a_{110} > 0, a_{220} > 0$;

$$
(a_{112}, \bar{x}_2) = \underset{\text{increasing-inflection source}}{\underbrace{(\text{SO}, \text{II})}} \quad \text{if } \bar{x}_2 \in (-\infty, a_{121}),
$$

$$
(a_{112}, \bar{x}_2) = \underset{\text{decreasing-inflection sink}}{\underbrace{(\text{SI}, \text{DI})}} \quad \text{if } \bar{x}_2 \in (a_{121}, \infty),
\qquad (4.60)
$$

for $a_{110} < 0, a_{220} > 0$;

$$
(a_{112}, \bar{x}_2) = \underset{\text{increasing-inflection sink}}{\underbrace{(\text{SI}, \text{II})}} \quad \text{if } \bar{x}_2 \in (-\infty, a_{121}),
$$

$$
(a_{112}, \bar{x}_2) = \underset{\text{decreasing-inflection source}}{\underbrace{(\text{SO}, \text{DI})}} \quad \text{if } \bar{x}_2 \in (a_{121}, \infty),
\qquad (4.61)
$$

for $a_{110} > 0, a_{220} < 0$;

$$
(a_{112}, \bar{x}_2) = \underset{\text{decreasing-inflection source}}{\underbrace{(\text{SO}, \text{DI})}} \quad \text{if } \bar{x}_2 \in (-\infty, a_{121}),
$$

$$
(a_{112}, \bar{x}_2) = \underset{\text{increasing-inflection sink}}{\underbrace{(\text{SI}, \text{II})}} \quad \text{if } \bar{x}_2 \in (a_{121}, \infty),
\qquad (4.62)
$$

for $a_{110} < 0, a_{220} < 0$.

The inflection-source and sink infinite-equilibriums of $x_1^* = a_{112} = a_{211}$ are presented in Fig. 4.12. The infinite-equilibriums are for the switching bifurcation for the equilibriums for $a_{112} < a_{211}$ and $a_{112} \in (a_{211}, a_{212})$. The parabola-saddles on the infinite-equilibriums are for the switching bifurcations of a saddle and centers with hyperbolic flows at $(x_2^*, x_1^*) = (a_{121}, a_{211})$. The parabola-saddles not on the infinite-equilibriums are for the appearing bifurcations of a saddle and center with hyperbolic flows at $(x_2^*, x_1^*) = (a_{121}, a_{212})$.

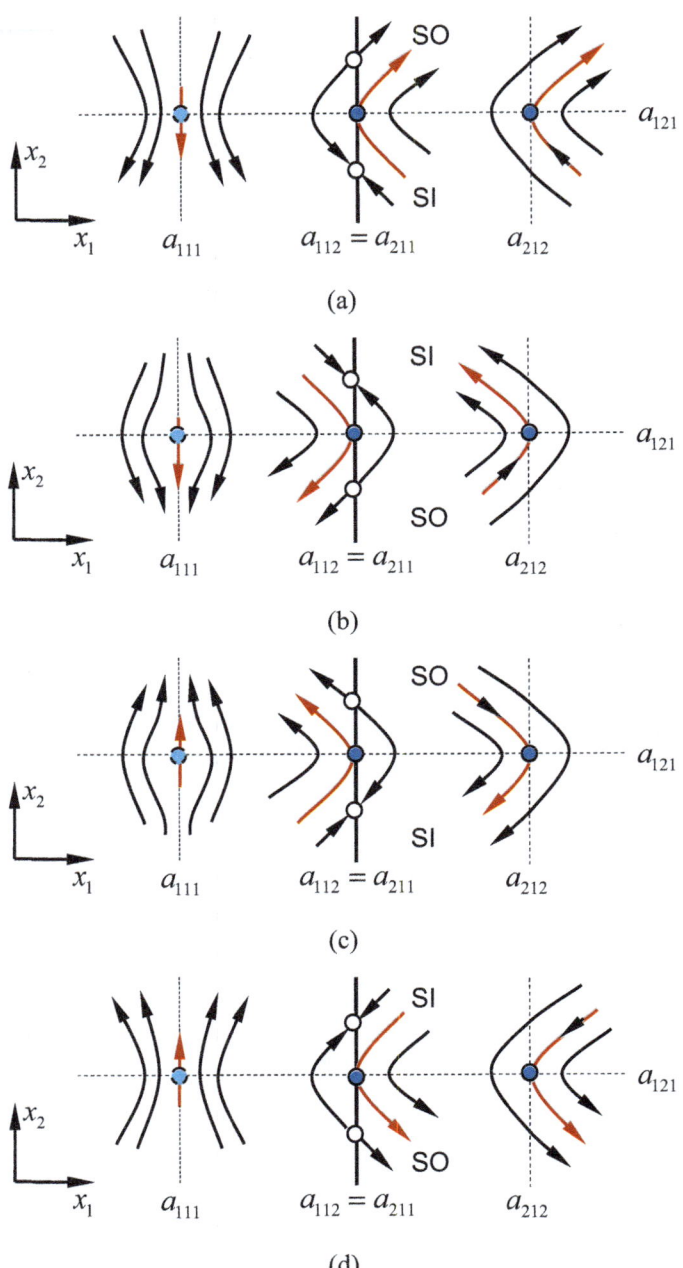

Fig. 4.12 Phase portraits ($a_{111} = a_{211}$) for two-dimensional systems on the x_1-direction with $x_1^* = a_{111}, a_{211}$ and on the x_2-direction with $x_2^* = a_{121}$. The equilibriums flows: (**a**) ($a_{110} > 0$, $a_{220} > 0$), (**b**) ($a_{110} < 0$, $a_{220} > 0$), (**c**) ($a_{110} > 0$, $a_{220} < 0$), (**d**) ($a_{110} < 0$, $a_{220} < 0$)

Consider a dynamical system as

$$\begin{aligned}
\dot{x}_1 &= a_{110}(x_1 - a_{111})(x_1 - a_{112})(x_2 - a_{121}), \\
\dot{x}_2 &= a_{220}(x_1 - a_{212})(x_1 - a_{211})^2.
\end{aligned} \tag{4.63}$$

For the infinite equilibriums of $x_1 = a_{212} = a_{11l_1}$ ($l_1 = 1, 2$), the corresponding dynamical behaviors and phase portraits can be developed. Readers can complete such analysis as an exercise for a better understanding of switching bifurcations.

4.2.2 Parabola-Sink and Source Infinite-Equilibriums

Consider a dynamical system with a parabola-sink and source infinite-equilibrium or $a_{212} = a_{11l_1}$ ($l_1 = 1, 2$) as

$$\begin{aligned}
\dot{x}_1 &= a_{110}(x_1 - a_{111})(x_1 - a_{112})(x_2 - a_{121}), \\
\dot{x}_2 &= a_{220}(x_1 - a_{211})(x_1 - a_{212})^2.
\end{aligned} \tag{4.64}$$

For $a_{212} = a_{11l_1}$ ($l_1, l_2 \in \{1, 2\}$, $l_1 \neq l_2$), the corresponding first integral manifold is

$$\begin{aligned}
&\frac{1}{2}\left[(x_1 - a_{11l_2})^2 - (x_{10} - a_{11l_2})^2\right] + (2a_{11l_2} - a_{211} - a_{212})(x_1 - x_{10}) \\
&+ (a_{11l_2} - a_{211})(a_{11l_2} - a_{212}) \ln \frac{|x_1 - a_{11l_2}|}{|x_{10} - a_{11l_2}|} \\
&= \frac{1}{2}\frac{a_{110}}{a_{220}}\left[(x_2 - a_{121})^2 - (x_{20} - a_{121})^2\right].
\end{aligned} \tag{4.65}$$

(i) For $a_{212} = a_{111} = a_{11l_1}$, the equilibriums of $x_1^* = a_{111}, a_{112}, a_{211}, a_{212}$ with $x_2^* = a_{121}$ have the following properties:

$$\left.\begin{array}{c}
(a_{112}, a_{121}) \\
(a_{121}, a_{212}) \\
(a_{121}, a_{211})
\end{array}\right\} = \left\{\begin{array}{c}
\underbrace{(\text{DP}:\text{UP}, \text{pF})}_{\text{hyperbolic flow } (+)} \\
\underbrace{({}_{\text{UP}}\text{SO}:{}_{\text{DP}}\text{SI}, \text{UP}_+:\text{DP}_+)}_{\text{circular source-to-sink}} \\
\underbrace{(\text{UP}_+, \text{UP}_+)}_{\text{positive saddle}}
\end{array}\right\} \quad \text{for } a_{110} > 0, a_{220} > 0, \tag{4.66}$$

$$\left.\begin{array}{c}
(a_{112}, a_{121}) \\
(a_{121}, a_{212}) \\
(a_{121}, a_{211})
\end{array}\right\} = \left\{\begin{array}{c}
\underbrace{(\text{UP}:\text{DP}, \text{pF})}_{\text{hyperbolic-secant flow } (+)} \\
\underbrace{({}_{\text{DP}}\text{SI}:{}_{\text{UP}}\text{SO}, \text{DP}_+:\text{UP}_+)}_{\text{hyperbolic sink-to-source}} \\
\underbrace{(\text{DP}_+, \text{DP}_-)}_{\text{CCW center}}
\end{array}\right\} \quad \text{for } a_{110} < 0, a_{220} > 0, \tag{4.67}$$

$$\left\{\begin{array}{c}(a_{112},a_{121})\\(a_{121},a_{212})\\(a_{121},a_{211})\end{array}\right\} = \left\{\begin{array}{c}\underbrace{(\mathrm{UP:DP,nF})}_{\text{hyperbolic-secant flow }(-)}\\\underbrace{(_{\mathrm{DP}}\mathrm{SO}:_{\mathrm{UP}}\mathrm{SI},\mathrm{DP}_-:\mathrm{UP}_-)}_{\text{hyperbolic source-to-sink}}\\\underbrace{(\mathrm{DP}_-,\mathrm{DP}_+)}_{\text{CW center}}\end{array}\right\} \quad \text{for } a_{110}>0, a_{220}<0, \quad (4.68)$$

$$\left\{\begin{array}{c}(a_{112},a_{121})\\(a_{121},a_{212})\\(a_{121},a_{211})\end{array}\right\} = \left\{\begin{array}{c}\underbrace{(\mathrm{DP:UP,nF})}_{\text{hyperbolic flow }(-)}\\\underbrace{(_{\mathrm{UP}}\mathrm{SI}:_{\mathrm{DP}}\mathrm{SO},\mathrm{UP}_-:\mathrm{DP}_-)}_{\text{circular sink-to-source}}\\\underbrace{(\mathrm{UP}_-,\mathrm{UP}_-)}_{\text{negative saddle}}\end{array}\right\} \quad \text{for } a_{110}<0, a_{220}<0, \quad (4.69)$$

and the infinite-equilibrium of $x_1^* = a_{111} = a_{212}$ is summarized as follows:

$$(a_{111},\bar{x}_2) = \underbrace{(\mathrm{SO,UP})}_{\text{up-parabola source}} \quad \text{if } \bar{x}_2 \in (-\infty, a_{121}),$$

$$(a_{111},\bar{x}_2) = \underbrace{(\mathrm{SI,DP})}_{\text{down-parabola sink}} \quad \text{if } \bar{x}_2 \in (a_{121},\infty), \quad (4.70)$$

for $a_{110}>0, a_{220}>0$;

$$(a_{111},\bar{x}_2) = \underbrace{(\mathrm{SI,DP})}_{\text{down-parabola sink}} \quad \text{if } \bar{x}_2 \in (-\infty, a_{121}),$$

$$(a_{111},\bar{x}_2) = \underbrace{(\mathrm{SO,UP})}_{\text{up-parabola source}} \quad \text{if } \bar{x}_2 \in (a_{121},\infty), \quad (4.71)$$

for $a_{110}<0, a_{220}>0$;

$$(a_{111},\bar{x}_2) = \underbrace{(\mathrm{SO,DP})}_{\text{down-parabola source}} \quad \text{if } \bar{x}_2 \in (-\infty, a_{121}),$$

$$(a_{111},\bar{x}_2) = \underbrace{(\mathrm{SI,UP})}_{\text{up-parabola sink}} \quad \text{if } \bar{x}_2 \in (a_{121},\infty), \quad (4.72)$$

for $a_{110}>0, a_{220}<0$;

$$(a_{111},\bar{x}_2) = \underbrace{(\mathrm{SI,UP})}_{\text{up-parabola sink}} \quad \text{if } \bar{x}_2 \in (-\infty, a_{121}),$$

$$(a_{111},\bar{x}_2) = \underbrace{(\mathrm{SO,DP})}_{\text{down-parabola source}} \quad \text{if } \bar{x}_2 \in (a_{121},\infty), \quad (4.73)$$

for $a_{110}<0, a_{220}<0$.

The infinite-equilibriums of $x_1^* = a_{111} = a_{212}$ are parabola-sink and source, as presented in Fig. 4.13. The infinite-equilibrium is for the switching bifurcation of equilibriums for $a_{111} \in (a_{211}, a_{212})$ with $a_{212} < a_{112}$ and $a_{212} < a_{111}$. The hyperbolic and circular sink-to-source and source-to-sink are presented.

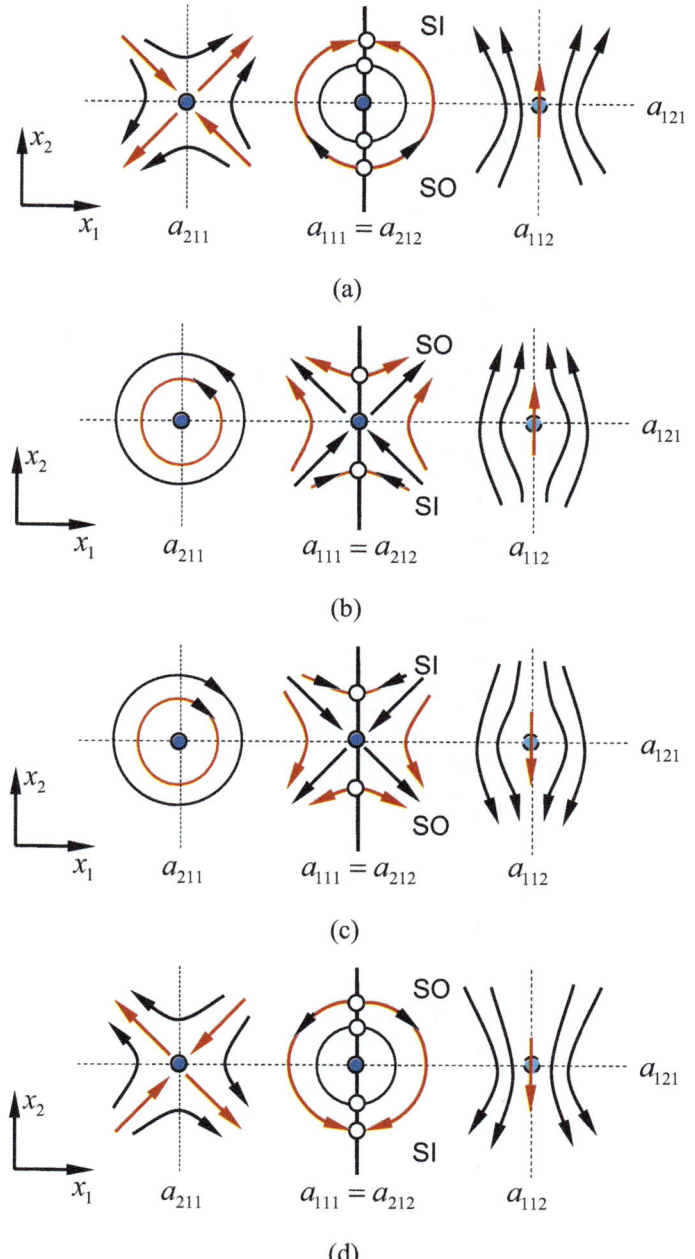

Fig. 4.13 Phase portraits ($a_{111} = a_{211}$) for two-dimensional systems on the x_1-direction with $x_1^* = a_{111}, a_{112}, a_{211}, a_{212}$ and on the x_2-direction with $x_2^* = a_{121}$. The equilibriums flows: (**a**) ($a_{110} > 0, a_{220} > 0$), (**b**) ($a_{110} < 0, a_{220} > 0$), (**c**) ($a_{110} > 0, a_{220} < 0$), (**d**) ($a_{110} < 0, a_{220} < 0$)

(ii) For $a_{212} = a_{112}$ with $a_{211} < a_{111}$, the equilibriums of $x_1^* = a_{111}, a_{112}, a_{211}, a_{212}$ with $x_2^* = a_{121}$ have the following properties:

$$
\left.\begin{array}{c}
(a_{121}, a_{212}) \\
(a_{111}, a_{121}) \\
(a_{121}, a_{211})
\end{array}\right\}
=
\left\{
\begin{array}{c}
\underbrace{({}_{DP}SI:{}_{UP}SO, DP_+:UP_+)}_{\text{hyperbolic sink-to-source}} \\
\underbrace{(UP:DP, pF)}_{\text{hyperbolic-secant flow }(+)} \\
\underbrace{(UP_+, UP_+)}_{\text{positive saddle}}
\end{array}
\right\}
\quad \text{for } a_{110} > 0, a_{220} > 0, \quad (4.74)
$$

$$
\left.\begin{array}{c}
(a_{121}, a_{212}) \\
(a_{111}, a_{121}) \\
(a_{121}, a_{211})
\end{array}\right\}
=
\left\{
\begin{array}{c}
\underbrace{({}_{UP}SO:{}_{DP}SI, UP_+:DP_+)}_{\text{circular source-to-sink}} \\
\underbrace{(DP:UP, pF)}_{\text{hyperbolic flow }(+)} \\
\underbrace{(DP_+, DP_-)}_{\text{CCW center}}
\end{array}
\right\}
\quad \text{for } a_{110} < 0, a_{220} > 0, \quad (4.75)
$$

$$
\left.\begin{array}{c}
(a_{121}, a_{212}) \\
(a_{111}, a_{121}) \\
(a_{121}, a_{211})
\end{array}\right\}
=
\left\{
\begin{array}{c}
\underbrace{({}_{UP}SI:{}_{DP}SO, UP_-:DP_-)}_{\text{circular sink-to-source}} \\
\underbrace{(DP:UP, nF)}_{\text{hyperbolic flow }(-)} \\
\underbrace{(DP_-, DP_-)}_{\text{CW center}}
\end{array}
\right\}
\quad \text{for } a_{110} > 0, a_{220} < 0, \quad (4.76)
$$

$$
\left.\begin{array}{c}
(a_{121}, a_{212}) \\
(a_{111}, a_{121}) \\
(a_{121}, a_{211})
\end{array}\right\}
=
\left\{
\begin{array}{c}
\underbrace{({}_{DP}SO:{}_{UP}SI, DP_-:UP_-)}_{\text{hyperbolic source-to-sink}} \\
\underbrace{(UP:DP, nF)}_{\text{hyperbolic-secant flow }(-)} \\
\underbrace{(UP_-, UP_-)}_{\text{negative saddle}}
\end{array}
\right\}
\quad \text{for } a_{110} < 0, a_{220} < 0, \quad (4.77)
$$

and the infinite-equilibrium of $x_1^* = a_{112} = a_{212}$ is summarized as follows:

$$
\begin{aligned}
(a_{112}, \bar{x}_2) &= \underbrace{(SI, DP)}_{\text{down-parabola sink}} \quad \text{if } \bar{x}_2 \in (-\infty, a_{121}), \\
(a_{112}, \bar{x}_2) &= \underbrace{(SO, UP)}_{\text{up-parabola source}} \quad \text{if } \bar{x}_2 \in (a_{121}, \infty), \quad (4.78)
\end{aligned}
$$

for $a_{110} > 0, a_{220} > 0;$

$$(a_{111}, \bar{x}_2) = \underbrace{(\text{SO}, \text{UP})}_{\text{up-parabola source}} \quad \text{if } \bar{x}_2 \in (-\infty, a_{121}),$$

$$(a_{111}, \bar{x}_2) = \underbrace{(\text{SI}, \text{DP})}_{\text{down-parabola sink}} \quad \text{if } \bar{x}_2 \in (a_{121}, \infty), \tag{4.79}$$

for $a_{110} < 0, a_{220} > 0$;

$$(a_{111}, \bar{x}_2) = \underbrace{(\text{SI}, \text{UP})}_{\text{up-parabola sink}} \quad \text{if } \bar{x}_2 \in (-\infty, a_{121}),$$

$$(a_{111}, \bar{x}_2) = \underbrace{(\text{SO}, \text{DP})}_{\text{down-parabola source}} \quad \text{if } \bar{x}_2 \in (a_{121}, \infty), \tag{4.80}$$

for $a_{110} > 0, a_{220} < 0$;

$$(a_{112}, \bar{x}_2) = \underbrace{(\text{SO}, \text{DP})}_{\text{down-parabola source}} \quad \text{if } \bar{x}_2 \in (-\infty, a_{121}),$$

$$(a_{112}, \bar{x}_2) = \underbrace{(\text{SI}, \text{UP})}_{\text{up-parabola sink}} \quad \text{if } \bar{x}_2 \in (a_{121}, \infty), \tag{4.81}$$

for $a_{110} < 0, a_{220} < 0$;

The infinite-equilibriums of $x_1^* = a_{111} = a_{212}$ are parabola-sink and source, as presented in Fig. 4.14. The infinite-equilibrium is for the switching bifurcation of equilibriums for $a_{112} < a_{212}$ and $a_{212} \in (a_{111}, a_{112})$ with $a_{211} < a_{111}$. The hyperbolic and circular sink-to-source and source-to-sink are also presented.

(iii) For $a_{212} = a_{112}$ with $a_{211} < a_{111}$, the equilibriums of $x_1^* = a_{111}, a_{112}, a_{211}, a_{212}$ with $x_2^* = a_{121}$ have the following properties:

$$\left\{ \begin{array}{c} (a_{121}, a_{212}) \\ (a_{121}, a_{211}) \\ (a_{111}, a_{121}) \end{array} \right\} = \left\{ \begin{array}{c} \underbrace{(_{\text{DP}}\text{SI} :_{\text{UP}}\text{SO}, \text{DP}_+ : \text{UP}_+)}_{\text{hyperbolic sink-to-source}} \\ \underbrace{(\text{DP}_+, \text{DP}_-)}_{\text{CCW center}} \\ \underbrace{(\text{DP} : \text{UP}, \text{nF})}_{\text{hyperbolic flow } (-)} \end{array} \right\} \quad \text{for } a_{110} > 0, a_{220} > 0, \tag{4.82}$$

$$\left\{ \begin{array}{c} (a_{121}, a_{212}) \\ (a_{121}, a_{211}) \\ (a_{111}, a_{121}) \end{array} \right\} = \left\{ \begin{array}{c} \underbrace{(_{\text{UP}}\text{SO} :_{\text{DP}}\text{SI}, \text{UP}_+ : \text{DP}_+)}_{\text{circular source-to-sink}} \\ \underbrace{(\text{UP}_+, \text{UP}_+)}_{\text{positive saddle}} \\ \underbrace{(\text{UP} : \text{DP}, \text{nF})}_{\text{hyperbolic-secant flow } (-)} \end{array} \right\} \quad \text{for } a_{110} < 0, a_{220} > 0, \tag{4.83}$$

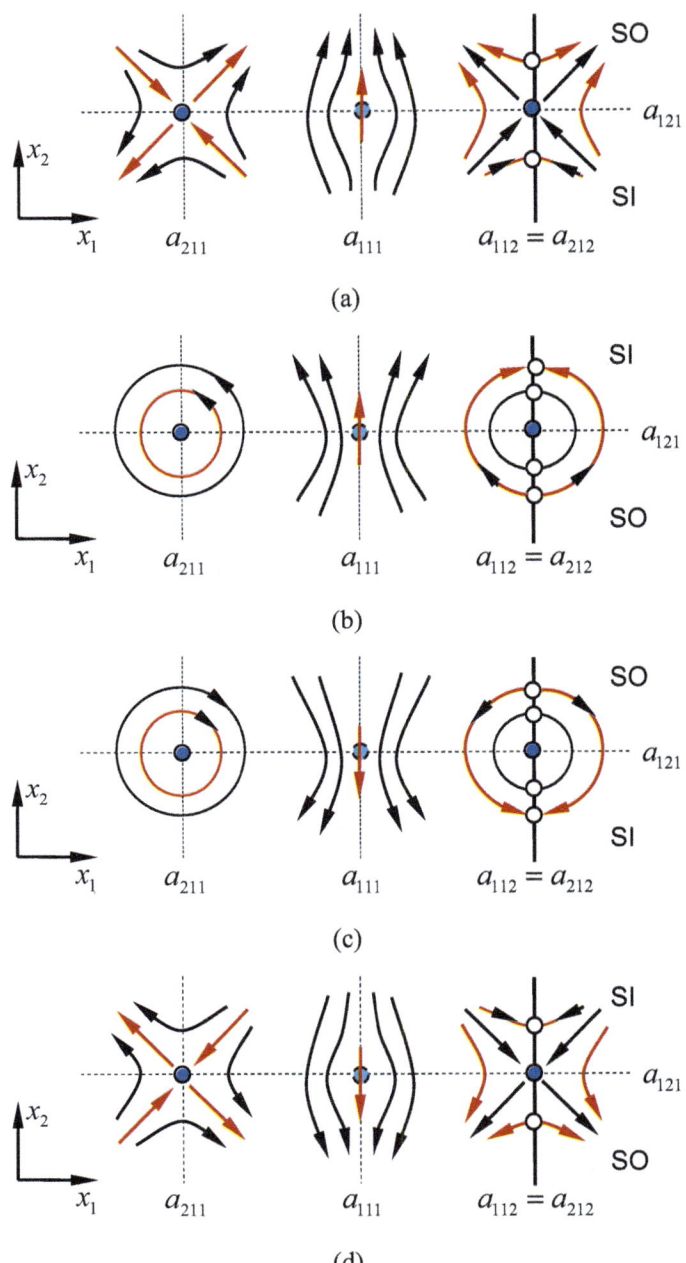

Fig. 4.14 Phase portraits ($a_{112} = a_{212}$) for two-dimensional systems on the x_1-direction with $x_1^* = a_{111}, a_{112}, a_{211}, a_{212}$ and on the x_2-direction with $x_2^* = a_{121}$. The equilibriums flows: (**a**) ($a_{110} > 0, a_{220} > 0$), (**b**) ($a_{110} < 0, a_{220} > 0$), (**c**) ($a_{110} > 0, a_{220} < 0$), (**d**) ($a_{110} < 0, a_{220} < 0$)

$$\left\{\begin{array}{c}(a_{121},a_{212})\\(a_{121},a_{211})\\(a_{111},a_{121})\end{array}\right\}=\left\{\begin{array}{c}\overbrace{(_{\mathrm{UP}}\mathrm{SI}:_{\mathrm{DP}}\mathrm{SO},\mathrm{UP}_-:\mathrm{DP}_-)}^{\text{circular sink-to-source}}\\\underbrace{(\mathrm{UP}_-,\mathrm{UP}_-)}_{\text{negative saddle}}\\\underbrace{(\mathrm{UP}:\mathrm{DP},\mathrm{pF})}_{\text{hyperbolic-secant flow }(+)}\end{array}\right\}\quad\text{for }a_{110}>0,a_{220}<0,\quad(4.84)$$

$$\left\{\begin{array}{c}(a_{121},a_{212})\\(a_{111},a_{121})\\(a_{121},a_{211})\end{array}\right\}=\left\{\begin{array}{c}\overbrace{(_{\mathrm{DP}}\mathrm{SO}:_{\mathrm{UP}}\mathrm{SI},\mathrm{DP}_-:\mathrm{UP}_-)}^{\text{hyperbolic source-to-sink}}\\\underbrace{(\mathrm{DP}_-,\mathrm{DP}_-)}_{\text{CW center}}\\\underbrace{(\mathrm{DP}:\mathrm{UP},\mathrm{pF})}_{\text{hyperbolic flow }(+)}\end{array}\right\}\quad\text{for }a_{110}<0,a_{220}<0,\quad(4.85)$$

and the infinite-equilibrium of $x_1^*=a_{112}=a_{212}$ has the same properties as in Eqs. (4.78)–(4.81). The infinite-equilibrium is for the switching bifurcation of equilibriums for $a_{112}\in(a_{211},a_{212})$ and $a_{212}<a_{112}$ with $a_{111}<a_{211}$. The hyperbolic and circular sink-to-source and source-to-sink are also presented in Fig. 4.15.

4.2.3 Third-Order Inflection-Sink and Source Infinite-Equilibria

Consider a dynamical system with a parabola sink and source infinite-equilibrium as

$$\dot{x}_1=a_{110}(x_1-a_{111})(x_1-a_{112})(x_2-a_{121}),$$
$$\dot{x}_2=a_{220}(x_1-a_{211})^3.\tag{4.86}$$

For $a_{211}=a_{11l_1}$ $(l_1,l_2\in\{1,2\},l_1\neq l_2)$, the corresponding first integral manifold is

$$\frac{1}{2}\left[(x_1-a_{11l_2})^2-(x_{10}-a_{11l_2})^2\right]+2(a_{11l_2}-a_{211})(x_1-x_{10})$$
$$+(a_{11l_2}-a_{211})^2\ln\left|\frac{x_1-a_{11l_2}}{x_{10}-a_{11l_2}}\right|=\frac{1}{2}\frac{a_{110}}{a_{220}}\left[(x_2-a_{121})^2-(x_{20}-a_{121})^2\right].\tag{4.87}$$

(i) For $a_{211}=a_{111}$, the equilibriums of $x_1^*=a_{111},a_{112},a_{211}$ with $x_2^*=a_{121}$ have the following properties:

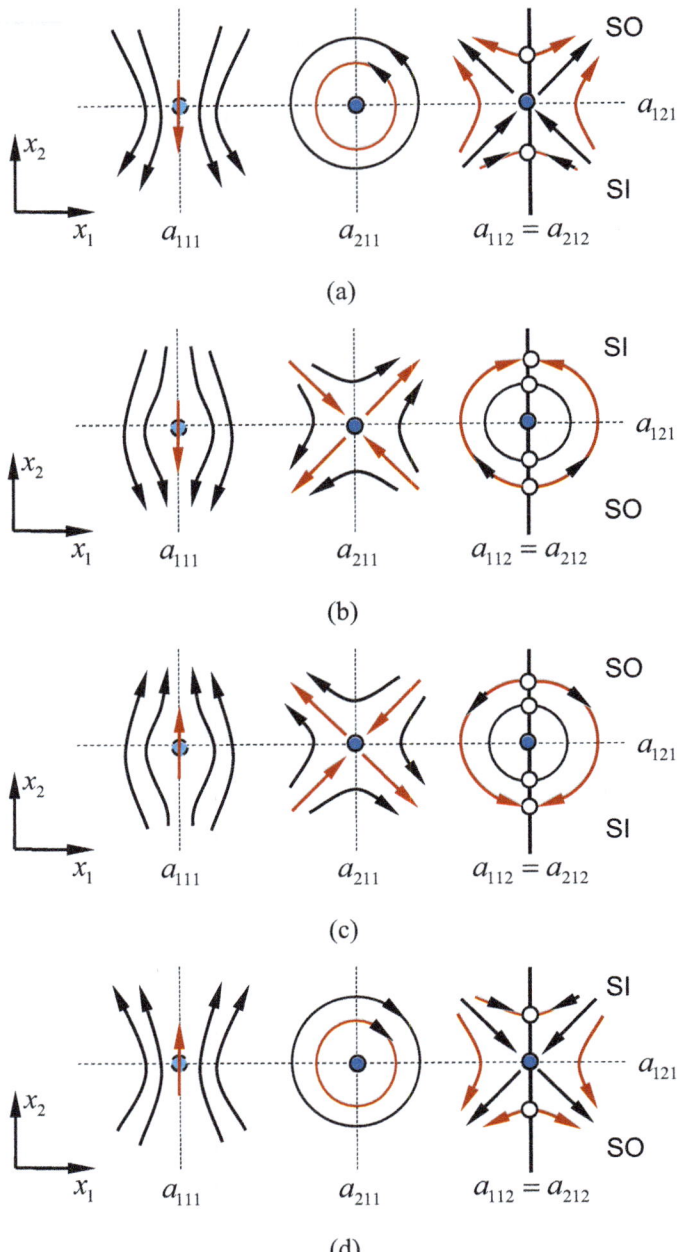

Fig. 4.15 Phase portraits ($a_{112} = a_{212}$) for two-dimensional systems on the x_1-direction with $x_1^* = a_{111}, a_{112}, a_{211}, a_{212}$ and on the x_2-direction with $x_2^* = a_{121}$. The equilibriums flows: (**a**) ($a_{110} > 0, a_{220} > 0$), (**b**) ($a_{110} < 0, a_{220} > 0$), (**c**) ($a_{110} > 0, a_{220} < 0$), (**d**) ($a_{110} < 0, a_{220} < 0$)

$$\left.\begin{array}{c} \{(a_{112}, a_{121})\} \\ \{(a_{211}, a_{121})\} \end{array}\right\} = \left\{ \begin{array}{c} \underbrace{(\text{DP}:\text{UP}, \text{pF})}_{\text{hyperbolic flow } (+)} \\ \underbrace{(_{2^{\text{nd}}(\text{II}:\text{DI})}\text{DP}, \text{SO}:\text{SI}\text{LS})}_{\text{second-order down-parabola lower-saddle}} \end{array} \right\} \quad \text{for } a_{110} > 0, a_{220} > 0, \quad (4.88)$$

$$\left.\begin{array}{c} \{(a_{112}, a_{121})\} \\ \{(a_{211}, a_{121})\} \end{array}\right\} = \left\{ \begin{array}{c} \underbrace{(\text{UP}:\text{DP}, \text{pF})}_{\text{hyperbolic-secant flow } (+)} \\ \underbrace{(_{2^{\text{nd}}(\text{DI}:\text{II})}\text{UP}, \text{SI}:\text{SO}\text{US})}_{\text{second-order up-parabola upper-saddle}} \end{array} \right\} \quad \text{for } a_{110} < 0, a_{220} > 0, \quad (4.89)$$

$$\left.\begin{array}{c} \{(a_{112}, a_{121})\} \\ \{(a_{211}, a_{121})\} \end{array}\right\} = \left\{ \begin{array}{c} \underbrace{(\text{UP}:\text{DP}, \text{nF})}_{\text{hyperbolic-secant flow } (-)} \\ \underbrace{(_{2^{\text{nd}}(\text{DI}:\text{II})}\text{UP}, \text{SO}:\text{SI}\text{LS})}_{\text{second-order up-parabola lower-saddle}} \end{array} \right\} \quad \text{for } a_{110} > 0, a_{220} < 0, \quad (4.90)$$

$$\left.\begin{array}{c} \{(a_{112}, a_{121})\} \\ \{(a_{211}, a_{121})\} \end{array}\right\} = \left\{ \begin{array}{c} \underbrace{(\text{DP}:\text{UP}, \text{nF})}_{\text{hyperbolic flow } (-)} \\ \underbrace{(_{2^{\text{nd}}(\text{II}:\text{DI})}\text{DP}, \text{SI}:\text{SO}\text{US})}_{\text{second-order down-parabola upper-saddle}} \end{array} \right\} \quad \text{for } a_{110} < 0, a_{220} < 0, \quad (4.91)$$

and the infinite-equilibrium of $x_1^* = a_{111} = a_{211}$ is summarized as follows:

$$(a_{111}, \bar{x}_2) = \underbrace{(\text{SO}, 2^{\text{nd}}\text{II})}_{\text{second-order increasing-inflection source}} \quad \text{if } \bar{x}_2 \in (-\infty, a_{121}),$$

$$(a_{111}, \bar{x}_2) = \underbrace{(\text{SI}, 2^{\text{nd}}\text{DI})}_{\text{second-order decreasing-inflection sink}} \quad \text{if } \bar{x}_2 \in (a_{121}, \infty), \quad (4.92)$$

for $a_{110} > 0, a_{220} > 0$;

$$(a_{111}, \bar{x}_2) = \underbrace{(\text{SI}, 2^{\text{nd}}\text{DI})}_{\text{second-order decreasing-inflection sink}} \quad \text{if } \bar{x}_2 \in (-\infty, a_{121}),$$

$$(a_{111}, \bar{x}_2) = \underbrace{(\text{SO}, 2^{\text{nd}}\text{II})}_{\text{second-order increasing-inflection source}} \quad \text{if } \bar{x}_2 \in (a_{121}, \infty), \quad (4.93)$$

for $a_{110} < 0, a_{220} > 0$;

$$(a_{111}, \bar{x}_2) = \underbrace{(\text{SO}, 2^{\text{nd}}\text{DI})}_{\text{second-order decreasing-inflection source}} \quad \text{if } \bar{x}_2 \in (-\infty, a_{121}),$$

$$(a_{111}, \bar{x}_2) = \underbrace{(\text{SI}, 2^{\text{nd}}\text{II})}_{\text{second-order increasing-inflection sink}} \quad \text{if } \bar{x}_2 \in (a_{121}, \infty), \quad (4.94)$$

for $a_{110} > 0, a_{220} < 0$;

$$(a_{111}, \bar{x}_2) = \underbrace{(\text{SI}, 2^{\text{nd}}\text{II})}_{\text{second-order increasing-inflection sink}} \quad \text{if } \bar{x}_2 \in (-\infty, a_{121}),$$

$$(a_{111}, \bar{x}_2) = \underbrace{(\text{SO}, 2^{\text{nd}}\text{DI})}_{\text{second-order decreasing-inflection source}} \quad \text{if } \bar{x}_2 \in (a_{121}, \infty), \quad (4.95)$$

for $a_{110} < 0, a_{220} < 0$;

The infinite-equilibriums of $x_1^* = a_{111} = a_{211}$ are second-order inflection sink and source, as presented in Fig. 4.16. The infinite-equilibrium is for the switching bifurcation of equilibriums for $a_{211} = a_{111}$. The parabola-saddles are based on the second-order inflection sink and source.

(ii) For $a_{211} = a_{112}$, the equilibriums of $x_1^* = a_{111}, a_{112}, a_{211}$ with $x_2^* = a_{121}$ have the following properties:

$$\left. \begin{matrix} (a_{211}, a_{121}) \\ (a_{111}, a_{121}) \end{matrix} \right\} = \left\{ \begin{matrix} \underbrace{(_{2^{\text{nd}}(\text{DI:II})}\text{UP}, \text{SI:SO}\text{US})}_{\text{second-order up-parabola upper-saddle}} \\ \underbrace{(\text{DP:UP, nF})}_{\text{hyperbolic flow } (-)} \end{matrix} \right\} \quad \text{for } a_{110} > 0, a_{220} > 0, \quad (4.96)$$

$$\left. \begin{matrix} (a_{211}, a_{121}) \\ (a_{111}, a_{121}) \end{matrix} \right\} = \left\{ \begin{matrix} \underbrace{(_{2^{\text{nd}}(\text{II:DI})}\text{DP}, \text{SO:SI}\text{LS})}_{\text{second-order down-parabola lower-saddle}} \\ \underbrace{(\text{UP:DP, nF})}_{\text{hyperbolic-secant flow } (-)} \end{matrix} \right\} \quad \text{for } a_{110} < 0, a_{220} > 0, \quad (4.97)$$

$$\left. \begin{matrix} (a_{211}, a_{121}) \\ (a_{111}, a_{121}) \end{matrix} \right\} = \left\{ \begin{matrix} \underbrace{(_{2^{\text{nd}}(\text{II:DI})}\text{DP}, \text{SI:SO}\text{US})}_{\text{second-order down-parabola upper-saddle}} \\ \underbrace{(\text{UP:DP, pF})}_{\text{hyperbolic-secant flow } (+)} \end{matrix} \right\} \quad \text{for } a_{110} > 0, a_{220} < 0, \quad (4.98)$$

$$\left. \begin{matrix} (a_{211}, a_{121}) \\ (a_{111}, a_{121}) \end{matrix} \right\} = \left\{ \begin{matrix} \underbrace{(_{2^{\text{nd}}(\text{DI:II})}\text{UP}, \text{SO:SI}\text{LS})}_{\text{second-order up-parabola lower-saddle}} \\ \underbrace{(\text{DP:UP, pF})}_{\text{hyperbolic flow } (+)} \end{matrix} \right\} \quad \text{for } a_{110} < 0, a_{220} < 0, \quad (4.99)$$

and the infinite equilibrium of $x_1^* = a_{111} = a_{211}$ is summarized as follows:

$$(a_{112}, \bar{x}_2) = \underbrace{(\text{SO}, 2^{\text{nd}}\text{II})}_{\text{second-order increasing-inflection source}} \quad \text{if } \bar{x}_2 \in (-\infty, a_{121}),$$

$$(a_{112}, \bar{x}_2) = \underbrace{(\text{SI}, 2^{\text{nd}}\text{DI})}_{\text{second-order decreasing-inflection sink}} \quad \text{if } \bar{x}_2 \in (a_{121}, \infty), \quad (4.100)$$

for $a_{110} > 0, a_{220} > 0$;

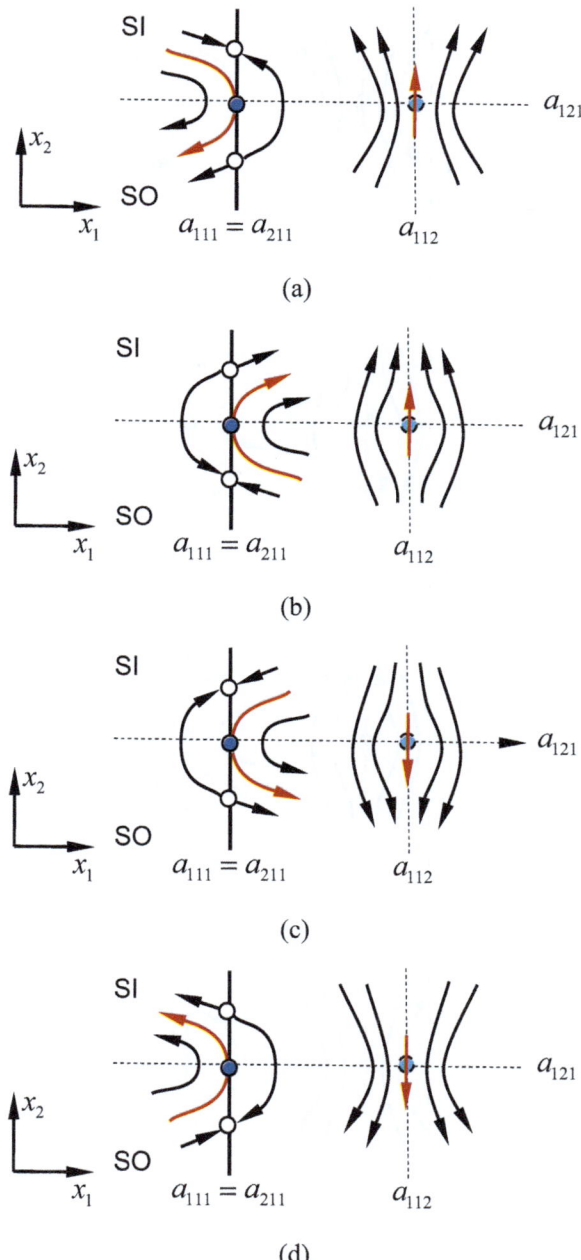

Fig. 4.16 Phase portraits ($a_{111} = a_{211}$) for two-dimensional systems on the x_1-direction with $x_1^* = a_{111}, a_{112}, a_{211}$ and on the x_2-direction with $x_2^* = a_{121}$. The equilibriums flows: (**a**) ($a_{110} > 0$, $a_{220} > 0$), (**b**) ($a_{110} < 0$, $a_{220} > 0$), (**c**) ($a_{110} > 0$, $a_{220} < 0$), (**d**) ($a_{110} < 0$, $a_{220} < 0$)

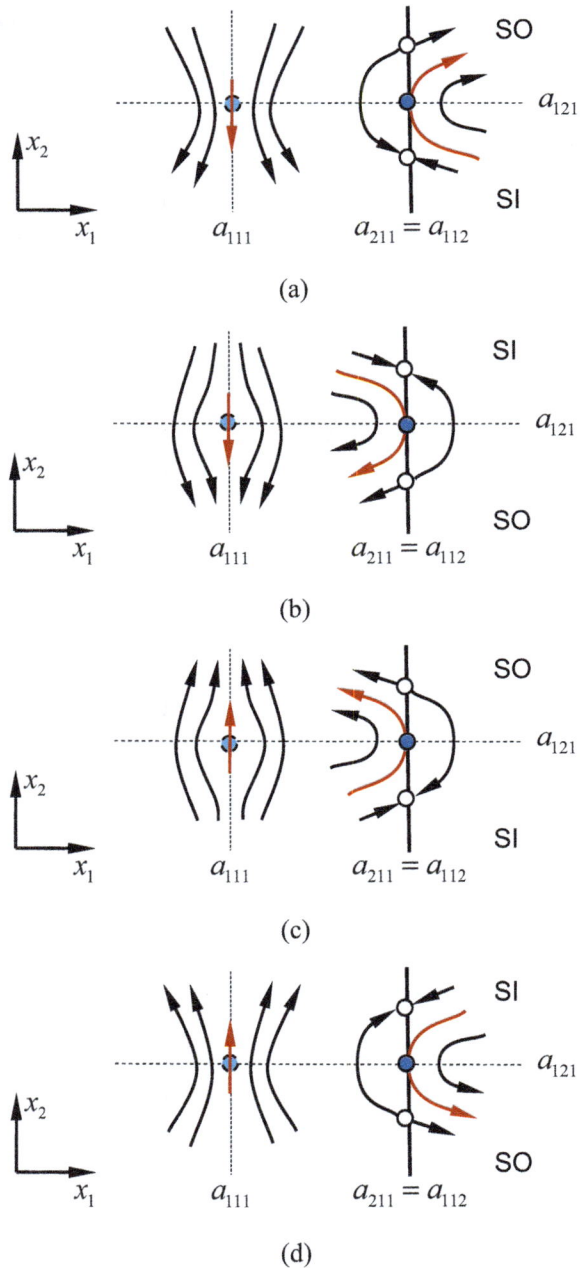

Fig. 4.17 Phase portraits ($a_{112} = a_{211}$) for two-dimensional systems on the x_1-direction with $x_1^* = a_{111}, a_{112}, a_{211}$ and on the x_2-direction with $x_2^* = a_{121}$. The equilibriums flows: (**a**) ($a_{110} > 0$, $a_{220} > 0$), (**b**) ($a_{110} < 0$, $a_{220} > 0$), (**c**) ($a_{110} > 0$, $a_{220} < 0$), (**d**) ($a_{110} < 0$, $a_{220} < 0$)

$$(a_{112}, \bar{x}_2) = \underbrace{(\text{SI}, 2^{\text{nd}}\text{DI})}_{\text{second-order decreasing-inflection sink}} \quad \text{if } \bar{x}_2 \in (-\infty, a_{121}),$$

$$(a_{112}, \bar{x}_2) = \underbrace{(\text{SO}, 2^{\text{nd}}\text{II})}_{\text{second-order increasing-inflection source}} \quad \text{if } \bar{x}_2 \in (a_{121}, \infty), \tag{4.101}$$

for $a_{110} < 0, a_{220} > 0$;

$$(a_{112}, \bar{x}_2) = \underbrace{(\text{SO}, 2^{\text{nd}}\text{DI})}_{\text{second-order decreasing-inflection source}} \quad \text{if } \bar{x}_2 \in (-\infty, a_{121}),$$

$$(a_{112}, \bar{x}_2) = \underbrace{(\text{SI}, 2^{\text{nd}}\text{II})}_{\text{second-order increasing-inflection sink}} \quad \text{if } \bar{x}_2 \in (a_{121}, \infty), \tag{4.102}$$

for $a_{110} > 0, a_{220} < 0$;

$$(a_{112}, \bar{x}_2) = \underbrace{(\text{SI}, 2^{\text{nd}}\text{II})}_{\text{second-order increasing-inflection sink}} \quad \text{if } \bar{x}_2 \in (-\infty, a_{121}),$$

$$(a_{111}, \bar{x}_2) = \underbrace{(\text{SO}, 2^{\text{nd}}\text{DI})}_{\text{second-order decreasing-inflection source}} \quad \text{if } \bar{x}_2 \in (a_{121}, \infty), \tag{4.103}$$

for $a_{110} < 0, a_{220} < 0$;

The infinite-equilibriums of $x_1^* = a_{112} = a_{211}$ are second-order inflection sinks and sources, as presented in Fig. 4.17. The infinite-equilibrium is for the switching bifurcation of equilibriums for $a_{211} = a_{112}$. The parabola-saddles are based on the second-order inflection-sink and source.

Chapter 5
Simple Equilibriums and Hyperbolic Flows

In this chapter, simple equilibriums and hyperbolic flows forming a series in the crossing and product cubic systems are presented, and the corresponding switching dynamics are discussed through the inflection-source and sink infinite-equilibriums. The parabola-saddles on single and double inflection-sources and sink infinite-equilibriums are discussed. Parabola-saddles are the switching bifurcations of a saddle and a hyperbolic-secant flow with a center and hyperbolic flow.

5.1   Simple Equilibrium Series with Hyperbolic Flows

Consider a dynamical system as

$$
\begin{aligned}
\dot{x}_1 &= a_{110}(x_1 - a_{111})(x_1 - a_{112})(x_2 - a_{121}), \\
\dot{x}_2 &= a_{220}(x_1 - a_{211})(x_1 - a_{212})(x_1 - a_{213}),
\end{aligned}
\tag{5.1}
$$

and the corresponding first integral manifold is

$$
\begin{aligned}
&\frac{1}{2}\Big[(x_1 - a_{11s_1})^2 - (x_{10} - a_{11s_1})^2\Big] \\
&+ \sum_{l_1=1, l_2 \ne l_1}^{2} \frac{\prod_{s_2=1, s_2 \ne s_1}^{3}(a_{11l_1} - a_{21s_2})}{a_{11l_1} - a_{11l_2}}(x_1 - x_{10}) \\
&+ \sum_{l_1=1, l_2 \ne l_1}^{2} \frac{\prod_{s_1=1}^{3}(a_{112} - a_{21s_1})}{a_{11l_1} - a_{11l_2}} \ln \frac{|x_1 - a_{11l_1}|}{|x_{10} - a_{11l_1}|} \\
&= \frac{1}{2}\frac{a_{110}}{a_{220}}\Big[(x_2 - a_{121})^2 - (x_{20} - a_{121})^2\Big].
\end{aligned}
\tag{5.2}
$$

(i) For $a_{112} < a_{211}$, the simple equilibrium series has the following properties:

© The Author(s), under exclusive license to Springer Nature Switzerland AG 2025
A. C. J. Luo, *Two-dimensional Crossing and Product Cubic Systems, Vol. II*,
https://doi.org/10.1007/978-3-031-57100-8_5

$$
\left\{
\begin{array}{ll}
(a_{121}, a_{213}) & - \\
(a_{121}, a_{211}) & (a_{121}, a_{212}) \\
(a_{111}, a_{121}) & (a_{112}, a_{121})
\end{array}
\right\}
=
\left\{
\begin{array}{ll}
\underbrace{(\mathrm{UP}_+, \mathrm{UP}_+)}_{\text{positive saddle}} & - \\[2mm]
\underbrace{(\mathrm{UP}_-, \mathrm{UP}_-)}_{\text{positive saddle}} & \underbrace{(\mathrm{DP}_-, \mathrm{DP}_+)}_{\text{CW center}} \\[2mm]
\underbrace{(\mathrm{DP}:\mathrm{UP}, \mathrm{nF})}_{\text{hyperbolic flow }(-)} & \underbrace{(\mathrm{UP}:\mathrm{DP}, \mathrm{nF})}_{\text{hyperbolic-secant flow }(-)}
\end{array}
\right\}
\tag{5.3}
$$

for $a_{110} > 0$ and $a_{220} > 0$;

$$
\left\{
\begin{array}{ll}
(a_{121}, a_{213}) & - \\
(a_{121}, a_{211}) & (a_{121}, a_{212}) \\
(a_{111}, a_{121}) & (a_{112}, a_{121})
\end{array}
\right\}
=
\left\{
\begin{array}{ll}
\underbrace{(\mathrm{DP}_+, \mathrm{DP}_-)}_{\text{CCW center}} & - \\[2mm]
\underbrace{(\mathrm{DP}_-, \mathrm{DP}_+)}_{\text{CCW center}} & \underbrace{(\mathrm{UP}_-, \mathrm{UP}_-)}_{\text{negative saddle}} \\[2mm]
\underbrace{(\mathrm{UP}:\mathrm{DP}, \mathrm{nF})}_{\text{hyperbolic-secant flow }(-)} & \underbrace{(\mathrm{DP}:\mathrm{UP}, \mathrm{nF})}_{\text{hyperbolic flow }(-)}
\end{array}
\right\}
\tag{5.4}
$$

for $a_{110} < 0$ and $a_{220} > 0$;

$$
\left\{
\begin{array}{ll}
(a_{121}, a_{213}) & - \\
(a_{121}, a_{211}) & (a_{121}, a_{212}) \\
(a_{111}, a_{121}) & (a_{112}, a_{121})
\end{array}
\right\}
=
\left\{
\begin{array}{ll}
\underbrace{(\mathrm{DP}_-, \mathrm{DP}_+)}_{\text{CW center}} & - \\[2mm]
\underbrace{(\mathrm{DP}_-, \mathrm{DP}_+)}_{\text{CW center}} & \underbrace{(\mathrm{UP}_+, \mathrm{UP}_+)}_{\text{positive saddle}} \\[2mm]
\underbrace{(\mathrm{UP}:\mathrm{DP}, \mathrm{pF})}_{\text{hyperbolic-secant flow }(+)} & \underbrace{(\mathrm{DP}:\mathrm{UP}, \mathrm{pF})}_{\text{hyperbolic flow }(+)}
\end{array}
\right\}
\tag{5.5}
$$

for $a_{110} > 0$ and $a_{220} < 0$;

$$
\left\{
\begin{array}{ll}
(a_{121}, a_{213}) & - \\
(a_{121}, a_{211}) & (a_{121}, a_{212}) \\
(a_{111}, a_{121}) & (a_{112}, a_{121})
\end{array}
\right\}
=
\left\{
\begin{array}{ll}
\underbrace{(\mathrm{UP}_-, \mathrm{UP}_-)}_{\text{negative saddle}} & - \\[2mm]
\underbrace{(\mathrm{UP}_-, \mathrm{UP}_-)}_{\text{negative saddle}} & \underbrace{(\mathrm{DP}_+, \mathrm{DP}_-)}_{\text{CCW center}} \\[2mm]
\underbrace{(\mathrm{DP}:\mathrm{UP}, \mathrm{pF})}_{\text{hyperbolic flow }(+)} & \underbrace{(\mathrm{UP}:\mathrm{DP}, \mathrm{pF})}_{\text{hyperbolic-secant flow }(+)}
\end{array}
\right\}
\tag{5.6}
$$

for $a_{110} < 0$ and $a_{220} < 0$;

as shown in Fig. 5.1. The directrix flows of $x_1 = a_{111}, a_{112}$ are positive and negative. The directrix flows are separatrix flows, which cannot be passed by the other flows. The saddle and center equilibriums are on the right side of the hyperbolic flows.

(ii) For $a_{211} \in (a_{111}, a_{112})$, the simple equilibrium series has the following properties:

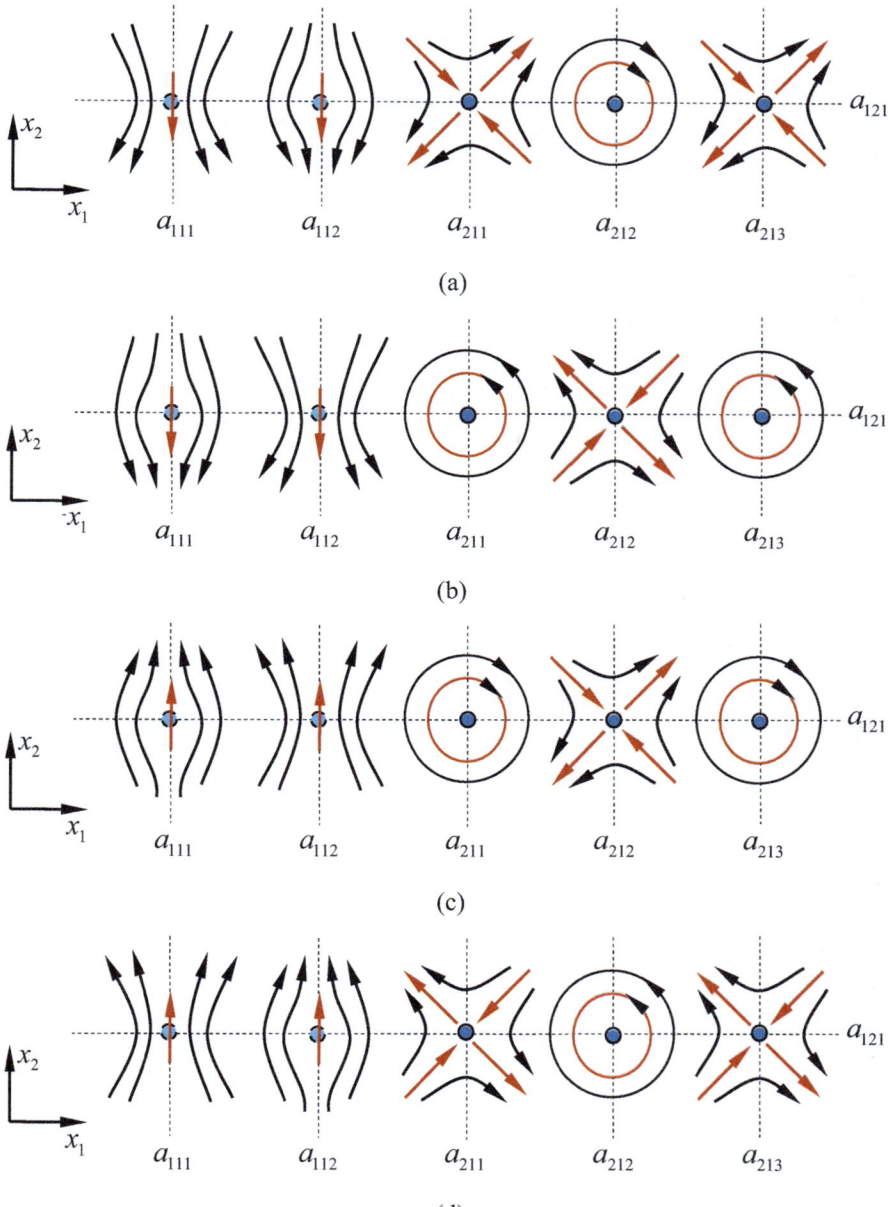

Fig. 5.1 Phase portraits ($a_{112} < a_{211}$) for two-dimensional systems on the x_1-direction with $x_1^* = a_{111}, a_{112}, a_{211}, a_{212}, a_{213}$ and on the x_2-direction with $x_2^* = a_{121}$. (**a**) ($a_{110} > 0, a_{220} > 0$), (**b**) ($a_{110} < 0, a_{220} > 0$), (**c**) ($a_{110} > 0, a_{220} < 0$), (**d**) ($a_{110} < 0, a_{220} < 0$)

$$
\left\{
\begin{array}{ll}
(a_{121}, a_{213}) & - \\
(a_{112}, a_{121}) & (a_{121}, a_{212}) \\
(a_{111}, a_{121}) & (a_{121}, a_{211})
\end{array}
\right\}
=
\left\{
\begin{array}{ll}
\underbrace{(\mathrm{UP}_+, \mathrm{UP}_+)}_{\text{positive saddle}} & - \\[2ex]
\underbrace{(\mathrm{DP}:\mathrm{UP}, \mathrm{pF})}_{\text{hyperbolic flow }(+)} & \underbrace{(\mathrm{DP}_-, \mathrm{DP}_+)}_{\text{CW center}} \\[2ex]
\underbrace{(\mathrm{DP}:\mathrm{UP}, \mathrm{nF})}_{\text{hyperbolic flow }(-)} & \underbrace{(\mathrm{DP}_+, \mathrm{DP}_-)}_{\text{CCW center}}
\end{array}
\right\}
\tag{5.7}
$$

for $a_{110} > 0$ and $a_{220} > 0$;

$$
\left\{
\begin{array}{ll}
(a_{121}, a_{213}) & - \\
(a_{112}, a_{121}) & (a_{121}, a_{212}) \\
(a_{111}, a_{121}) & (a_{121}, a_{211})
\end{array}
\right\}
=
\left\{
\begin{array}{ll}
\underbrace{(\mathrm{DP}_+, \mathrm{DP}_-)}_{\text{CCW center}} & - \\[2ex]
\underbrace{(\mathrm{UP}:\mathrm{DP}, \mathrm{pF})}_{\text{hyperbolic-secant flow }(+)} & \underbrace{(\mathrm{UP}_-, \mathrm{UP}_-)}_{\text{negative saddle}} \\[2ex]
\underbrace{(\mathrm{UP}:\mathrm{DP}, \mathrm{nF})}_{\text{hyperbolic-secant flow }(-)} & \underbrace{(\mathrm{UP}_+, \mathrm{UP}_+)}_{\text{positive saddle}}
\end{array}
\right\}
\tag{5.8}
$$

for $a_{110} < 0$ and $a_{220} > 0$;

$$
\left\{
\begin{array}{ll}
(a_{121}, a_{213}) & - \\
(a_{112}, a_{121}) & (a_{121}, a_{212}) \\
(a_{111}, a_{121}) & (a_{121}, a_{211})
\end{array}
\right\}
=
\left\{
\begin{array}{ll}
\underbrace{(\mathrm{DP}_-, \mathrm{DP}_+)}_{\text{CW center}} & - \\[2ex]
\underbrace{(\mathrm{UP}:\mathrm{DP}, \mathrm{nF})}_{\text{hyperbolic-secant flow }(-)} & \underbrace{(\mathrm{UP}_+, \mathrm{UP}_+)}_{\text{positive saddle}} \\[2ex]
\underbrace{(\mathrm{UP}:\mathrm{DP}, \mathrm{pF})}_{\text{hyperbolic-secant flow }(+)} & \underbrace{(\mathrm{UP}_-, \mathrm{UP}_-)}_{\text{negative saddle}}
\end{array}
\right\}
\tag{5.9}
$$

for $a_{110} > 0$ and $a_{220} < 0$;

$$
\left\{
\begin{array}{ll}
(a_{121}, a_{213}) & - \\
(a_{112}, a_{121}) & (a_{121}, a_{212}) \\
(a_{111}, a_{121}) & (a_{121}, a_{211})
\end{array}
\right\}
=
\left\{
\begin{array}{ll}
\underbrace{(\mathrm{UP}_-, \mathrm{UP}_-)}_{\text{negative saddle}} & - \\[2ex]
\underbrace{(\mathrm{DP}:\mathrm{UP}, \mathrm{nF})}_{\text{hyperbolic flow }(-)} & \underbrace{(\mathrm{DP}_+, \mathrm{DP}_-)}_{\text{CCW center}} \\[2ex]
\underbrace{(\mathrm{DP}:\mathrm{UP}, \mathrm{pF})}_{\text{hyperbolic flow }(+)} & \underbrace{(\mathrm{DP}_-, \mathrm{DP}_+)}_{\text{CW center}}
\end{array}
\right\}
\tag{5.10}
$$

for $a_{110} < 0$ and $a_{220} < 0$;

as shown in Fig. 5.2. The directrix flows of $x_1 = a_{111}, a_{112}$ are positive and negative. The directrix flows are separatrix flows, which cannot be passed by the other flows. The two centers are separated by a hyperbolic flow, and the two saddles are separated by a hyperbolic-secant flow.

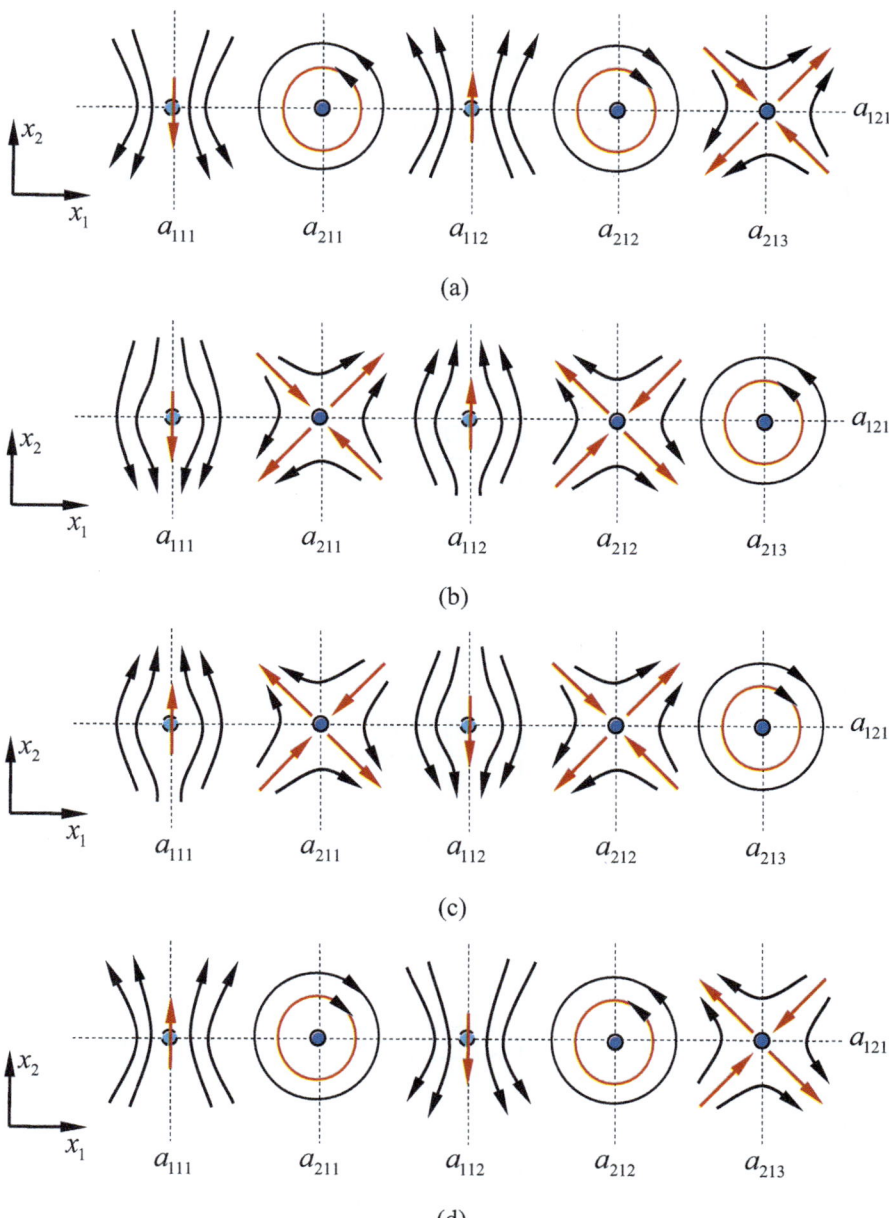

Fig. 5.2 Phase portraits ($a_{211} \in (a_{111}, a_{112})$) for two-dimensional systems on the x_1-direction with $x_1^* = a_{111}, a_{211}$ and on the x_2-direction with $x_2^* = a_{121}$. (**a**) ($a_{110} > 0, a_{220} > 0$), (**b**) ($a_{110} < 0$, $a_{220} > 0$), (**c**) ($a_{110} > 0, a_{220} < 0$), (**d**) ($a_{110} < 0, a_{220} < 0$)

(iii) For a_{211}, $a_{212} \in (a_{211}, a_{112})$, simple equilibriums of $x_1^* = a_{111}, a_{112}, a_{211},$ a_{212}, a_{213} with $x_2^* = a_{121}$ have the following properties:

$$
\left.\begin{array}{cc}
(a_{121}, a_{213}) & - \\
(a_{121}, a_{212}) & (a_{112}, a_{121}) \\
(a_{111}, a_{121}) & (a_{121}, a_{211})
\end{array}\right\} = \left\{\begin{array}{cc}
\underbrace{(\mathrm{UP}_+, \mathrm{UP}_+)}_{\text{positive saddle}} & - \\
\underbrace{(\mathrm{UP}_-, \mathrm{UP}_-)}_{\text{negative saddle}} & \underbrace{(\mathrm{UP}:\mathrm{DP},\mathrm{nF})}_{\text{hyperbolic-secant flow } (-)} \\
\underbrace{(\mathrm{DP}:\mathrm{UP},\mathrm{nF})}_{\text{hyperbolic flow } (-)} & \underbrace{(\mathrm{DP}_+, \mathrm{DP}_-)}_{\text{CCW center}}
\end{array}\right\} \quad (5.11)
$$

for $a_{110} > 0$ and $a_{220} > 0$;

$$
\left.\begin{array}{cc}
(a_{121}, a_{213}) & - \\
(a_{121}, a_{212}) & (a_{112}, a_{121}) \\
(a_{111}, a_{121}) & (a_{121}, a_{211})
\end{array}\right\} = \left\{\begin{array}{cc}
\underbrace{(\mathrm{DP}_+, \mathrm{DP}_-)}_{\text{CCW center}} & - \\
\underbrace{(\mathrm{DP}_-, \mathrm{DP}_-)}_{\text{CW center}} & \underbrace{(\mathrm{DP}:\mathrm{UP},\mathrm{nF})}_{\text{hyperbolic flow } (-)} \\
\underbrace{(\mathrm{UP}:\mathrm{DP},\mathrm{nF})}_{\text{hyperbolic-secant flow } (-)} & \underbrace{(\mathrm{UP}_+, \mathrm{UP}_+)}_{\text{positive saddle}}
\end{array}\right\} \quad (5.12)
$$

for $a_{110} < 0$ and $a_{220} > 0$;

$$
\left.\begin{array}{cc}
(a_{121}, a_{213}) & - \\
(a_{121}, a_{212}) & (a_{112}, a_{121}) \\
(a_{111}, a_{121}) & (a_{121}, a_{211})
\end{array}\right\} = \left\{\begin{array}{cc}
\underbrace{(\mathrm{DP}_-, \mathrm{DP}_+)}_{\text{CW center}} & - \\
\underbrace{(\mathrm{DP}_+, \mathrm{DP}_-)}_{\text{CCW center}} & \underbrace{(\mathrm{DP}:\mathrm{UP},\mathrm{pF})}_{\text{hyperbolic flow } (+)} \\
\underbrace{(\mathrm{UP}:\mathrm{DP},\mathrm{pF})}_{\text{hyperbolic-secant flow } (+)} & \underbrace{(\mathrm{UP}_-, \mathrm{UP}_-)}_{\text{negative saddle}}
\end{array}\right\} \quad (5.13)
$$

for $a_{110} > 0$ and $a_{220} < 0$;

$$
\left.\begin{array}{cc}
(a_{121}, a_{213}) & - \\
(a_{121}, a_{212}) & (a_{112}, a_{121}) \\
(a_{111}, a_{121}) & (a_{121}, a_{211})
\end{array}\right\} = \left\{\begin{array}{cc}
\underbrace{(\mathrm{UP}_-, \mathrm{UP}_-)}_{\text{negative saddle}} & - \\
\underbrace{(\mathrm{UP}_+, \mathrm{UP}_+)}_{\text{positive saddle}} & \underbrace{(\mathrm{UP}:\mathrm{DP},\mathrm{pF})}_{\text{hyperbolic-secant flow } (+)} \\
\underbrace{(\mathrm{DP}:\mathrm{UP},\mathrm{pF})}_{\text{hyperbolic flow } (+)} & \underbrace{(\mathrm{DP}_-, \mathrm{DP}_+)}_{\text{CW center}}
\end{array}\right\} \quad (5.14)
$$

for $a_{110} < 0$ and $a_{220} < 0$;

as shown in Fig. 5.3. The directrix flows of $x_1 = a_{111}, a_{112}$ are positive and negative. The directrix flows are separatrix flows, which cannot be passed by the other flows. The two centers are also separated by a hyperbolic flow, and the two saddles are separated by a hyperbolic-secant flow.

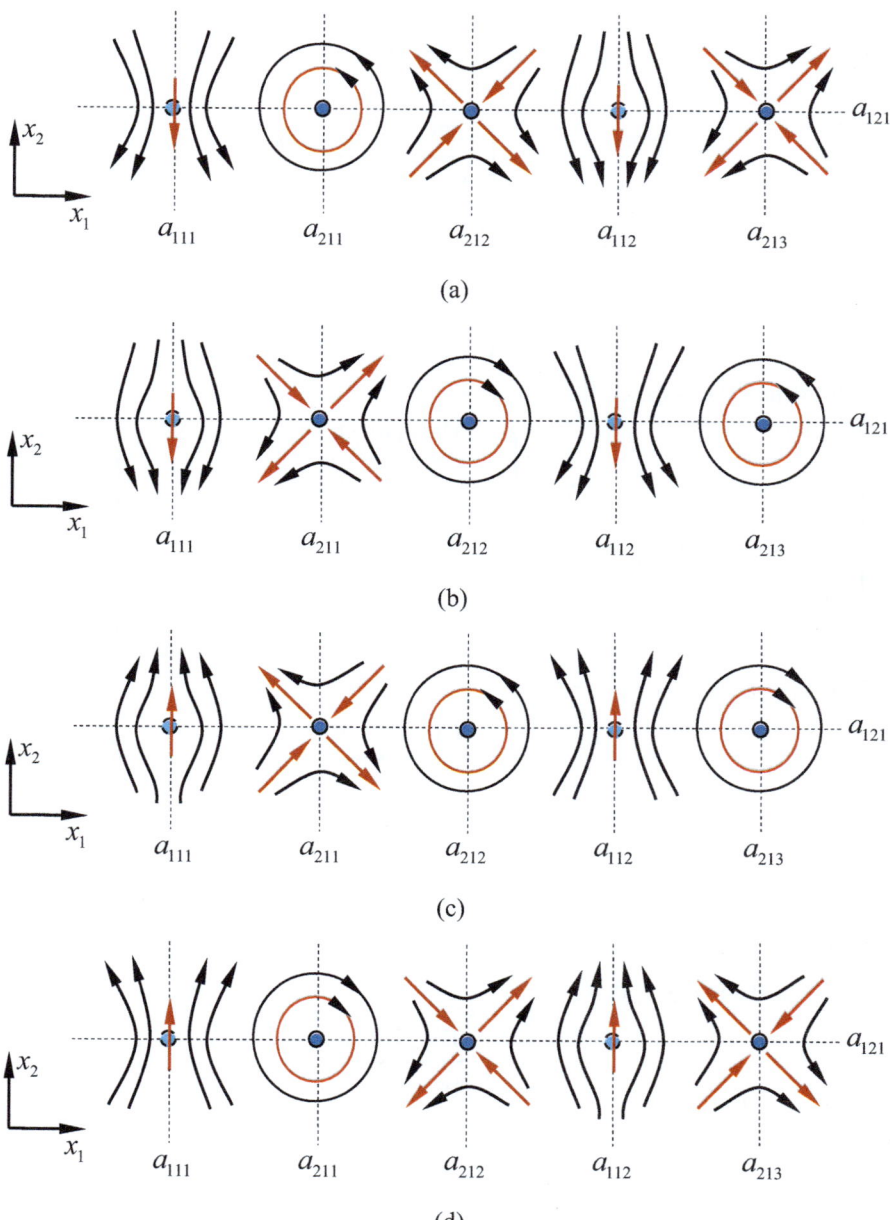

Fig. 5.3 Phase portraits ($a_{211}, a_{212} \in (a_{111}, a_{112})$) for two-dimensional systems on the x_1-direction with $x_1^* = a_{111}, a_{112}, a_{211}, a_{212}, a_{213}$ and on the x_2-direction with $x_2^* = a_{121}$. (**a**) ($a_{110} > 0, a_{220} > 0$), (**b**) ($a_{110} < 0, a_{220} > 0$), (**c**) ($a_{110} > 0, a_{220} < 0$), (**d**) ($a_{110} < 0, a_{220} < 0$)

(iv) For the case of $a_{211}, a_{212}, a_{213} \in (a_{111}, a_{112})$, the simple equilibriums of $x_1^* = a_{111}, a_{112}, a_{211}, a_{212}, a_{213}$ with $x_2^* = a_{121}$ have the following properties:

$$
\left\{
\begin{array}{ll}
(a_{112}, a_{121}) & - \\
(a_{121}, a_{212}) & (a_{121}, a_{213}) \\
(a_{111}, a_{121}) & (a_{121}, a_{211})
\end{array}
\right\}
=
\left\{
\begin{array}{ll}
\underbrace{(\mathrm{DP:UP, pF})}_{\text{hyperbolic flow }(+)} & - \\
\underbrace{(\mathrm{UP_-, UP_-})}_{\text{negative saddle}} & \underbrace{(\mathrm{DP_+, DP_-})}_{\text{CCW center}} \\
\underbrace{(\mathrm{DP:UP, nF})}_{\text{hyperbolic flow }(-)} & \underbrace{(\mathrm{DP_+, DP_-})}_{\text{CCW center}}
\end{array}
\right\}
\tag{5.15}
$$

for $a_{110} > 0$ and $a_{220} > 0$;

$$
\left\{
\begin{array}{ll}
(a_{112}, a_{121}) & - \\
(a_{121}, a_{212}) & (a_{121}, a_{213}) \\
(a_{111}, a_{121}) & (a_{121}, a_{211})
\end{array}
\right\}
=
\left\{
\begin{array}{ll}
\underbrace{(\mathrm{UP:DP, pF})}_{\text{hyperbolic-secant flow }(+)} & - \\
\underbrace{(\mathrm{DP_-, DP_+})}_{\text{CW center}} & \underbrace{(\mathrm{UP_+, UP_+})}_{\text{positive saddle}} \\
\underbrace{(\mathrm{UP:DP, nF})}_{\text{hyperbolic-secant flow }(-)} & \underbrace{(\mathrm{UP_+, UP_+})}_{\text{positive saddle}}
\end{array}
\right\}
\tag{5.16}
$$

for $a_{110} < 0$ and $a_{220} > 0$;

$$
\left\{
\begin{array}{ll}
(a_{112}, a_{121}) & - \\
(a_{121}, a_{212}) & (a_{121}, a_{213}) \\
(a_{111}, a_{121}) & (a_{121}, a_{211})
\end{array}
\right\}
=
\left\{
\begin{array}{ll}
\underbrace{(\mathrm{UP:DP, nF})}_{\text{hyperbolic-secant flow }(-)} & - \\
\underbrace{(\mathrm{DP_+, DP_-})}_{\text{CCW center}} & \underbrace{(\mathrm{UP_-, UP_-})}_{\text{negative saddle}} \\
\underbrace{(\mathrm{UP:DP, pF})}_{\text{hyperbolic-secant flow }(+)} & \underbrace{(\mathrm{UP_-, UP_-})}_{\text{negative saddle}}
\end{array}
\right\}
\tag{5.17}
$$

for $a_{110} > 0$ and $a_{220} < 0$;

$$
\left\{
\begin{array}{ll}
(a_{112}, a_{121}) & - \\
(a_{121}, a_{212}) & (a_{121}, a_{213}) \\
(a_{111}, a_{121}) & (a_{121}, a_{211})
\end{array}
\right\}
=
\left\{
\begin{array}{ll}
\underbrace{(\mathrm{DP:UP, nF})}_{\text{hyperbolic flow }(-)} & - \\
\underbrace{(\mathrm{UP_+, UP_+})}_{\text{positive saddle}} & \underbrace{(\mathrm{DP_-, DP_+})}_{\text{CW center}} \\
\underbrace{(\mathrm{DP:UP, pF})}_{\text{hyperbolic flow }(+)} & \underbrace{(\mathrm{DP_-, DP_+})}_{\text{CW center}}
\end{array}
\right\}
\tag{5.18}
$$

for $a_{110} < 0$ and $a_{220} < 0$;

as shown in Fig. 5.4. The directrix flows of $x_1 = a_{111}, a_{112}$ are positive and negative. The directrix flows are separatrix flows, which cannot be passed by the other flows. The two centers separated by a saddle are the same properties, and the two saddles separated by a center have the same properties.

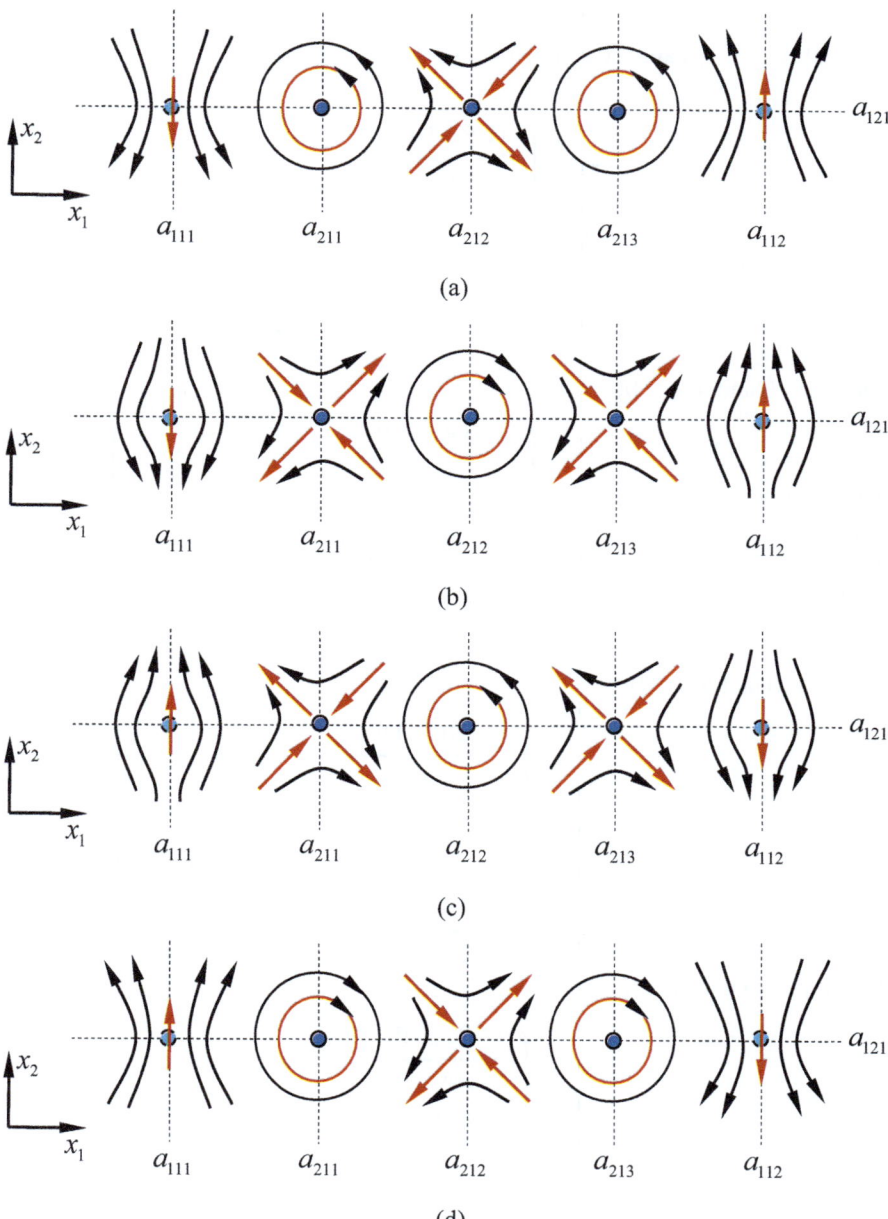

Fig. 5.4 Phase portraits ($a_{211}, a_{212}, a_{213} \in (a_{111}, a_{112})$) for two-dimensional systems on the x_1-direction with $x_1^* = a_{111}, a_{112}, a_{211}, a_{212}, a_{213}$ and on the x_2-direction with $x_2^* = a_{121}$. (**a**) ($a_{110} > 0$, $a_{220} > 0$), (**b**) ($a_{110} < 0, a_{220} > 0$), (**c**) ($a_{110} > 0, a_{220} < 0$), (**d**) ($a_{110} < 0, a_{220} < 0$)

(v) For $a_{111}, a_{112} \in (a_{211}, a_{212})$, simple equilibriums of $x_1^* = a_{111}, a_{112}, a_{211}, a_{212}, a_{213}$ with $x_2^* = a_{121}$ have the following properties:

$$
\left\{
\begin{array}{ll}
(a_{121}, a_{213}) & - \\
(a_{112}, a_{121}) & (a_{121}, a_{212}) \\
(a_{121}, a_{211}) & (a_{111}, a_{121})
\end{array}
\right\}
=
\left\{
\begin{array}{ll}
\underbrace{(\mathrm{UP_+}, \mathrm{UP_+})}_{\text{positive saddle}} & - \\[2ex]
\underbrace{(\mathrm{DP} : \mathrm{UP}, \mathrm{pF})}_{\text{hyperbolic flow (+)}} & \underbrace{(\mathrm{DP_-}, \mathrm{DP_+})}_{\text{CW center}} \\[2ex]
\underbrace{(\mathrm{UP_+}, \mathrm{UP_+})}_{\text{positive saddle}} & \underbrace{(\mathrm{UP} : \mathrm{DP}, \mathrm{pF})}_{\text{hyperbolic-secant flow (+)}}
\end{array}
\right\}
\quad (5.19)
$$

for $a_{110} > 0$ and $a_{220} > 0$;

$$
\left\{
\begin{array}{ll}
(a_{121}, a_{213}) & - \\
(a_{112}, a_{121}) & (a_{121}, a_{212}) \\
(a_{121}, a_{211}) & (a_{111}, a_{121})
\end{array}
\right\}
=
\left\{
\begin{array}{ll}
\underbrace{(\mathrm{DP_+}, \mathrm{DP_-})}_{\text{CCW center}} & - \\[2ex]
\underbrace{(\mathrm{UP} : \mathrm{DP}, \mathrm{pF})}_{\text{hyperbolic-secant flow (+)}} & \underbrace{(\mathrm{UP_-}, \mathrm{UP_-})}_{\text{negative saddle}} \\[2ex]
\underbrace{(\mathrm{DP_+}, \mathrm{DP_-})}_{\text{CCW center}} & \underbrace{(\mathrm{DP} : \mathrm{UP}, \mathrm{pF})}_{\text{hyperbolic flow (+)}}
\end{array}
\right\}
\quad (5.20)
$$

for $a_{110} < 0$ and $a_{220} > 0$;

$$
\left\{
\begin{array}{ll}
(a_{121}, a_{213}) & - \\
(a_{112}, a_{121}) & (a_{121}, a_{212}) \\
(a_{121}, a_{211}) & (a_{111}, a_{121})
\end{array}
\right\}
=
\left\{
\begin{array}{ll}
\underbrace{(\mathrm{DP_-}, \mathrm{DP_+})}_{\text{CW center}} & - \\[2ex]
\underbrace{(\mathrm{UP} : \mathrm{DP}, \mathrm{nF})}_{\text{hyperbolic-secant flow (-)}} & \underbrace{(\mathrm{UP_+}, \mathrm{UP_+})}_{\text{positive saddle}} \\[2ex]
\underbrace{(\mathrm{DP_-}, \mathrm{DP_+})}_{\text{CW center}} & \underbrace{(\mathrm{DP} : \mathrm{UP}, \mathrm{nF})}_{\text{hyperbolic flow (-)}}
\end{array}
\right\}
\quad (5.21)
$$

for $a_{110} > 0$ and $a_{220} < 0$;

$$
\left\{
\begin{array}{ll}
(a_{121}, a_{213}) & - \\
(a_{112}, a_{121}) & (a_{121}, a_{212}) \\
(a_{121}, a_{211}) & (a_{111}, a_{121})
\end{array}
\right\}
=
\left\{
\begin{array}{ll}
\underbrace{(\mathrm{UP_-}, \mathrm{UP_-})}_{\text{negative saddle}} & - \\[2ex]
\underbrace{(\mathrm{DP} : \mathrm{UP}, \mathrm{nF})}_{\text{hyperbolic flow (-)}} & \underbrace{(\mathrm{DP_+}, \mathrm{DP_-})}_{\text{CCW center}} \\[2ex]
\underbrace{(\mathrm{UP_-}, \mathrm{UP_-})}_{\text{negative saddle}} & \underbrace{(\mathrm{UP} : \mathrm{DP}, \mathrm{nF})}_{\text{hyperbolic-secant flow (-)}}
\end{array}
\right\}
\quad (5.22)
$$

for $a_{110} < 0$ and $a_{220} < 0$;

as shown in Fig. 5.5. The directrix flows of $x_1 = a_{111}, a_{112}$ are positive and negative. The directrix flows are separatrix flows, which cannot be passed by the other flows. The two centers on both ends are the same properties, and the two saddles on both

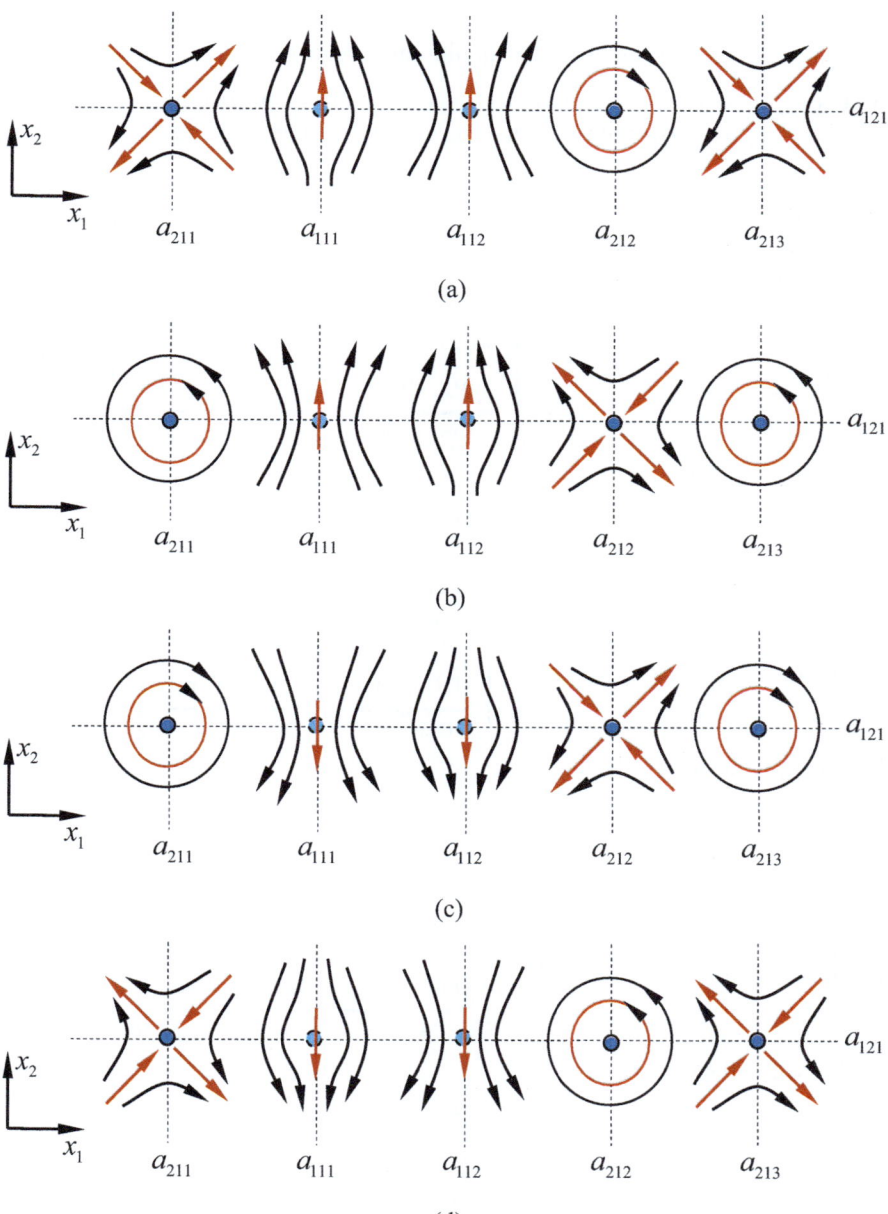

Fig. 5.5 Phase portraits $(a_{111}, a_{112} \in (a_{211}, a_{212}))$ for two-dimensional systems on the x_1-direction with $x_1^* = a_{111}, a_{112}, a_{211}, a_{212}, a_{213}$ and on the x_2-direction with $x_2^* = a_{121}$. (**a**) $(a_{110} > 0, a_{220} > 0)$, (**b**) $(a_{110} < 0, a_{220} > 0)$, (**c**) $(a_{110} > 0, a_{220} < 0)$, (**d**) $(a_{110} < 0, a_{220} < 0)$

ends are the same properties. The properties of the simple equilibriums are the same as case (i). Only the locations of the equilibriums are different.

(vi) For $a_{111} \in (a_{211}, a_{212})$, and $a_{112} \in (a_{212}, a_{213})$, the simple equilibriums of $x_1^* = a_{111}, a_{112}, a_{211}, a_{212}, a_{213}$ with $x_2^* = a_{121}$ have the following properties:

$$
\left\{
\begin{array}{ll}
(a_{121}, a_{213}) & - \\
(a_{121}, a_{212}) & (a_{112}, a_{121}) \\
(a_{121}, a_{211}) & (a_{111}, a_{121})
\end{array}
\right\}
=
\left\{
\begin{array}{ll}
\underbrace{(\mathrm{UP}_+, \mathrm{UP}_+)}_{\text{positive saddle}} & - \\
\underbrace{(\mathrm{UP}_-, \mathrm{UP}_-)}_{\text{negative saddle}} & \underbrace{(\mathrm{UP}:\mathrm{DP}, \mathrm{nF})}_{\text{hyperbolic-secant flow }(-)} \\
\underbrace{(\mathrm{UP}_+, \mathrm{UP}_+)}_{\text{positive saddle}} & \underbrace{(\mathrm{UP}:\mathrm{DP}, \mathrm{pF})}_{\text{hyperbolic-secant flow }(+)}
\end{array}
\right\}
\quad (5.23)
$$

for $a_{110} > 0$ and $a_{220} > 0$;

$$
\left\{
\begin{array}{ll}
(a_{121}, a_{213}) & - \\
(a_{121}, a_{212}) & (a_{112}, a_{121}) \\
(a_{121}, a_{211}) & (a_{111}, a_{121})
\end{array}
\right\}
=
\left\{
\begin{array}{ll}
\underbrace{(\mathrm{DP}_+, \mathrm{DP}_-)}_{\text{CCW center}} & - \\
\underbrace{(\mathrm{DP}_-, \mathrm{DP}_+)}_{\text{CW center}} & \underbrace{(\mathrm{DP}:\mathrm{UP}, \mathrm{nF})}_{\text{hyperbolic flow }(-)} \\
\underbrace{(\mathrm{DP}_+, \mathrm{DP}_-)}_{\text{CCW center}} & \underbrace{(\mathrm{DP}:\mathrm{UP}, \mathrm{pF})}_{\text{hyperbolic flow }(+)}
\end{array}
\right\}
\quad (5.24)
$$

for $a_{110} < 0$ and $a_{220} > 0$;

$$
\left\{
\begin{array}{ll}
(a_{121}, a_{213}) & - \\
(a_{121}, a_{212}) & (a_{112}, a_{121}) \\
(a_{121}, a_{211}) & (a_{111}, a_{121})
\end{array}
\right\}
=
\left\{
\begin{array}{ll}
\underbrace{(\mathrm{DP}_-, \mathrm{DP}_+)}_{\text{CW center}} & - \\
\underbrace{(\mathrm{DP}_+, \mathrm{DP}_-)}_{\text{CW center}} & \underbrace{(\mathrm{DP}:\mathrm{UP}, \mathrm{pF})}_{\text{hyperbolic flow }(+)} \\
\underbrace{(\mathrm{DP}_-, \mathrm{DP}_+)}_{\text{CW center}} & \underbrace{(\mathrm{DP}:\mathrm{UP}, \mathrm{nF})}_{\text{hyperbolic flow }(-)}
\end{array}
\right\}
\quad (5.25)
$$

for $a_{110} > 0$ and $a_{220} < 0$;

$$
\left\{
\begin{array}{ll}
(a_{121}, a_{213}) & - \\
(a_{121}, a_{212}) & (a_{112}, a_{121}) \\
(a_{121}, a_{211}) & (a_{111}, a_{121})
\end{array}
\right\}
=
\left\{
\begin{array}{ll}
\underbrace{(\mathrm{UP}_-, \mathrm{UP}_-)}_{\text{negative saddle}} & - \\
\underbrace{(\mathrm{UP}_+, \mathrm{UP}_+)}_{\text{positive saddle}} & \underbrace{(\mathrm{UP}:\mathrm{DP}, \mathrm{pF})}_{\text{hyperbolic-secant flow }(+)} \\
\underbrace{(\mathrm{UP}_-, \mathrm{UP}_-)}_{\text{negative saddle}} & \underbrace{(\mathrm{UP}:\mathrm{DP}, \mathrm{nF})}_{\text{hyperbolic-secant flow }(-)}
\end{array}
\right\}
\quad (5.26)
$$

for $a_{110} < 0$ and $a_{220} < 0$;

as shown in Fig. 5.6. The directrix flows of $x_1 = a_{111}, a_{112}$ are positive and negative. The directrix flows are separatrix flows, which cannot be passed by the other flows.

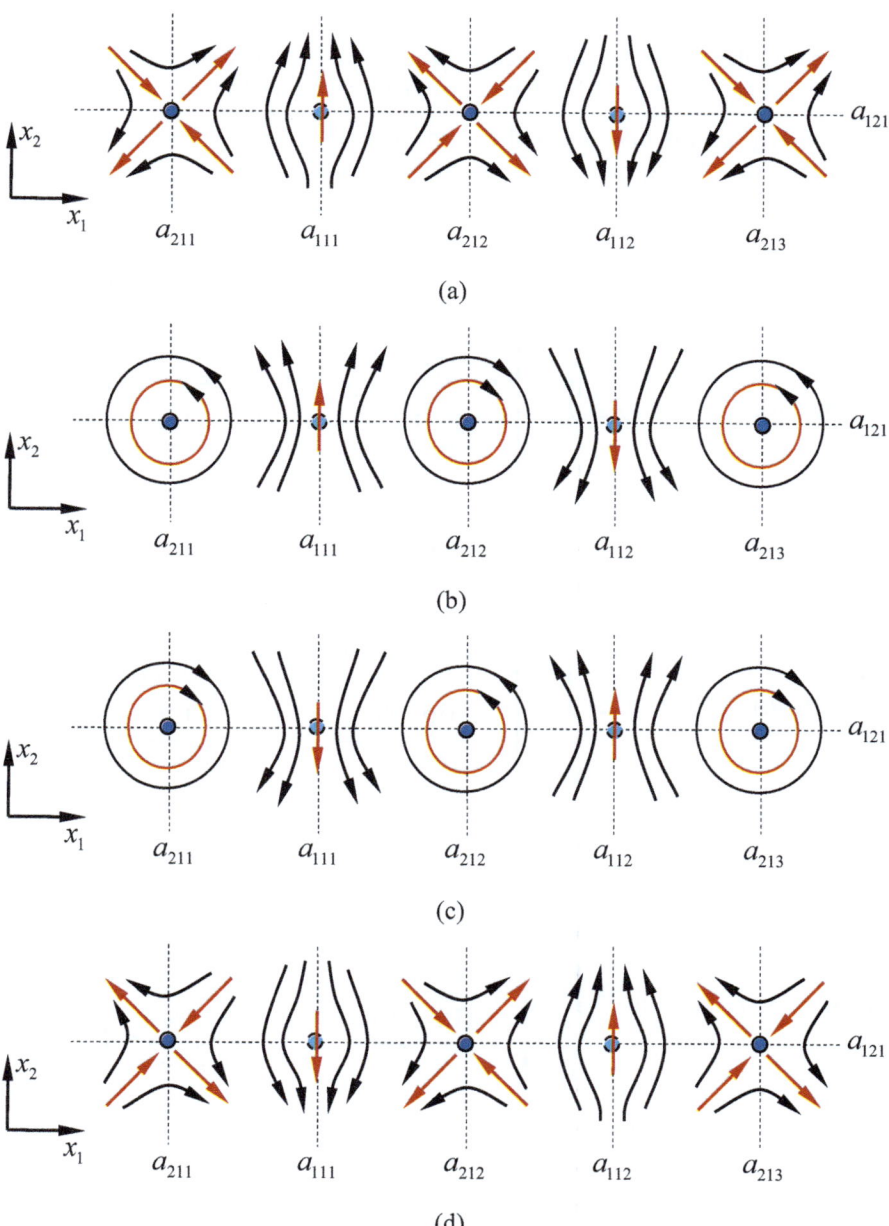

Fig. 5.6 Phase portraits ($a_{111} \in (a_{211}, a_{212})$, $a_{112} \in (a_{212}, a_{213})$) for two-dimensional systems on the x_1-direction with $x_1^* = a_{111}, a_{112}, a_{211}, a_{212}, a_{213}$ and on the x_2-direction with $x_2^* = a_{121}$. (**a**) ($a_{110} > 0, a_{220} > 0$), (**b**) ($a_{110} < 0, a_{220} > 0$), (**c**) ($a_{110} > 0, a_{220} < 0$), (**d**) ($a_{110} < 0, a_{220} < 0$)

The three centers are separated by the two hyperbolic flows, and the three saddle are separated by the two hyperbolic-secant flows. Since the directrix flows exist, the three centers are independent, and the three saddles are independent.

(vii) For $a_{111}, a_{112} \in (a_{212}, a_{213})$, the simple equilibriums of $x_1^* = a_{111}, a_{112}, a_{211}, a_{212}, a_{213}$ with $x_2^* = a_{121}$ have the following properties:

$$
\left\{
\begin{array}{cc}
(a_{121}, a_{213}) & - \\
(a_{111}, a_{121}) & (a_{112}, a_{121}) \\
(a_{121}, a_{211}) & (a_{121}, a_{212})
\end{array}
\right\}
=
\left\{
\begin{array}{cc}
\underbrace{(\mathrm{UP}_+, \mathrm{UP}_+)}_{\text{positive saddle}} & - \\
\underbrace{(\mathrm{DP}:\mathrm{UP}, \mathrm{nF})}_{\text{hyperbolic flow }(-)} & \underbrace{(\mathrm{UP}:\mathrm{DP}, \mathrm{nF})}_{\text{hyperbolic-secant flow }(-)} \\
\underbrace{(\mathrm{UP}_+, \mathrm{UP}_+)}_{\text{positive saddle}} & \underbrace{(\mathrm{DP}_-, \mathrm{DP}_+)}_{\text{CW center}}
\end{array}
\right\}
\quad (5.27)
$$

for $a_{110} > 0$ and $a_{220} > 0$;

$$
\left\{
\begin{array}{cc}
(a_{121}, a_{213}) & - \\
(a_{111}, a_{121}) & (a_{112}, a_{121}) \\
(a_{121}, a_{211}) & (a_{121}, a_{212})
\end{array}
\right\}
=
\left\{
\begin{array}{cc}
\underbrace{(\mathrm{DP}_+, \mathrm{DP}_-)}_{\text{CCW center}} & - \\
\underbrace{(\mathrm{UP}:\mathrm{DP}, \mathrm{nF})}_{\text{hyperbolic-secant flow }(-)} & \underbrace{(\mathrm{DP}:\mathrm{UP}, \mathrm{nF})}_{\text{hyperbolic flow }(-)} \\
\underbrace{(\mathrm{DP}_+, \mathrm{DP}_-)}_{\text{CCW center}} & \underbrace{(\mathrm{UP}_-, \mathrm{UP}_-)}_{\text{negative saddle}}
\end{array}
\right\}
\quad (5.28)
$$

for $a_{110} < 0$ and $a_{220} > 0$;

$$
\left\{
\begin{array}{cc}
(a_{121}, a_{213}) & - \\
(a_{111}, a_{121}) & (a_{112}, a_{121}) \\
(a_{121}, a_{211}) & (a_{121}, a_{212})
\end{array}
\right\}
=
\left\{
\begin{array}{cc}
\underbrace{(\mathrm{DP}_-, \mathrm{DP}_+)}_{\text{CW center}} & - \\
\underbrace{(\mathrm{UP}:\mathrm{DP}, \mathrm{pF})}_{\text{hyperbolic-secant flow }(+)} & \underbrace{(\mathrm{DP}:\mathrm{UP}, \mathrm{pF})}_{\text{hyperbolic flow }(+)} \\
\underbrace{(\mathrm{DP}_-, \mathrm{DP}_+)}_{\text{CW center}} & \underbrace{(\mathrm{UP}_+, \mathrm{UP}_+)}_{\text{positive saddle}}
\end{array}
\right\}
\quad (5.29)
$$

for $a_{110} > 0$ and $a_{220} < 0$;

$$
\left\{
\begin{array}{cc}
(a_{121}, a_{213}) & - \\
(a_{111}, a_{121}) & (a_{112}, a_{121}) \\
(a_{121}, a_{211}) & (a_{121}, a_{212})
\end{array}
\right\}
=
\left\{
\begin{array}{cc}
\underbrace{(\mathrm{UP}_-, \mathrm{UP}_-)}_{\text{negative saddle}} & - \\
\underbrace{(\mathrm{DP}:\mathrm{UP}, \mathrm{pF})}_{\text{hyperbolic flow }(+)} & \underbrace{(\mathrm{UP}:\mathrm{DP}, \mathrm{pF})}_{\text{hyperbolic-secant flow }(+)} \\
\underbrace{(\mathrm{UP}_-, \mathrm{UP}_-)}_{\text{negative saddle}} & \underbrace{(\mathrm{DP}_+, \mathrm{DP}_-)}_{\text{CCW saddle}}
\end{array}
\right\}
\quad (5.30)
$$

for $a_{110} < 0$ and $a_{220} < 0$;

as shown in Fig. 5.7. The directrix flows of $x_1 = a_{111}, a_{112}$ are positive and negative. The directrix flows are separatrix flows, which cannot be passed by the other flows.

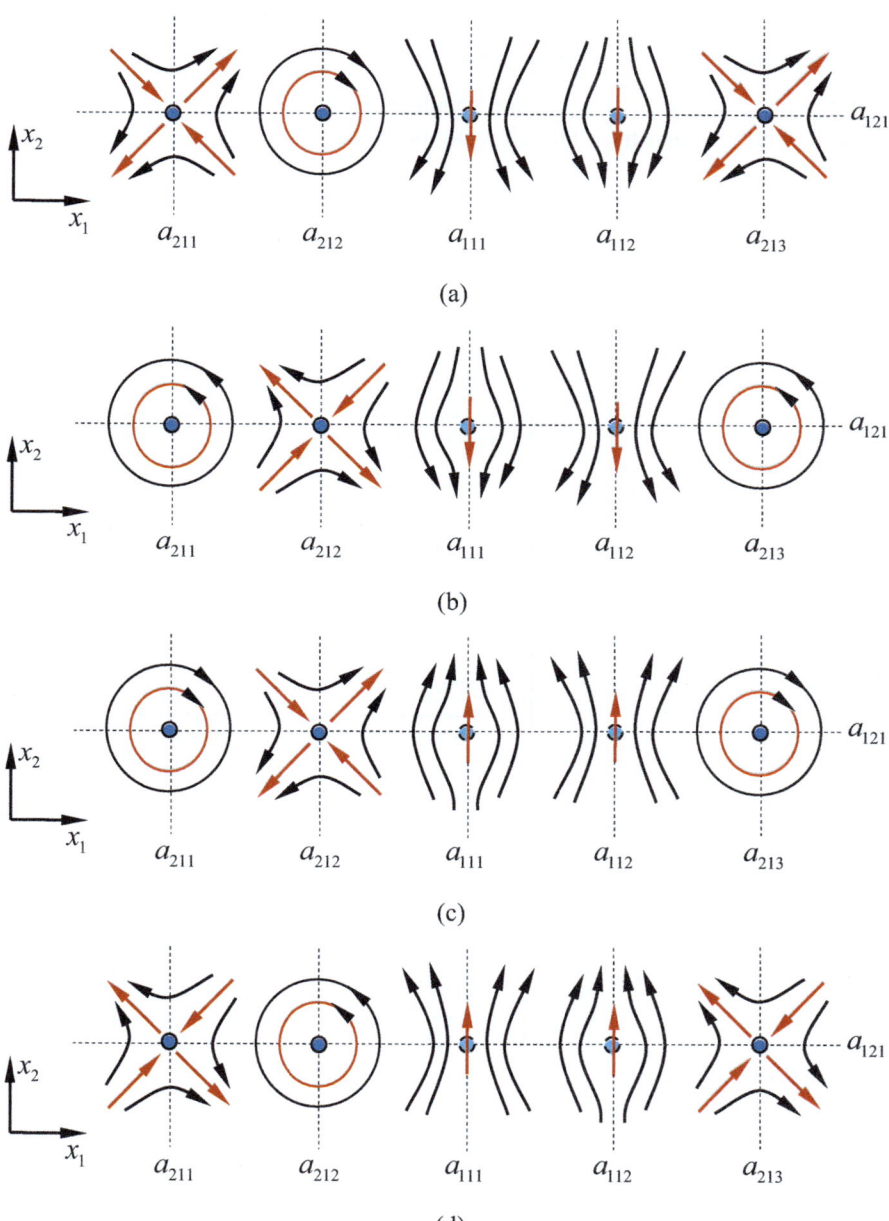

Fig. 5.7   Phase portraits $(a_{111}, a_{112} \in (a_{212}, a_{213}))$ for two-dimensional systems on the x_1-direction wit $x_1^* = a_{111}, a_{112}, a_{211}, a_{212}, a_{213}$ and on the x_2-direction with $x_2^* = a_{121}$. (**a**) $(a_{110} > 0, a_{220} > 0)$, (**b**) $(a_{110} < 0, a_{220} > 0)$, (**c**) $(a_{110} > 0, a_{220} < 0)$, (**d**) $(a_{110} < 0, a_{220} < 0)$

The properties of the simple equilibriums are the same as cases (i) and (v). Only the locations of the equilibriums are different.

(viii) For $a_{212}, a_{213} \in (a_{111}, a_{112})$, the simple equilibriums of $x_1^* = a_{111}, a_{112}, a_{211}, a_{212}, a_{213}$ with $x_2^* = a_{121}$ have the following properties:

$$
\left\{
\begin{array}{ll}
(a_{112}, a_{121}) & - \\
(a_{121}, a_{212}) & (a_{121}, a_{213}) \\
(a_{121}, a_{211}) & (a_{111}, a_{121})
\end{array}
\right\}
=
\left\{
\begin{array}{ll}
\underbrace{(\mathrm{DP} : \mathrm{UP}, \mathrm{pF})}_{\text{hyperbolic flow }(+)} & - \\
\underbrace{(\mathrm{UP}_-, \mathrm{UP}_-)}_{\text{negative saddle}} & \underbrace{(\mathrm{DP}_+, \mathrm{DP}_-)}_{\text{CCW center}} \\
\underbrace{(\mathrm{UP}_+, \mathrm{UP}_+)}_{\text{positive saddle}} & \underbrace{(\mathrm{UP} : \mathrm{DP}, \mathrm{pF})}_{\text{hyperbolic-secant flow }(+)}
\end{array}
\right\}
\tag{5.31}
$$

for $a_{110} > 0$ and $a_{220} > 0$;

$$
\left\{
\begin{array}{ll}
(a_{112}, a_{121}) & - \\
(a_{121}, a_{212}) & (a_{121}, a_{213}) \\
(a_{121}, a_{211}) & (a_{111}, a_{121})
\end{array}
\right\}
=
\left\{
\begin{array}{ll}
\underbrace{(\mathrm{UP} : \mathrm{DP}, \mathrm{pF})}_{\text{hyperbolic-secant flow }(+)} & - \\
\underbrace{(\mathrm{DP}_-, \mathrm{DP}_+)}_{\text{CW center}} & \underbrace{(\mathrm{UP}_+, \mathrm{UP}_+)}_{\text{positive saddle}} \\
\underbrace{(\mathrm{DP}_+, \mathrm{DP}_-)}_{\text{CCW center}} & \underbrace{(\mathrm{DP} : \mathrm{UP}, \mathrm{pF})}_{\text{hyperbolic flow }(+)}
\end{array}
\right\}
\tag{5.32}
$$

for $a_{110} < 0$ and $a_{220} > 0$;

$$
\left\{
\begin{array}{ll}
(a_{112}, a_{121}) & - \\
(a_{121}, a_{212}) & (a_{121}, a_{213}) \\
(a_{121}, a_{211}) & (a_{111}, a_{121})
\end{array}
\right\}
=
\left\{
\begin{array}{ll}
\underbrace{(\mathrm{UP} : \mathrm{DP}, \mathrm{nF})}_{\text{hyperbolic-secant flow }(-)} & - \\
\underbrace{(\mathrm{DP}_+, \mathrm{DP}_-)}_{\text{CCW center}} & \underbrace{(\mathrm{UP}_-, \mathrm{UP}_-)}_{\text{negative saddle}} \\
\underbrace{(\mathrm{DP}_-, \mathrm{DP}_+)}_{\text{CW center}} & \underbrace{(\mathrm{DP} : \mathrm{UP}, \mathrm{nF})}_{\text{hyperbolic flow }(-)}
\end{array}
\right\}
\tag{5.33}
$$

for $a_{110} > 0$ and $a_{220} < 0$;

$$
\left\{
\begin{array}{ll}
(a_{112}, a_{121}) & - \\
(a_{121}, a_{212}) & (a_{121}, a_{213}) \\
(a_{121}, a_{211}) & (a_{111}, a_{121})
\end{array}
\right\}
=
\left\{
\begin{array}{ll}
\underbrace{(\mathrm{DP} : \mathrm{UP}, \mathrm{nF})}_{\text{hyperbolic flow }(-)} & - \\
\underbrace{(\mathrm{UP}_+, \mathrm{UP}_+)}_{\text{positive saddle}} & \underbrace{(\mathrm{DP}_-, \mathrm{DP}_+)}_{\text{CW saddle}} \\
\underbrace{(\mathrm{UP}_-, \mathrm{UP}_-)}_{\text{negative saddle}} & \underbrace{(\mathrm{UP} : \mathrm{DP}, \mathrm{nF})}_{\text{hyperbolic-secant flow }(-)}
\end{array}
\right\}
\tag{5.34}
$$

for $a_{110} < 0$ and $a_{220} < 0$;

as shown in Fig. 5.8. The directrix flows of $x_1 = a_{111}, a_{112}$ are positive and negative. The directrix flows are separatrix flows, which cannot be passed by the other flows.

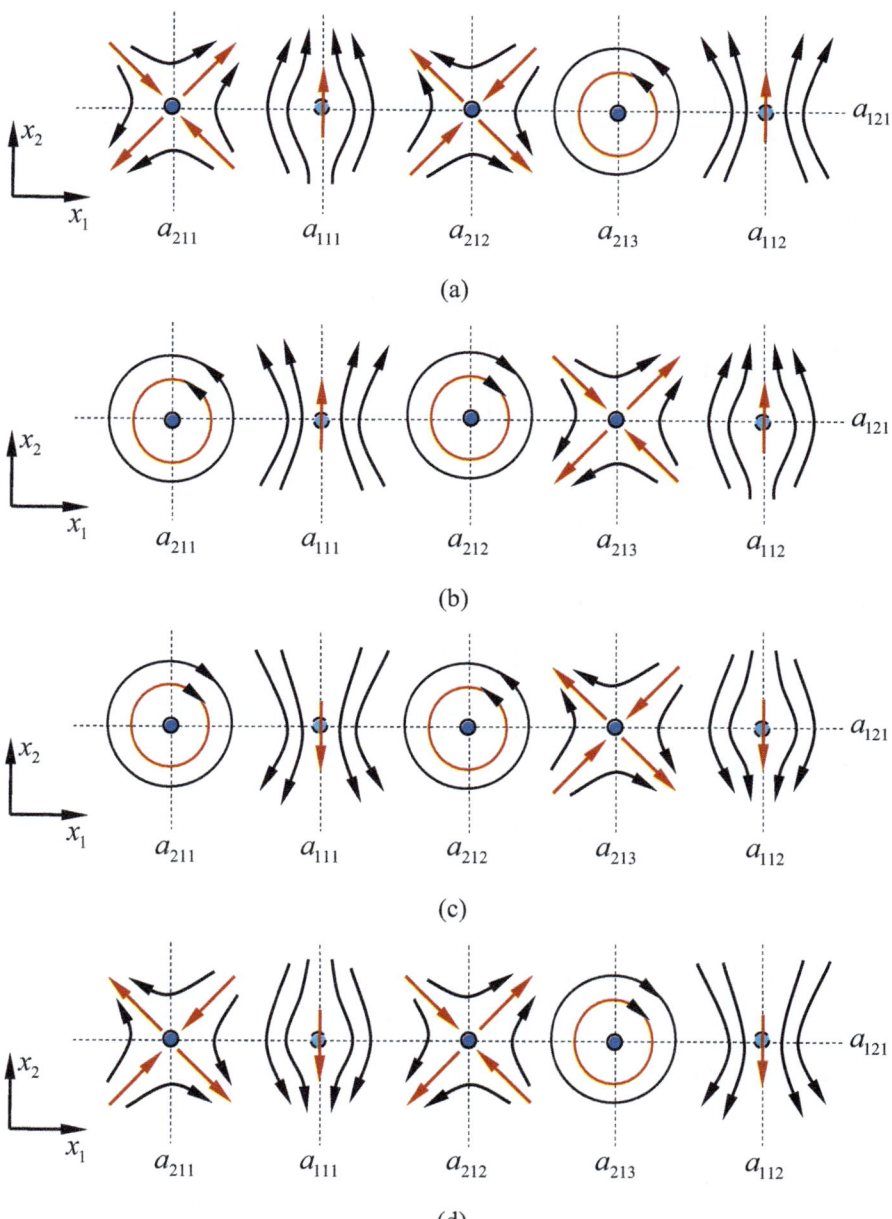

Fig. 5.8 Phase portraits ($a_{212}, a_{213} \in (a_{111}, a_{112})$) for two-dimensional systems on the x_1-direction with $x_1^* = a_{111}, a_{112}, a_{211}, a_{212}, a_{213}$ and on the x_2-direction with $x_2^* = a_{121}$. (**a**) ($a_{110} > 0, a_{220} > 0$), (**b**) ($a_{110} < 0, a_{220} > 0$), (**c**) ($a_{110} > 0, a_{220} < 0$), (**d**) ($a_{110} < 0, a_{220} < 0$)

The two saddles are separated by a hyperbolic-secant flow, and the two centers are separated by a hyperbolic flow.

(ix) For $a_{213} \in (a_{111}, a_{112})$, the simple equilibriums of $x_1^* = a_{111}, a_{112}, a_{211}, a_{212}, a_{213}$ with $x_2^* = a_{121}$ have the following properties:

$$
\left\{
\begin{array}{ll}
(a_{112}, a_{121}) & - \\
(a_{111}, a_{121}) & (a_{121}, a_{213}) \\
(a_{121}, a_{211}) & (a_{121}, a_{212})
\end{array}
\right\}
=
\left\{
\begin{array}{ll}
\underbrace{(\mathrm{DP}:\mathrm{UP},\mathrm{pF})}_{\text{hyperbolic flow }(+)} & - \\
\underbrace{(\mathrm{DP}:\mathrm{UP},\mathrm{nF})}_{\text{hyperbolic flow }(-)} & \underbrace{(\mathrm{DP}_+,\mathrm{DP}_-)}_{\text{CCW center}} \\
\underbrace{(\mathrm{UP}_+,\mathrm{UP}_+)}_{\text{positive saddle}} & \underbrace{(\mathrm{DP}_-,\mathrm{DP}_+)}_{\text{CW center}}
\end{array}
\right\}
\tag{5.35}
$$

for $a_{110} > 0$ and $a_{220} > 0$;

$$
\left\{
\begin{array}{ll}
(a_{112}, a_{121}) & - \\
(a_{111}, a_{121}) & (a_{121}, a_{213}) \\
(a_{121}, a_{211}) & (a_{121}, a_{212})
\end{array}
\right\}
=
\left\{
\begin{array}{ll}
\underbrace{(\mathrm{UP}:\mathrm{DP},\mathrm{pF})}_{\text{hyperbolic-secant flow }(+)} & - \\
\underbrace{(\mathrm{UP}:\mathrm{DP},\mathrm{nF})}_{\text{hyperbolic-secant flow }(-)} & \underbrace{(\mathrm{UP}_+,\mathrm{UP}_+)}_{\text{positive saddle}} \\
\underbrace{(\mathrm{DP}_+,\mathrm{DP}_-)}_{\text{CCW center}} & \underbrace{(\mathrm{UP}_-,\mathrm{UP}_-)}_{\text{negative saddle}}
\end{array}
\right\}
\tag{5.36}
$$

for $a_{110} < 0$ and $a_{220} > 0$;

$$
\left\{
\begin{array}{ll}
(a_{112}, a_{121}) & - \\
(a_{111}, a_{121}) & (a_{121}, a_{213}) \\
(a_{121}, a_{211}) & (a_{121}, a_{212})
\end{array}
\right\}
=
\left\{
\begin{array}{ll}
\underbrace{(\mathrm{UP}:\mathrm{DP},\mathrm{nF})}_{\text{hyperbolic-secant flow }(-)} & - \\
\underbrace{(\mathrm{UP}:\mathrm{DP},\mathrm{pF})}_{\text{hyperbolic-secant flow }(+)} & \underbrace{(\mathrm{UP}_-,\mathrm{UP}_-)}_{\text{negative saddle}} \\
\underbrace{(\mathrm{DP}_-,\mathrm{DP}_+)}_{\text{CW center}} & \underbrace{(\mathrm{UP}_+,\mathrm{UP}_+)}_{\text{positive saddle}}
\end{array}
\right\}
\tag{5.37}
$$

for $a_{110} > 0$ and $a_{220} < 0$;

$$
\left\{
\begin{array}{ll}
(a_{112}, a_{121}) & - \\
(a_{111}, a_{121}) & (a_{121}, a_{213}) \\
(a_{121}, a_{211}) & (a_{121}, a_{212})
\end{array}
\right\}
=
\left\{
\begin{array}{ll}
\underbrace{(\mathrm{DP}:\mathrm{UP},\mathrm{nF})}_{\text{hyperbolic flow }(-)} & - \\
\underbrace{(\mathrm{DP}:\mathrm{UP},\mathrm{pF})}_{\text{hyperbolic flow }(+)} & \underbrace{(\mathrm{DP}_-,\mathrm{DP}_+)}_{\text{CW saddle}} \\
\underbrace{(\mathrm{UP}_-,\mathrm{UP}_-)}_{\text{negative saddle}} & \underbrace{(\mathrm{DP}_+,\mathrm{DP}_-)}_{\text{CCW saddle}}
\end{array}
\right\}
\tag{5.38}
$$

for $a_{110} < 0$ and $a_{220} < 0$;

as shown in Fig. 5.9. The directrix flows of $x_1 = a_{111}, a_{112}$ are positive and negative. The directrix flows are separatrix flows, which cannot be passed by the other flows.

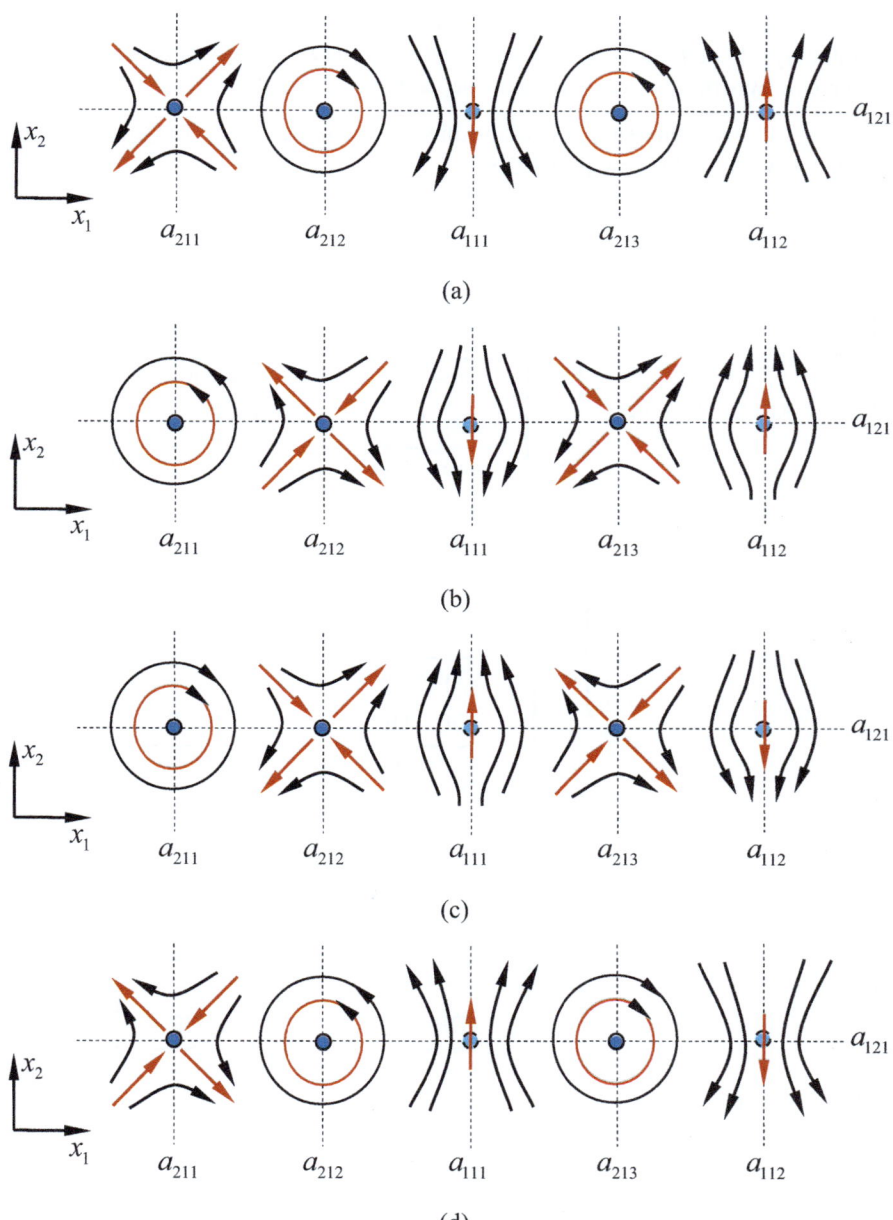

Fig. 5.9 Phase portraits ($a_{213} \in (a_{111}, a_{112})$) for two-dimensional systems on the x_1 direction with $x_1^* = a_{111}, a_{112}, a_{211}, a_{212}, a_{213}$ and on the x_2 direction with $x_2^* = a_{121}$. (**a**) ($a_{110} > 0, a_{220} > 0$), (**b**) ($a_{110} < 0, a_{220} > 0$), (**c**) ($a_{110} > 0, a_{220} < 0$), (**d**) ($a_{110} < 0, a_{220} < 0$)

The two saddles are separated by a hyperbolic-secant flow, and the two centers are separated by a hyperbolic flow. The two hyperbolic flows are separated by a center, and the two hyperbolic-secant flows are separated by a saddle.

(x) For $a_{213} < a_{111}$, the simple equilibriums of $x_1^* = a_{111}, a_{112}, a_{211}, a_{212}, a_{213}$ with $x_2^* = a_{121}$ have the following properties:

$$
\left\{
\begin{array}{ll}
(a_{112}, a_{121}) & - \\
(a_{121}, a_{213}) & (a_{111}, a_{121}) \\
(a_{121}, a_{211}) & (a_{121}, a_{212})
\end{array}
\right\}
=
\left\{
\begin{array}{ll}
\underbrace{(\mathrm{DP}:\mathrm{UP},\mathrm{pF})}_{\text{hyperbolic flow }(+)} & - \\[1em]
\underbrace{(\mathrm{UP}_+,\mathrm{UP}_+)}_{\text{positive saddle}} & \underbrace{(\mathrm{UP}:\mathrm{DP},\mathrm{pF})}_{\text{hyperbolic-secnt flow }(+)} \\[1em]
\underbrace{(\mathrm{UP}_+,\mathrm{UP}_+)}_{\text{positive saddle}} & \underbrace{(\mathrm{DP}_-,\mathrm{DP}_+)}_{\text{CW center}}
\end{array}
\right\}
\quad (5.39)
$$

for $a_{110} > 0$ and $a_{220} > 0$;

$$
\left\{
\begin{array}{ll}
(a_{112}, a_{121}) & - \\
(a_{121}, a_{213}) & (a_{111}, a_{121}) \\
(a_{121}, a_{211}) & (a_{121}, a_{212})
\end{array}
\right\}
=
\left\{
\begin{array}{ll}
\underbrace{(\mathrm{UP}:\mathrm{DP},\mathrm{pF})}_{\text{hyperbolic-secant flow }(+)} & - \\[1em]
\underbrace{(\mathrm{DP}_+,\mathrm{DP}_-)}_{\text{CCW center}} & \underbrace{(\mathrm{DP}:\mathrm{UP},\mathrm{pF})}_{\text{hyperbolic flow }(+)} \\[1em]
\underbrace{(\mathrm{DP}_+,\mathrm{DP}_-)}_{\text{CCW center}} & \underbrace{(\mathrm{UP}_-,\mathrm{UP}_-)}_{\text{negative saddle}}
\end{array}
\right\}
\quad (5.40)
$$

for $a_{110} < 0$ and $a_{220} > 0$;

$$
\left\{
\begin{array}{ll}
(a_{112}, a_{121}) & - \\
(a_{121}, a_{213}) & (a_{111}, a_{121}) \\
(a_{121}, a_{211}) & (a_{121}, a_{212})
\end{array}
\right\}
=
\left\{
\begin{array}{ll}
\underbrace{(\mathrm{UP}:\mathrm{DP},\mathrm{nF})}_{\text{hyperbolic-secant flow }(-)} & - \\[1em]
\underbrace{(\mathrm{DP}_-,\mathrm{DP}_+)}_{\text{CW center}} & \underbrace{(\mathrm{DP}:\mathrm{UP},\mathrm{nF})}_{\text{hyperbolic flow }(-)} \\[1em]
\underbrace{(\mathrm{DP}_-,\mathrm{DP}_+)}_{\text{CW center}} & \underbrace{(\mathrm{UP}_+,\mathrm{UP}_+)}_{\text{positive saddle}}
\end{array}
\right\}
\quad (5.41)
$$

for $a_{110} > 0$ and $a_{220} < 0$;

$$
\left\{
\begin{array}{ll}
(a_{112}, a_{121}) & - \\
(a_{111}, a_{121}) & (a_{121}, a_{213}) \\
(a_{121}, a_{211}) & (a_{121}, a_{212})
\end{array}
\right\}
=
\left\{
\begin{array}{ll}
\underbrace{(\mathrm{DP}:\mathrm{UP},\mathrm{nF})}_{\text{hyperbolic flow }(-)} & - \\[1em]
\underbrace{(\mathrm{UP}_-,\mathrm{UP}_-)}_{\text{negative saddle}} & \underbrace{(\mathrm{UP}:\mathrm{DP},\mathrm{nF})}_{\text{hyperbolic-secant flow }(-)} \\[1em]
\underbrace{(\mathrm{UP}_-,\mathrm{UP}_-)}_{\text{negative saddle}} & \underbrace{(\mathrm{DP}_+,\mathrm{DP}_-)}_{\text{CCW saddle}}
\end{array}
\right\}
\quad (5.42)
$$

for $a_{110} < 0$ and $a_{220} < 0$;

as shown in Fig. 5.10. The directrix flows of $x_1 = a_{111}, a_{112}$ are positive and negative. The directrix flows are separatrix flows, which cannot be passed by the

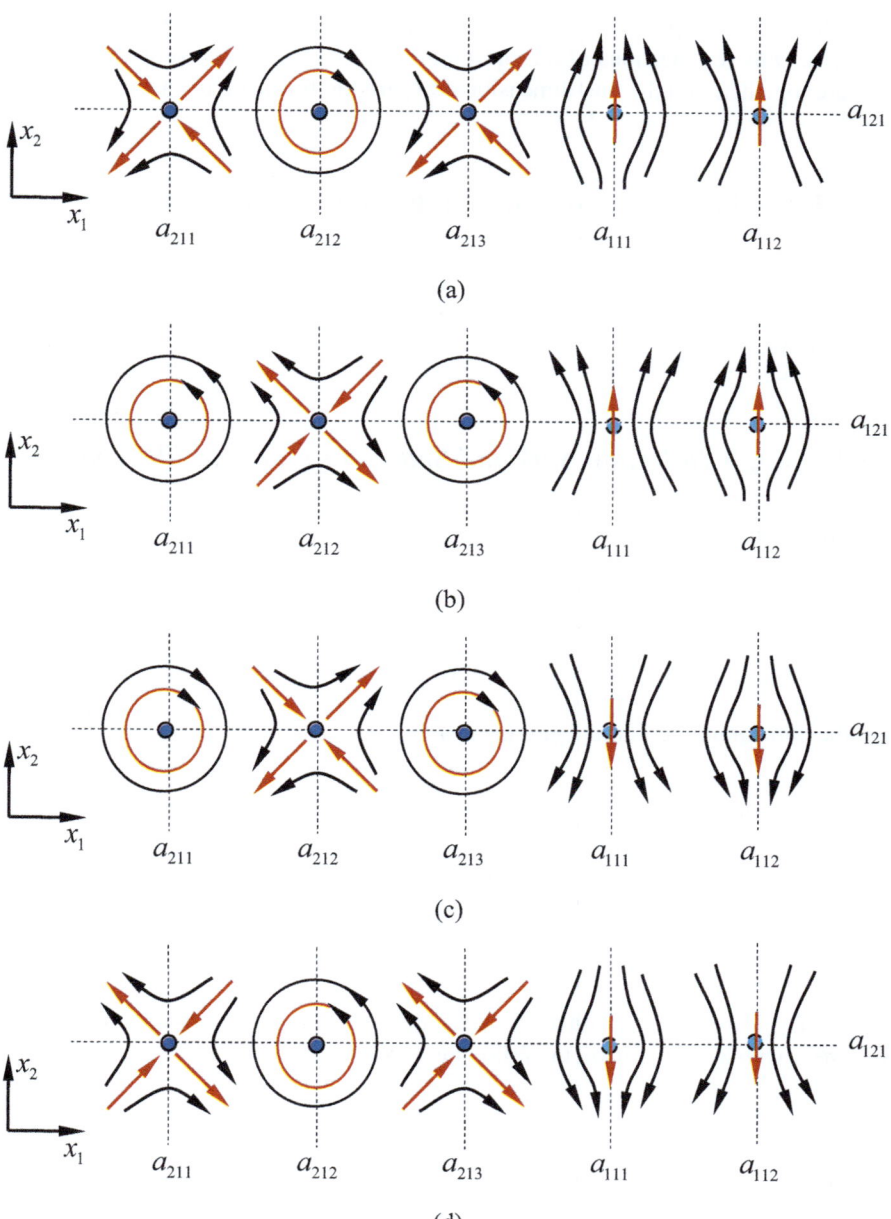

Fig. 5.10 Phase portraits ($a_{213} < a_{111}$) for two-dimensional systems on the x_1-direction with $x_1^* = a_{111}, a_{112}, a_{211}, a_{212}, a_{213}$ and on the x_2-direction with $x_2^* = a_{121}$. (**a**) ($a_{110} > 0, a_{220} > 0$), (**b**) ($a_{110} < 0, a_{220} > 0$), (**c**) ($a_{110} > 0, a_{220} < 0$), (**d**) ($a_{110} < 0, a_{220} < 0$)

other flows. The properties of the simple equilibriums are the same as cases (i), (v), and (vii). Only the locations of the equilibriums are different. The simple equilibriums are together, and the one-dimensional hyperbolic flows are together.

5.2   Inflection-Sink and Source Infinite-Equilibriums

In this section, inflection-sink and source infinite-equilibriums will be presented for the switching bifurcations of saddles and centers with hyperbolic and hyperbolic-secant flows.

5.2.1   Single Inflection Sink and Source Infinite Equilibriums

Consider a dynamical system for $a_{11s_1} = a_{21l_1}$ and $a_{11s_2} \neq a_{21l_2}$ ($s_1, s_2 \in \{1, 2, 3\}$; $s_1 \neq s_2$; $l_1, l_2 \in \{1, 2\}$, $l_1 \neq l_2$) as

$$\begin{aligned}
\dot{x}_1 &= a_{110}(x_1 - a_{11l_1})(x_1 - a_{11l_2})(x_2 - a_{121}), \\
\dot{x}_2 &= a_{220}(x_1 - a_{21s_1})(x_1 - a_{21s_2})(x_1 - a_{21s_3}),
\end{aligned} \tag{5.43}$$

and the corresponding first integral manifold is

$$\begin{aligned}
&\frac{1}{2}\left[(x_1 - a_{11l_2})^2 - (x_{10} - a_{11l_2})^2\right] + (2a_{11l_2} - a_{21s_2} - a_{21s_3})(x_1 - x_{10}) \\
&+ \prod_{s_2 = 1, s_2 \neq s_1}^{3}(a_{11l_2} - a_{21s_2}) \ln \frac{|x_1 - a_{11l_2}|}{|x_{10} - a_{11l_2}|} \\
&= \frac{1}{2}\frac{a_{110}}{a_{220}}\left[(x_2 - a_{121})^2 - (x_{20} - a_{121})^2\right].
\end{aligned} \tag{5.44}$$

(i) For $a_{211} = a_{111} = a_{11l_1}$ with $a_{112} < a_{212}$, the equilibriums of $x_1^* = a_{111}, a_{112}, a_{211}, a_{212}, a_{213}$ with $x_2^* = a_{121}$ have the following properties:

$$\left\{\begin{matrix} (a_{121}, a_{212}) & (a_{121}, a_{213}) \\ (a_{121}, a_{211}) & (a_{112}, a_{121}) \end{matrix}\right\} = \left\{\begin{matrix} \overbrace{(\text{DP}_-, \text{DP}_+)}^{\text{CW center}} & \overbrace{(\text{UP}_+, \text{UP}_+)}^{\text{postive saddle}} \\ \underbrace{(_{\text{II:DI}}\text{DP}, _{\text{SO:SI}}\text{LS})}_{\text{down-parabola lower-saddle}} & \underbrace{(\text{DP} : \text{UP}, \text{pF})}_{\text{hyperbolic flow } (+)} \end{matrix}\right\} \tag{5.45}$$

for $a_{110} > 0, a_{220} > 0$,

$$\left\{ \begin{array}{ll} (a_{121}, a_{212}) & (a_{121}, a_{213}) \\ (a_{121}, a_{211}) & (a_{112}, a_{121}) \end{array} \right\} = \left\{ \begin{array}{ll} \underbrace{(\text{UP}_-, \text{UP}_-)}_{\text{negative saddle}} & \underbrace{(\text{DP}_+, \text{DP}_-)}_{\text{CCW center}} \\ \underbrace{(_{\text{DI:II}}\text{UP}, _{\text{SI:SO}}\text{US})}_{\text{up-parabola upper-saddle}} & \underbrace{(\text{UP}:\text{DP}, \text{pF})}_{\text{hyperbolic-secant flow }(+)} \end{array} \right\}$$

for $a_{110} < 0, a_{220} > 0$,

$$(5.46)$$

$$\left\{ \begin{array}{ll} (a_{121}, a_{212}) & (a_{121}, a_{213}) \\ (a_{121}, a_{211}) & (a_{112}, a_{121}) \end{array} \right\} = \left\{ \begin{array}{ll} \underbrace{(\text{UP}_+, \text{UP}_+)}_{\text{psotive saddle}} & \underbrace{(\text{DP}_-, \text{DP}_+)}_{\text{CW center}} \\ \underbrace{(_{\text{DI:II}}\text{UP}, _{\text{SO:SI}}\text{LS})}_{\text{up-parabola lower-saddle}} & \underbrace{(\text{UP}:\text{DP}, \text{nF})}_{\text{hyperbolic-secant flow }(-)} \end{array} \right\}$$

for $a_{110} > 0, a_{220} < 0$,

$$(5.47)$$

$$\left\{ \begin{array}{ll} (a_{121}, a_{212}) & (a_{121}, a_{213}) \\ (a_{121}, a_{211}) & (a_{112}, a_{121}) \end{array} \right\} = \left\{ \begin{array}{ll} \underbrace{(\text{DP}_+, \text{DP}_-)}_{\text{CCW center}} & \underbrace{(\text{UP}_-, \text{UP}_-)}_{\text{postive saddle}} \\ \underbrace{(_{\text{II:DI}}\text{DP}, _{\text{SI:SO}}\text{US})}_{\text{down-parabola upper-saddle}} & \underbrace{(\text{DP}:\text{UP}, \text{nF})}_{\text{hyperbolic flow }(-)} \end{array} \right\} \quad (5.48)$$

for $a_{110} < 0, a_{220} < 0$;

and the infinite equilibrium of $x_1^* = a_{111} = a_{211}$ is summarized in Table 5.1.

The inflection-source and sink infinite-equilibriums of $x_1^* = a_{111} = a_{211}$ are presented in Fig. 5.11. The infinite-equilibriums are for the switching bifurcation for the equilibriums for $a_{111} < a_{211}$ and $a_{111} \in (a_{211}, a_{212})$ with $a_{112} < a_{212}$. The parabola-saddles on the infinite-equilibriums are for the switching bifurcations of saddle and center with hyperbolic flows at $(x_2^*, x_1^*) = (a_{121}, a_{211})$. The saddle and center with the infinite-equilibrium are separated by a hyperbolic or hyperbolic-secant flow.

(ii) For $a_{211} = a_{111} = a_{11l_1}$ with $a_{112} \in (a_{212}, a_{213})$, the equilibriums of $x_1^* = a_{111}$, a_{112}, a_{211}, a_{212}, a_{213} with $x_2^* = a_{121}$ have the following properties:

Table 5.1 Infinite equilibriums of $x_1^* = a_{111} = a_{211}$

$(a_{111}, \bar{x}_2)$	$\bar{x}_2$	
(a_{110}, a_{220})	$(-\infty, a_{121})$	(a_{121}, ∞)
$(+,+)$	(SO,II)	(SI,DI)
$(-,+)$	(SI,DI)	(SO,II)
$(+,-)$	(SO,DI)	(SI,II)
$(-,-)$	(SI,II)	(SO,DI)

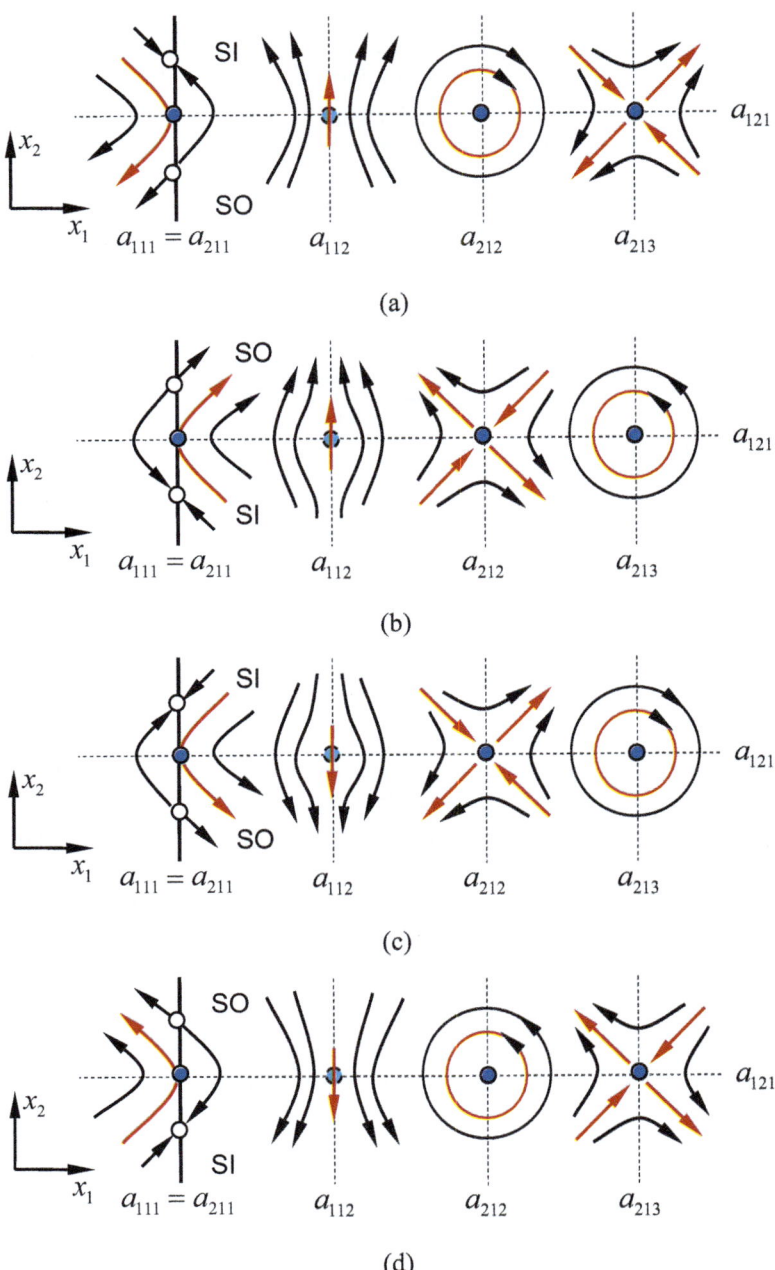

Fig. 5.11 Phase portraits ($a_{111} = a_{211}$ and $a_{112} < a_{212}$) for two-dimensional systems on the x_1-direction with $x_1^* = a_{111}, a_{112}, a_{211}, a_{212}, a_{213}$ and on the x_2-direction with $x_2^* = a_{121}$. (**a**) ($a_{110} > 0$, $a_{220} > 0$), (**b**) ($a_{110} < 0, a_{220} > 0$), (**c**) ($a_{110} > 0, a_{220} < 0$), (**d**) ($a_{110} < 0, a_{220} < 0$)

$$\left\{ \begin{matrix} (a_{112}, a_{121}) & (a_{121}, a_{213}) \\ (a_{121}, a_{211}) & (a_{121}, a_{212}) \end{matrix} \right\} = \left\{ \begin{matrix} \underbrace{(UP:DP, nF)}_{\text{hyperbolic-secant flow }(-)} & \underbrace{(UP_+, UP_+)}_{\text{postive saddle}} \\ \underbrace{(_{II:DI}DP, _{SO:SI}LS)}_{\text{down-parabola lower-saddle}} & \underbrace{(UP_-, UP_-)}_{\text{negative saddle}} \end{matrix} \right\} \quad (5.49)$$

for $a_{110} > 0, a_{220} > 0$,

$$\left\{ \begin{matrix} (a_{112}, a_{121}) & (a_{121}, a_{213}) \\ (a_{121}, a_{211}) & (a_{121}, a_{212}) \end{matrix} \right\} = \left\{ \begin{matrix} \underbrace{(DP:UP, nF)}_{\text{hyperbolic flow }(-)} & \underbrace{(DP_+, DP_-)}_{\text{CCW center}} \\ \underbrace{(_{DI:II}UP, _{SI:SO}US)}_{\text{up-parabola upper-saddle}} & \underbrace{(DP_-, DP_+)}_{\text{CW center}} \end{matrix} \right\} \quad (5.50)$$

for $a_{110} < 0, a_{220} > 0$,

$$\left\{ \begin{matrix} (a_{112}, a_{121}) & (a_{121}, a_{213}) \\ (a_{121}, a_{211}) & (a_{121}, a_{212}) \end{matrix} \right\} = \left\{ \begin{matrix} \underbrace{(DP:UP, pF)}_{\text{hyperbolic flow }(+)} & \underbrace{(DP_-, DP_+)}_{\text{CW center}} \\ \underbrace{(_{DI:II}UP, _{SO:SI}LS)}_{\text{up-parabola lower-saddle}} & \underbrace{(DP_+, DP_-)}_{\text{CCW center}} \end{matrix} \right\} \quad (5.51)$$

for $a_{110} > 0, a_{220} < 0$,

$$\left\{ \begin{matrix} (a_{112}, a_{121}) & (a_{121}, a_{213}) \\ (a_{121}, a_{211}) & (a_{121}, a_{212}) \end{matrix} \right\} = \left\{ \begin{matrix} \underbrace{(UP:DP, pF)}_{\text{hyperbolic-secant flow }(+)} & \underbrace{(UP_-, UP_-)}_{\text{negative saddle}} \\ \underbrace{(_{II:DI}DP, _{SI:SO}US)}_{\text{down-parabola upper-saddle}} & \underbrace{(UP_+, UP_+)}_{\text{postive saddle}} \end{matrix} \right\} \quad (5.52)$$

for $a_{110} < 0, a_{220} < 0$;

and the infinite-equilibrium of $x_1^* = a_{111} = a_{211}$ is summarized, which is the same as in Table 5.1. The inflection-source and sink infinite-equilibriums of $x_1^* = a_{111} = a_{211}$ are presented in Fig. 5.12. The infinite-equilibriums are for the switching bifurcation for the equilibriums for $a_{111} < a_{211}$ and $a_{111} \in (a_{211}, a_{212})$ with $a_{112} \in (a_{212}, a_{213})$. The parabola-saddles on the infinite-equilibriums are for the switching bifurcations of a saddle and centers with hyperbolic flows at $(x_2^*, x_1^*) = (a_{121}, a_{211})$. The infinite-equilibriums with a center and saddle are observed, and the infinite-equilibrium with simple equilibriums are separated by the hyperbolic or hyperbolic-secant flow with the corresponding directrix flow.

(iii) For $a_{211} = a_{111} = a_{11l_1}$ with $a_{213} < a_{112}$, the equilibriums of $x_1^* = a_{111}, a_{112}, a_{211}, a_{212}, a_{213}$ with $x_2^* = a_{121}$ have the following properties:

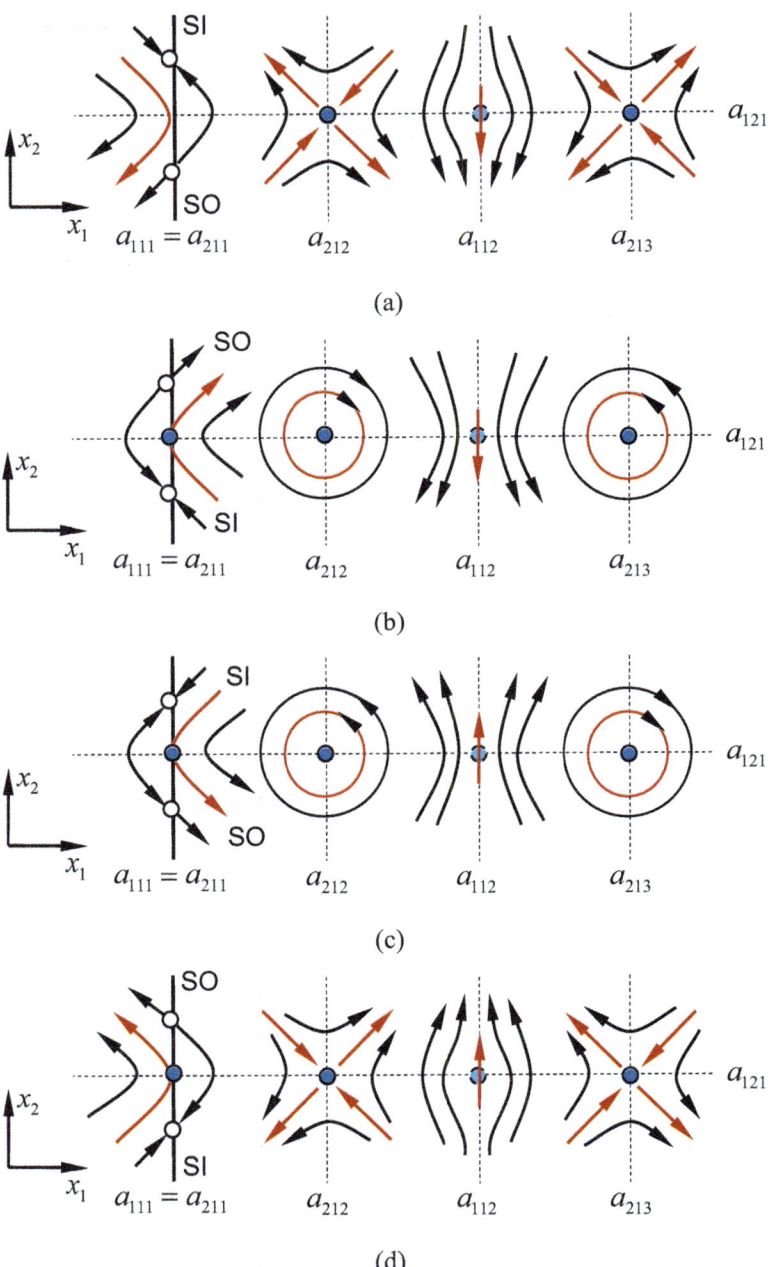

Fig. 5.12 Phase portraits ($a_{111} = a_{211}$ and $a_{212} < a_{112}$) for two-dimensional systems on the x_1-direction with $x_1^* = a_{111}, a_{112}, a_{211}, a_{212}, a_{213}$ and on the x_2-direction with $x_2^* = a_{121}$. (**a**) ($a_{110} > 0$, $a_{220} > 0$), (**b**) ($a_{110} < 0, a_{220} > 0$), (**c**) ($a_{110} > 0, a_{220} < 0$), (**d**) ($a_{110} < 0, a_{220} < 0$)

$$
\left\{
\begin{array}{ll}
(a_{121}, a_{213}) & (a_{112}, a_{121}) \\
(a_{121}, a_{211}) & (a_{121}, a_{212})
\end{array}
\right\}
=
\left\{
\begin{array}{ll}
\underbrace{(DP_+, DP_-)}_{\text{CCW center}} & \underbrace{(DP : UP, pF)}_{\text{hyperbolic flow (+)}} \\[4mm]
\underbrace{(_{\text{II:DI}}DP, _{\text{SO:SI}}LS)}_{\text{down-parabola lower-saddle}} & \underbrace{(UP_-, UP_-)}_{\text{negative saddle}}
\end{array}
\right\}
\tag{5.53}
$$

for $a_{110} > 0, a_{220} > 0$,

$$
\left\{
\begin{array}{ll}
(a_{121}, a_{213}) & (a_{112}, a_{121}) \\
(a_{121}, a_{211}) & (a_{121}, a_{212})
\end{array}
\right\}
=
\left\{
\begin{array}{ll}
\underbrace{(UP_+, UP_+)}_{\text{positive saddle}} & \underbrace{(UP : DP, pF)}_{\text{hyperbolic-seant flow (+)}} \\[4mm]
\underbrace{(_{\text{DI:II}}UP, _{\text{SI:SO}}US)}_{\text{up-parabola upper-saddle}} & \underbrace{(DP_-, DP_+)}_{\text{CW center}}
\end{array}
\right\}
\tag{5.54}
$$

for $a_{110} < 0, a_{220} > 0$,

$$
\left\{
\begin{array}{ll}
(a_{121}, a_{213}) & (a_{112}, a_{121}) \\
(a_{121}, a_{211}) & (a_{121}, a_{212})
\end{array}
\right\}
=
\left\{
\begin{array}{ll}
\underbrace{(UP_-, UP_-)}_{\text{negative saddle}} & \underbrace{(UP : DP, nF)}_{\text{hyperbolic-seant flow (−)}} \\[4mm]
\underbrace{(_{\text{DI:II}}UP, _{\text{SO:SI}}LS)}_{\text{up-parabola lower-saddle}} & \underbrace{(DP_+, DP_-)}_{\text{CCW center}}
\end{array}
\right\}
\tag{5.55}
$$

for $a_{110} > 0, a_{220} < 0$,

$$
\left\{
\begin{array}{ll}
(a_{121}, a_{213}) & (a_{112}, a_{121}) \\
(a_{121}, a_{211}) & (a_{121}, a_{212})
\end{array}
\right\}
=
\left\{
\begin{array}{ll}
\underbrace{(DP_-, DP_+)}_{\text{CW center}} & \underbrace{(DP : UP, nF)}_{\text{hyperbolic flow (−)}} \\[4mm]
\underbrace{(_{\text{II:DI}}DP, _{\text{SO:SI}}US)}_{\text{down-parabola upper-saddle}} & \underbrace{(UP_+, UP_+)}_{\text{postive saddle}}
\end{array}
\right\}
\tag{5.56}
$$

for $a_{110} < 0, a_{220} < 0$;

and the infinite-equilibrium of $x_1^* = a_{111} = a_{211}$ is summarized, which is the same as in Table 5.1. The inflection-source and sink infinite-equilibriums of $x_1^* = a_{111} = a_{211}$ are presented in Fig. 5.13. The infinite-equilibriums are for the switching bifurcation for the equilibriums for $a_{111} < a_{211}$ and $a_{111} \in (a_{211}, a_{212})$ with $a_{213} < a_{112}$. The parabola-saddles on the infinite-equilibriums are for the switching bifurcations of saddle and center with hyperbolic flows at $(x_2^*, x_1^*) = (a_{121}, a_{211})$. The infinite-equilibriums with two centers or two saddles are observed. However, two centers are separated by a hyperbolic flow, and two saddles are separated by a hyperbolic-secant flow.

(iv) For $a_{212} = a_{111} = a_{11l_1}$ with $a_{112} < a_{213}$, the equilibriums of $x_1^* = a_{111}, a_{112}, a_{211}$, a_{212}, a_{213} with $x_2^* = a_{121}$ have the following properties:

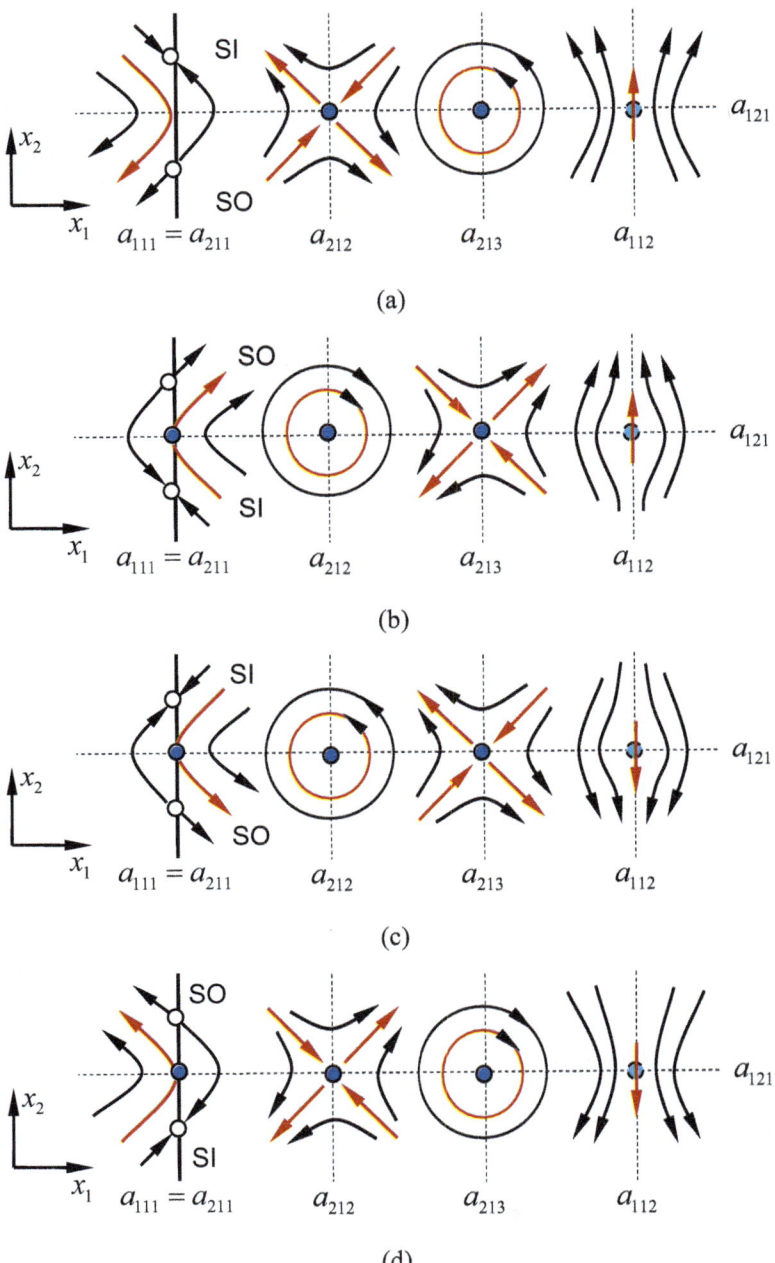

Fig. 5.13 Phase portraits ($a_{111} = a_{211}$ and $a_{213} < a_{112}$) for two-dimensional systems on the x_1-direction with $x_1^* = a_{111}, a_{112}, a_{211}, a_{212}, a_{213}$ and on the x_2-direction with $x_2^* = a_{121}$. (**a**) ($a_{110} > 0$, $a_{220} > 0$), (**b**) ($a_{110} < 0, a_{220} > 0$), (**c**) ($a_{110} > 0, a_{220} < 0$), (**d**) ($a_{110} < 0, a_{220} < 0$)

$$\left\{ \begin{array}{ll} (a_{112}, a_{121}) & (a_{121}, a_{213}) \\ (a_{121}, a_{211}) & (a_{121}, a_{212}) \end{array} \right\} = \left\{ \begin{array}{cc} \underbrace{(\text{UP} : \text{DP}, \text{nF})}_{\text{hyperbolic-secant flow } (-)} & \underbrace{(\text{UP}_+, \text{UP}_+)}_{\text{positive saddle}} \\ \underbrace{(\text{UP}_+, \text{UP}_+)}_{\text{positive saddle}} & \underbrace{\left(_{\text{DI:II}}\text{UP}, _{\text{SO:SI}}\text{LS}\right)}_{\text{up-parabola lower-saddle}} \end{array} \right\} \quad (5.57)$$

for $a_{110} > 0, a_{220} > 0$,

$$\left\{ \begin{array}{ll} (a_{112}, a_{121}) & (a_{121}, a_{213}) \\ (a_{121}, a_{211}) & (a_{121}, a_{212}) \end{array} \right\} = \left\{ \begin{array}{cc} \underbrace{(\text{DP} : \text{UP}, \text{nF})}_{\text{hyperbolic flow } (-)} & \underbrace{(\text{DP}_+, \text{DP}_-)}_{\text{CCW center}} \\ \underbrace{(\text{DP}_+, \text{DP}_-)}_{\text{CCW center}} & \underbrace{\left(_{\text{II:DI}}\text{DP}, _{\text{SI:SO}}\text{US}\right)}_{\text{down-parabola upper-saddle}} \end{array} \right\} \quad (5.58)$$

for $a_{110} < 0, a_{220} > 0$,

$$\left\{ \begin{array}{ll} (a_{112}, a_{121}) & (a_{121}, a_{213}) \\ (a_{121}, a_{211}) & (a_{121}, a_{212}) \end{array} \right\} = \left\{ \begin{array}{cc} \underbrace{(\text{DP} : \text{UP}, \text{pF})}_{\text{hyperbolic flow } (+)} & \underbrace{(\text{DP}_-, \text{DP}_+)}_{\text{CW center}} \\ \underbrace{(\text{DP}_-, \text{DP}_+)}_{\text{CW center}} & \underbrace{\left(_{\text{II:DI}}\text{DP}, _{\text{SO:SI}}\text{LS}\right)}_{\text{down-parabola lower-saddle}} \end{array} \right\} \quad (5.59)$$

for $a_{110} > 0, a_{220} < 0$,

$$\left\{ \begin{array}{ll} (a_{112}, a_{121}) & (a_{121}, a_{213}) \\ (a_{121}, a_{211}) & (a_{121}, a_{212}) \end{array} \right\} = \left\{ \begin{array}{cc} \underbrace{(\text{UP} : \text{DP}, \text{pF})}_{\text{hyperbolic-secant flow } (+)} & \underbrace{(\text{UP}_-, \text{UP}_-)}_{\text{negative saddle}} \\ \underbrace{(\text{UP}_-, \text{UP}_-)}_{\text{negative saddle}} & \underbrace{\left(_{\text{DI:II}}\text{UP}, _{\text{SI:SO}}\text{US}\right)}_{\text{up-parabola upper-saddle}} \end{array} \right\} \quad (5.60)$$

for $a_{110} < 0, a_{220} < 0$;

and the infinite-equilibrium of $x_1^* = a_{111} = a_{212}$ is summarized in Table 5.2.

The inflection-source and sink infinite-equilibriums of $x_1^* = a_{111} = a_{212}$ are presented in Fig. 5.14. The infinite-equilibriums are for the switching bifurcation for the equilibriums for $a_{111} \in (a_{211}, a_{212})$ and $a_{112} < a_{213}$ with $a_{211} < a_{111}$. Two centers are separated by the infinite-equilibrium with a hyperbolic or hyperbolic-secant flow.

Table 5.2 Infinite-equilibriums of $x_1^* = a_{111} = a_{212}$

$(a_{111}, \bar{x}_2)$	$\bar{x}_2$	
(a_{110}, a_{220})	$(-\infty, a_{121})$	(a_{121}, ∞)
$(+, +)$	(SO,DI)	(SI,II)
$(-, +)$	(SI,II)	(SO,DI)
$(+, -)$	(SO,II)	(SI,DI)
$(-, -)$	(SI,DI)	(SO,II)

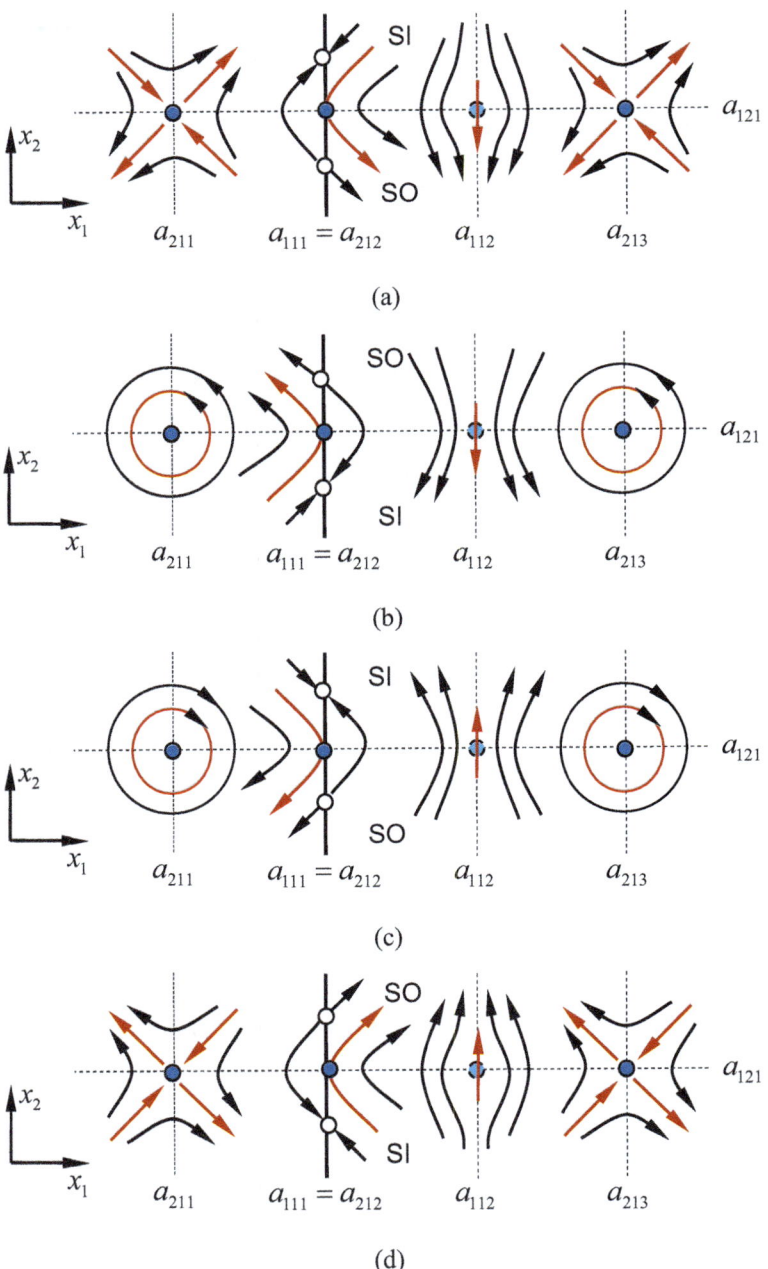

Fig. 5.14 Phase portraits ($a_{111} = a_{211}$ and $a_{212} < a_{112}$) for two-dimensional systems on the x_1-direction with $x_1^* = a_{111}, a_{112}, a_{211}, a_{212}, a_{213}$ and on the x_2-direction with $x_2^* = a_{121}$. (**a**) ($a_{110} > 0$, $a_{220} > 0$), (**b**) ($a_{110} < 0, a_{220} > 0$), (**c**) ($a_{110} > 0, a_{220} < 0$), (**d**) ($a_{110} < 0, a_{220} < 0$)

(v) For $a_{212} = a_{111} = a_{11l_1}$ with $a_{213} < a_{112}$, the equilibriums of $x_1^* = a_{111}, a_{112}, a_{211},$ a_{212}, a_{213} with $x_2^* = a_{121}$ have the following properties:

$$\left\{\begin{array}{ll} (a_{121}, a_{213}) & (a_{112}, a_{121}) \\ (a_{121}, a_{211}) & (a_{121}, a_{212}) \end{array}\right\} = \left\{\begin{array}{ll} \underbrace{(DP_+, DP_-)}_{\text{CCW center}} & \underbrace{(DP : UP, pF)}_{\text{hyperbolic flow (+)}} \\ \underbrace{(UP_+, UP_+)}_{\text{positive saddle}} & \underbrace{(_{DI:II}UP, _{SO:SI}LS)}_{\text{up-parabola lower-saddle}} \end{array}\right\} \quad (5.61)$$

for $a_{110} > 0, a_{220} > 0,$

$$\left\{\begin{array}{ll} (a_{121}, a_{213}) & (a_{112}, a_{121}) \\ (a_{121}, a_{211}) & (a_{121}, a_{212}) \end{array}\right\} = \left\{\begin{array}{ll} \underbrace{(UP_+, UP_+)}_{\text{positive saddle}} & \underbrace{(UP : DP, pF)}_{\text{hyperbolic-secant flow (+)}} \\ \underbrace{(DP_+, DP_-)}_{\text{CCW center}} & \underbrace{(_{II:DI}DP, _{SI:SO}US)}_{\text{down-parabola upper-saddle}} \end{array}\right\} \quad (5.62)$$

for $a_{110} < 0, a_{220} > 0,$

$$\left\{\begin{array}{ll} (a_{121}, a_{213}) & (a_{112}, a_{121}) \\ (a_{121}, a_{211}) & (a_{121}, a_{212}) \end{array}\right\} = \left\{\begin{array}{ll} \underbrace{(UP_-, UP_-)}_{\text{negative saddle}} & \underbrace{(UP : DP, nF)}_{\text{hyperbolic-secant flow (-)}} \\ \underbrace{(DP_-, DP_+)}_{\text{CW center}} & \underbrace{(_{II:DI}DP, _{SO:SI}LS)}_{\text{down-parabola lower-saddle}} \end{array}\right\} \quad (5.63)$$

for $a_{110} > 0, a_{220} < 0,$

$$\left\{\begin{array}{ll} (a_{121}, a_{213}) & (a_{112}, a_{121}) \\ (a_{121}, a_{211}) & (a_{121}, a_{212}) \end{array}\right\} = \left\{\begin{array}{ll} \underbrace{(DP_-, DP_+)}_{\text{CW center}} & \underbrace{(DP : UP, nF)}_{\text{hyperbolic flow (-)}} \\ \underbrace{(UP_-, UP_-)}_{\text{negative saddle}} & \underbrace{(_{DI:II}UP, _{SI:SO}US)}_{\text{up-parabola upper-saddle}} \end{array}\right\} \quad (5.64)$$

for $a_{110} < 0, a_{220} < 0$;

and the infinite-equilibrium of $x_1^* = a_{111} = a_{212}$ is summarized in Table 5.2. The inflection-source and sink infinite-equilibriums of $x_1^* = a_{111} = a_{212}$ are presented in Fig. 5.15. The infinite-equilibriums are for the switching bifurcation for the equilibriums for $a_{111} \in (a_{211}, a_{212})$ and $a_{111} \in (a_{212}, a_{213})$ with $a_{213} < a_{112}$. The center and saddle are separated by an infinite-equilibrium.

(vi) For $a_{213} = a_{111} = a_{11l_1}$, the equilibriums of $x_1^* = a_{111}, a_{112}, a_{211}, a_{212}, a_{213}$ with $x_2^* = a_{121}$ have the following properties:

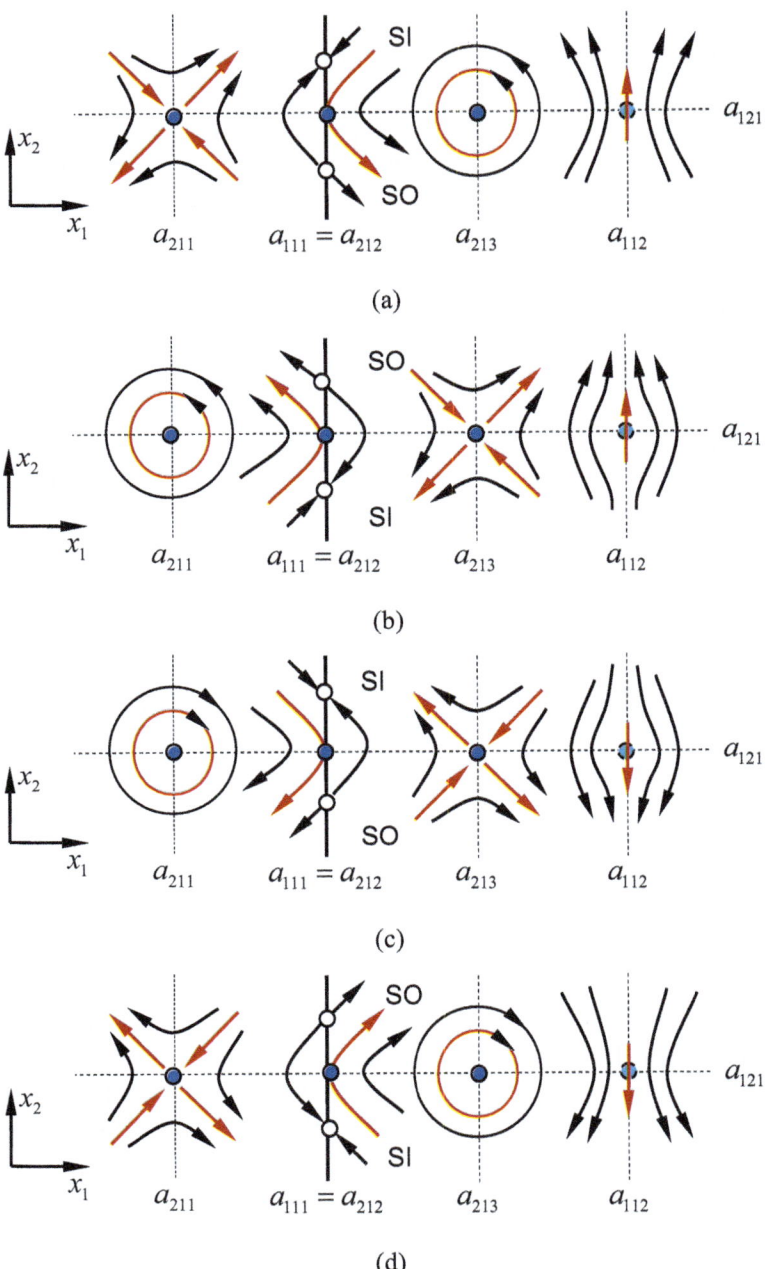

Fig. 5.15 Phase portraits ($a_{111} = a_{211}$ and $a_{213} < a_{112}$) for two-dimensional systems on the x_1-direction with $x_1^* = a_{111}, a_{112}, a_{211}, a_{212}, a_{213}$ and on the x_2-direction with $x_2^* = a_{121}$. (**a**) ($a_{110} > 0$, $a_{220} > 0$), (**b**) ($a_{110} < 0, a_{220} > 0$), (**c**) ($a_{110} > 0, a_{220} < 0$), (**d**) ($a_{110} < 0, a_{220} < 0$)

$$\left\{ \begin{array}{ll} (a_{121}, a_{213}) & (a_{112}, a_{121}) \\ (a_{121}, a_{211}) & (a_{121}, a_{212}) \end{array} \right\} = \left\{ \begin{array}{ll} \underbrace{(_{\text{II:DI}}\text{DP, }_{\text{SO:SI}}\text{LS})}_{\text{down-parabola lower-saddle}} & \underbrace{(\text{DP : UP, pF})}_{\text{hyperbolic flow (+)}} \\ \underbrace{(\text{UP}_+, \text{UP}_+)}_{\text{positive saddle}} & \underbrace{(\text{DP}_-, \text{DP}_+)}_{\text{CW center}} \end{array} \right\} \quad (5.65)$$

for $a_{110} > 0, a_{220} > 0,$

$$\left\{ \begin{array}{ll} (a_{121}, a_{213}) & (a_{112}, a_{121}) \\ (a_{121}, a_{211}) & (a_{121}, a_{212}) \end{array} \right\} = \left\{ \begin{array}{ll} \underbrace{(_{\text{DI:II}}\text{UP, }_{\text{SI:SO}}\text{US})}_{\text{up-parabola upper-saddle}} & \underbrace{(\text{UP : DP, pF})}_{\text{hyperbolic-secant flow (+)}} \\ \underbrace{(\text{DP}_+, \text{DP}_-)}_{\text{CCW center}} & \underbrace{(\text{UP}_-, \text{UP}_-)}_{\text{negative saddle}} \end{array} \right\} \quad (5.66)$$

for $a_{110} < 0, a_{220} > 0,$

$$\left\{ \begin{array}{ll} (a_{121}, a_{213}) & (a_{112}, a_{121}) \\ (a_{121}, a_{211}) & (a_{121}, a_{212}) \end{array} \right\} = \left\{ \begin{array}{ll} \underbrace{(_{\text{DI:II}}\text{UP, }_{\text{SO:SI}}\text{LS})}_{\text{up-parabola lower-saddle}} & \underbrace{(\text{UP : DP, nF})}_{\text{hyperbolic-secant flow (−)}} \\ \underbrace{(\text{DP}_-, \text{DP}_+)}_{\text{CW center}} & \underbrace{(\text{UP}_+, \text{UP}_+)}_{\text{positive saddle}} \end{array} \right\} \quad (5.67)$$

for $a_{110} > 0, a_{220} < 0,$

$$\left\{ \begin{array}{ll} (a_{121}, a_{213}) & (a_{112}, a_{121}) \\ (a_{121}, a_{211}) & (a_{121}, a_{212}) \end{array} \right\} = \left\{ \begin{array}{ll} \underbrace{(_{\text{II:DI}}\text{DP, }_{\text{SI:SO}}\text{US})}_{\text{down-parabola upper-saddle}} & \underbrace{(\text{DP : UP, nF})}_{\text{hyperbolic flow (−)}} \\ \underbrace{(\text{UP}_-, \text{UP}_-)}_{\text{negative saddle}} & \underbrace{(\text{DP}_+, \text{DP}_-)}_{\text{CCW center}} \end{array} \right\} \quad (5.68)$$

for $a_{110} < 0, a_{220} < 0;$

and the infinite-equilibrium of $x_1^* = a_{111} = a_{213}$ is summarized in Table 5.3.

The inflection-source and sink infinite-equilibriums of $x_1^* = a_{111} = a_{213}$ are presented in Fig. 5.16. The infinite-equilibriums are for the switching bifurcation for the equilibriums for $a_{111} \in (a_{212}, a_{213})$ and $a_{213} < a_{111}$ with $a_{212} < a_{111}$. The simple equilibrium and hyperbolic or hyperbolic-secant flow are separated by an infinite-equilibrium.

(vii) For $a_{211} = a_{112} = a_{11l_1}$, the equilibriums of $x_1^* = a_{111}, a_{112}, a_{211}, a_{212}, a_{213}$ with $x_2^* = a_{121}$ have the following properties:

Table 5.3 Infinite-equilibriums of $x_1^* = a_{111} = a_{213}$	$(a_{111}, \bar{x}_2)$	$\bar{x}_2$	
	(a_{110}, a_{220})	$(-\infty, a_{121})$	(a_{121}, ∞)
	$(+,+)$	(SO,II)	(SI,DI)
	$(-,+)$	(SI,DI)	(SO,II)
	$(+,-)$	(SO,DI)	(SI,II)
	$(-,-)$	(SI,II)	(SO,DI)

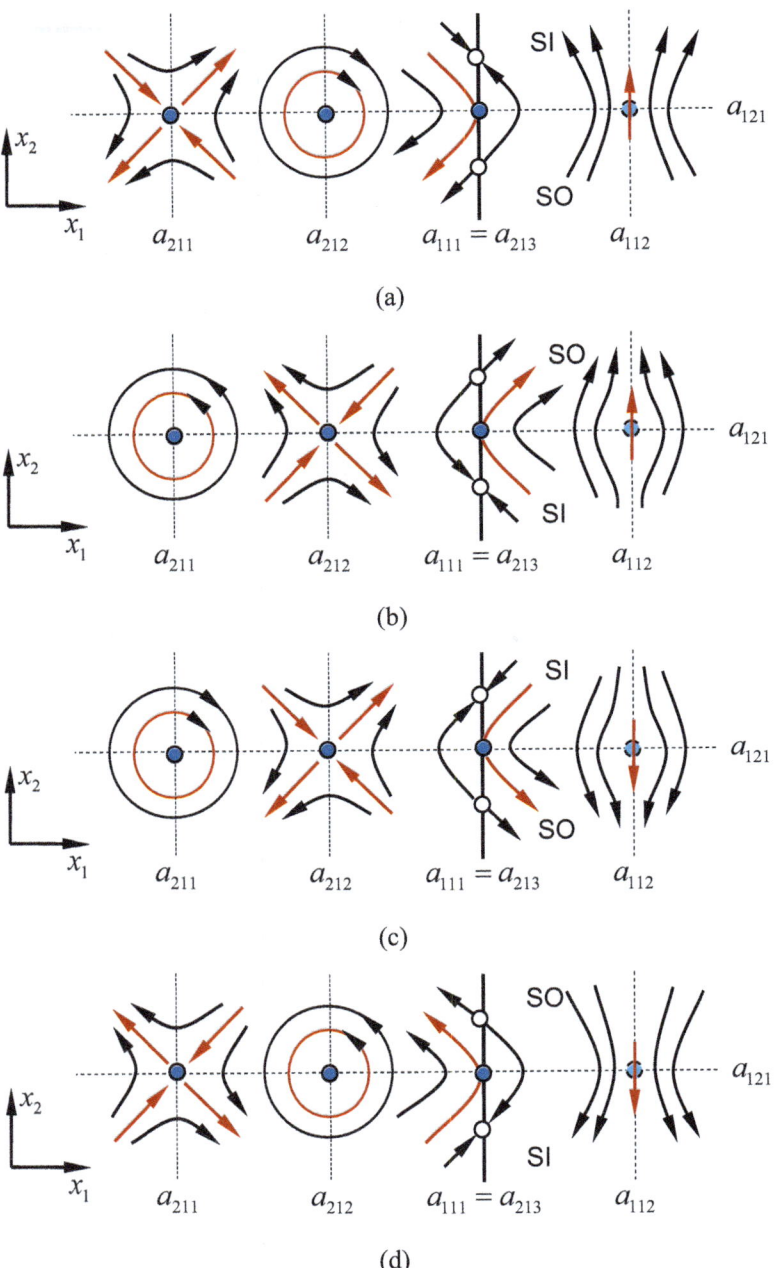

Fig. 5.16 Phase portraits ($a_{111} = a_{213}$) for two-dimensional systems on the x_1-direction with $x_1^* = a_{111}, a_{112}, a_{211}, a_{212}, a_{213}$ and on the x_2-direction with $x_2^* = a_{121}$. (**a**) ($a_{110} > 0, a_{220} > 0$), (**b**) ($a_{110} < 0, a_{220} > 0$), (**c**) ($a_{110} > 0, a_{220} < 0$), (**d**) ($a_{110} < 0, a_{220} < 0$)

$$\left\{ \begin{matrix} (a_{121}, a_{212}) & (a_{121}, a_{213}) \\ (a_{111}, a_{121}) & (a_{121}, a_{211}) \end{matrix} \right\} = \left\{ \begin{array}{cc} \underbrace{(DP_-, DP_+)}_{CW\ center} & \underbrace{(UP_+, UP_+)}_{positive\ saddle} \\ \underbrace{(DP:UP, nF)}_{hyperbolic\ flow\ (-)} & \underbrace{({}_{DI:II}UP, {}_{SI:SO}US)}_{up\text{-}parabola\ upper\text{-}saddle} \end{array} \right\} \quad (5.69)$$

for $a_{110} > 0, a_{220} > 0,$

$$\left\{ \begin{matrix} (a_{121}, a_{212}) & (a_{121}, a_{213}) \\ (a_{111}, a_{121}) & (a_{121}, a_{211}) \end{matrix} \right\} = \left\{ \begin{array}{cc} \underbrace{(UP_-, UP_-)}_{negative\ saddle} & \underbrace{(DP_+, DP_-)}_{CCW\ center} \\ \underbrace{(UP:DP, nF)}_{hyperbolic\text{-}secant\ flow\ (-)} & \underbrace{({}_{II:DI}DP, {}_{SO:SI}LS)}_{down\text{-}parabola\ lower\text{-}saddle} \end{array} \right\} \quad (5.70)$$

for $a_{110} < 0, a_{220} > 0,$

$$\left\{ \begin{matrix} (a_{121}, a_{213}) & (a_{112}, a_{121}) \\ (a_{121}, a_{211}) & (a_{121}, a_{212}) \end{matrix} \right\} = \left\{ \begin{array}{cc} \underbrace{(UP_+, UP_+)}_{positive\ saddle} & \underbrace{(DP_-, DP_+)}_{CW\ center} \\ \underbrace{(UP:DP, pF)}_{hyperbolic\text{-}secant\ flow\ (+)} & \underbrace{({}_{II:DI}DP, {}_{SI:SO}US)}_{down\text{-}parabola\ upper\text{-}saddle} \end{array} \right\} \quad (5.71)$$

for $a_{110} > 0, a_{220} < 0,$

$$\left\{ \begin{matrix} (a_{121}, a_{212}) & (a_{121}, a_{213}) \\ (a_{111}, a_{121}) & (a_{121}, a_{211}) \end{matrix} \right\} = \left\{ \begin{array}{cc} \underbrace{(DP_+, DP_-)}_{CCW\ center} & \underbrace{(UP_-, UP_-)}_{positive\ saddle} \\ \underbrace{(DP:UP, pF)}_{hyperbolic\ flow\ (+)} & \underbrace{({}_{DI:II}UP, {}_{SO:SI}LS)}_{up\text{-}parabola\ lower\text{-}saddle} \end{array} \right\} \quad (5.72)$$

for $a_{110} < 0, a_{220} < 0;$

and the infinite-equilibrium of $x_1^* = a_{112} = a_{211}$ is summarized in Table 5.4.

The inflection-source and sink infinite-equilibriums of $x_1^* = a_{112} = a_{211}$ are presented in Fig. 5.17. The infinite-equilibriums are for the switching bifurcation for the equilibriums for $a_{112} \in (a_{211}, a_{212})$ and $a_{112} < a_{211}$. Simple saddle and center equilibriums with a hyperbolic or hyperbolic-secant flow are separated by an infinite-equilibrium.

(viii) For $a_{212} = a_{112} = a_{11l_1}$ with $a_{111} < a_{211}$, the equilibriums of $x_1^* = a_{111}, a_{112},$ $a_{211}, a_{212}, a_{213}$ with $x_2^* = a_{121}$ have the following properties:

Table 5.4 Infinite-equilibriums of $x_1^* = a_{211} = a_{112}$

$(a_{112}, \bar{x}_2)$	$\bar{x}_2$	
(a_{110}, a_{220})	$(-\infty, a_{121})$	(a_{121}, ∞)
$(+,+)$	(SI,DI)	(SO,II)
$(-,+)$	(SO,II)	(SI,DI)
$(+,-)$	(SI,II)	(SO,DI)
$(-,-)$	(SO,DI)	(SI,II)

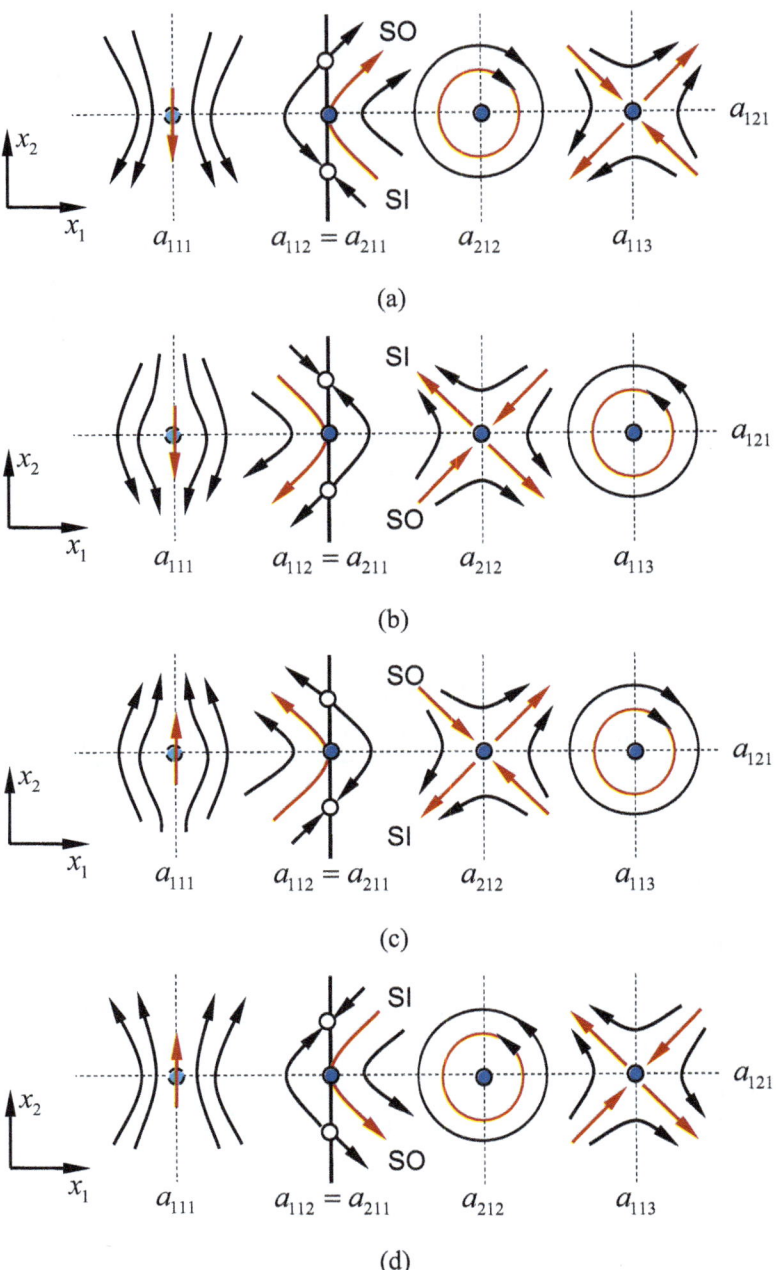

Fig. 5.17 Phase portraits ($a_{112} = a_{211}$) for two-dimensional systems on the x_1-direction with $x_1^* = a_{111}, a_{112}, a_{211}, a_{212}, a_{213}$ and on the x_2-direction with $x_2^* = a_{121}$. (**a**) ($a_{110} > 0, a_{220} > 0$), (**b**) ($a_{110} < 0, a_{220} > 0$), (**c**) ($a_{110} > 0, a_{220} < 0$), (**d**) ($a_{110} < 0, a_{220} < 0$)

$$\left\{ \begin{array}{cc} (a_{121}, a_{212}) & (a_{121}, a_{213}) \\ (a_{111}, a_{121}) & (a_{121}, a_{211}) \end{array} \right\} = \left\{ \begin{array}{cc} \underbrace{(_{\text{II:DI}}DP, \text{ }_{\text{SI:SO}}US)}_{\text{down-parabola upper-saddle}} & \underbrace{(UP_+, UP_+)}_{\text{positive saddle}} \\ \underbrace{(DP : UP, nF)}_{\text{hyperbolic flow } (-)} & \underbrace{(DP_+, DP_-)}_{\text{CCW center}} \end{array} \right\} \quad (5.73)$$

for $a_{110} > 0, a_{220} > 0,$

$$\left\{ \begin{array}{cc} (a_{121}, a_{212}) & (a_{121}, a_{213}) \\ (a_{111}, a_{121}) & (a_{121}, a_{211}) \end{array} \right\} = \left\{ \begin{array}{cc} \underbrace{(_{\text{DI:II}}UP, \text{ }_{\text{SO:SI}}LS)}_{\text{up-parabola lower-saddle}} & \underbrace{(DP_+, DP_-)}_{\text{CCW center}} \\ \underbrace{(UP : DP, nF)}_{\text{hyperbolic-secant flow } (-)} & \underbrace{(UP_+, UP_+)}_{\text{positive saddle}} \end{array} \right\} \quad (5.74)$$

for $a_{110} < 0, a_{220} > 0,$

$$\left\{ \begin{array}{cc} (a_{121}, a_{212}) & (a_{121}, a_{213}) \\ (a_{111}, a_{121}) & (a_{121}, a_{211}) \end{array} \right\} = \left\{ \begin{array}{cc} \underbrace{(_{\text{DI:II}}UP, \text{ }_{\text{SI:SO}}US)}_{\text{up-parabola upper-saddle}} & \underbrace{(DP_-, DP_+)}_{\text{CW center}} \\ \underbrace{(UP : DP, pF)}_{\text{hyperbolic-secant flow } (+)} & \underbrace{(UP_-, UP_-)}_{\text{negative saddle}} \end{array} \right\} \quad (5.75)$$

for $a_{110} > 0, a_{220} < 0,$

$$\left\{ \begin{array}{cc} (a_{121}, a_{212}) & (a_{121}, a_{213}) \\ (a_{111}, a_{121}) & (a_{121}, a_{211}) \end{array} \right\} = \left\{ \begin{array}{cc} \underbrace{(_{\text{II:DI}}DP, \text{ }_{\text{SO:SI}}LS)}_{\text{down-parabola lower-saddle}} & \underbrace{(UP_-, UP_-)}_{\text{negative saddle}} \\ \underbrace{(DP : UP, pF)}_{\text{hyperbolic flow } (+)} & \underbrace{(DP_-, DP_+)}_{\text{CW center}} \end{array} \right\} \quad (5.76)$$

for $a_{110} < 0, a_{220} < 0;$

and the infinite-equilibrium of $x_1^* = a_{112} = a_{212}$ is summarized in Table 5.5.

The inflection-source and sink infinite-equilibriums of $x_1^* = a_{112} = a_{212}$ are presented in Fig. 5.18. The infinite-equilibriums are for the switching bifurcation for the equilibriums for $a_{112} \in (a_{211}, a_{212})$ and $a_{112} \in (a_{212}, a_{213})$ with $a_{111} < a_{211}$. The simple saddle and center equilibriums are separated by an infinite-equilibrium.

(ix) For $a_{212} = a_{112} = a_{11l_1}$ with $a_{211} < a_{111}$, the equilibriums of $x_1^* = a_{111}, a_{112}$, $a_{211}, a_{212}, a_{213}$ with $x_2^* = a_{121}$ have the following properties:

Table 5.5 Infinite-equilibriums of $x_1^* = a_{212} = a_{112}$

$(a_{112}, \bar{x}_2)$	$\bar{x}_2$	
(a_{110}, a_{220})	$(-\infty, a_{121})$	(a_{121}, ∞)
$(+,+)$	(SI,II)	(SO,DI)
$(-,+)$	(SO,DI)	(SI,II)
$(+,-)$	(SI,DI)	(SO,II)
$(-,-)$	(SO,II)	(SI,DI)

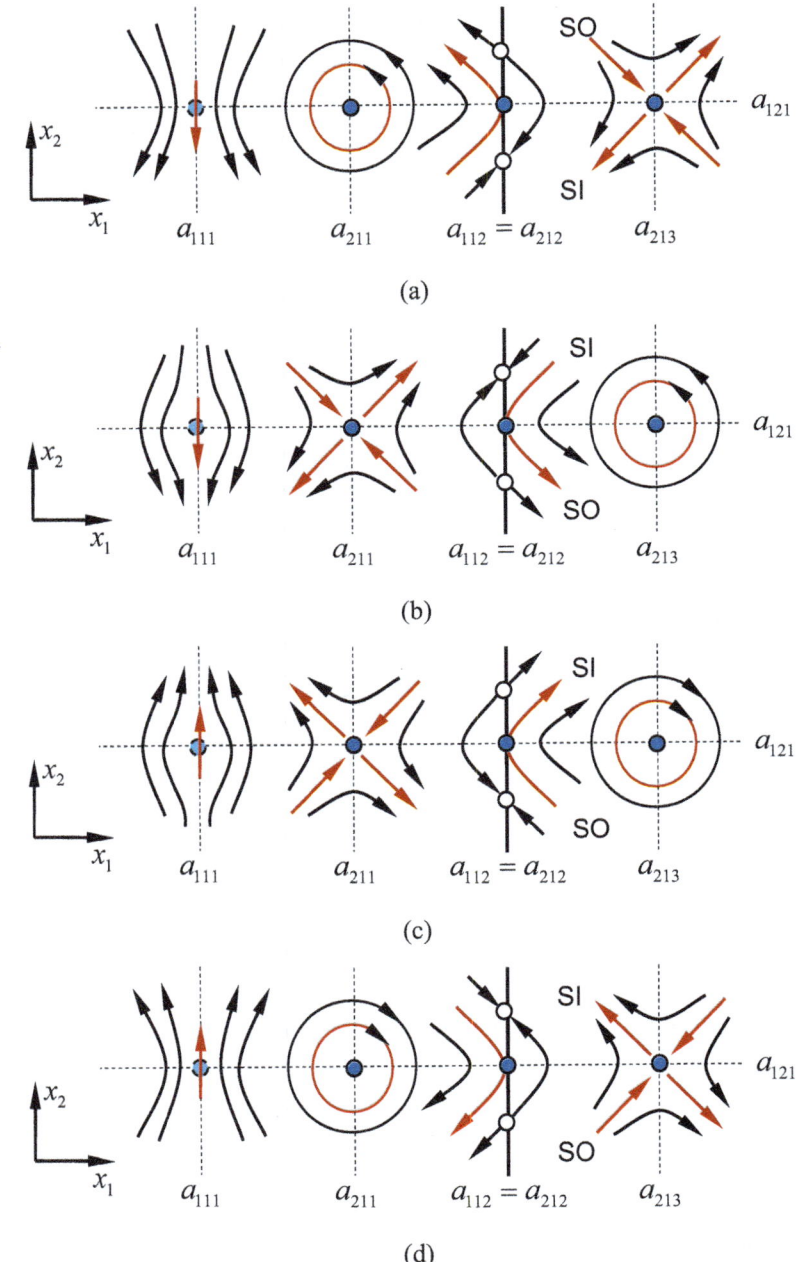

Fig. 5.18 Phase portraits ($a_{112} = a_{212}$ and $a_{111} < a_{211}$) for two-dimensional systems on the x_1-direction with $x_1^* = a_{111}, a_{112}, a_{211}, a_{212}, a_{213}$ and on the x_2-direction with $x_2^* = a_{121}$. (**a**) ($a_{110} > 0$, $a_{220} > 0$), (**b**) ($a_{110} < 0, a_{220} > 0$), (**c**) ($a_{110} > 0, a_{220} < 0$), (**d**) ($a_{110} < 0, a_{220} < 0$)

$$
\left\{ \begin{matrix} (a_{121}, a_{212}) & (a_{121}, a_{213}) \\ (a_{121}, a_{211}) & (a_{111}, a_{121}) \end{matrix} \right\} = \left\{ \begin{matrix} \underbrace{(_{\text{II:DI}}\text{DP}, _{\text{SI:SO}}\text{US})}_{\text{down-parabola upper-saddle}} & \underbrace{(\text{UP}_+, \text{UP}_+)}_{\text{positive saddle}} \\ \underbrace{(\text{UP}_+, \text{UP}_+)}_{\text{positive saddle}} & \underbrace{(\text{UP}:\text{DP}, \text{pF})}_{\text{hyperbolic-secant flow }(+)} \end{matrix} \right\} \tag{5.77}
$$

for $a_{110} > 0, a_{220} > 0,$

$$
\left\{ \begin{matrix} (a_{121}, a_{212}) & (a_{121}, a_{213}) \\ (a_{121}, a_{211}) & (a_{111}, a_{121}) \end{matrix} \right\} = \left\{ \begin{matrix} \underbrace{(_{\text{DI:II}}\text{UP}, _{\text{SO:SI}}\text{LS})}_{\text{up-parabola lower-saddle}} & \underbrace{(\text{DP}_+, \text{DP}_-)}_{\text{CCW center}} \\ \underbrace{(\text{DP}_+, \text{DP}_-)}_{\text{CCW center}} & \underbrace{(\text{DP}:\text{UP}, \text{pF})}_{\text{hyperbolic flow }(+)} \end{matrix} \right\} \tag{5.78}
$$

for $a_{110} < 0, a_{220} > 0,$

$$
\left\{ \begin{matrix} (a_{121}, a_{212}) & (a_{121}, a_{213}) \\ (a_{121}, a_{211}) & (a_{111}, a_{121}) \end{matrix} \right\} = \left\{ \begin{matrix} \underbrace{(_{\text{DI:II}}\text{UP}, _{\text{SI:SO}}\text{US})}_{\text{up-parabola upper-saddle}} & \underbrace{(\text{DP}_-, \text{DP}_+)}_{\text{CW center}} \\ \underbrace{(\text{DP}_-, \text{DP}_+)}_{\text{CW center}} & \underbrace{(\text{DP}:\text{UP}, \text{nF})}_{\text{hyperbolic flow }(-)} \end{matrix} \right\} \tag{5.79}
$$

for $a_{110} > 0, a_{220} < 0,$

$$
\left\{ \begin{matrix} (a_{121}, a_{212}) & (a_{121}, a_{213}) \\ (a_{121}, a_{211}) & (a_{111}, a_{121}) \end{matrix} \right\} = \left\{ \begin{matrix} \underbrace{(_{\text{II:DI}}\text{DP}, _{\text{SO:SI}}\text{LS})}_{\text{down-parabola lower-saddle}} & \underbrace{(\text{UP}_-, \text{UP}_-)}_{\text{positive saddle}} \\ \underbrace{(\text{UP}_-, \text{UP}_-)}_{\text{positive saddle}} & \underbrace{(\text{UP}:\text{DP}, \text{nF})}_{\text{hyperbolic-secant flow }(-)} \end{matrix} \right\} \tag{5.80}
$$

for $a_{110} < 0, a_{220} < 0;$

and the infinite-equilibrium of $x_1^* = a_{112} = a_{212}$ is summarized as in Table 5.5. The inflection-source and sink infinite-equilibriums of $x_1^* = a_{112} = a_{212}$ are presented in Fig. 5.19. The infinite-equilibriums are for the switching bifurcation for the equilibriums for $a_{112} \in (a_{211}, a_{212})$ and $a_{112} \in (a_{212}, a_{213})$ with $a_{211} < a_{111}$. The simple saddle and center equilibriums are separated by an infinite-equilibrium.

(x) For $a_{213} = a_{112} = a_{11l_1}$ with $a_{111} < a_{211}$, the equilibriums of $x_1^* = a_{111}, a_{112}, a_{211},$ a_{212}, a_{213} with $x_2^* = a_{121}$ have the following properties:

$$
\left\{ \begin{matrix} (a_{121}, a_{212}) & (a_{121}, a_{213}) \\ (a_{111}, a_{121}) & (a_{121}, a_{211}) \end{matrix} \right\} = \left\{ \begin{matrix} \underbrace{(\text{UP}_-, \text{UP}_-)}_{\text{negative saddle}} & \underbrace{(_{\text{DI:II}}\text{UP}, _{\text{SI:SO}}\text{US})}_{\text{up-parabola upper-saddle}} \\ \underbrace{(\text{DP}:\text{UP}, \text{nF})}_{\text{hyperbolic flow }(-)} & \underbrace{(\text{DP}_+, \text{DP}_-)}_{\text{CCW center}} \end{matrix} \right\} \tag{5.81}
$$

for $a_{110} > 0, a_{220} > 0,$

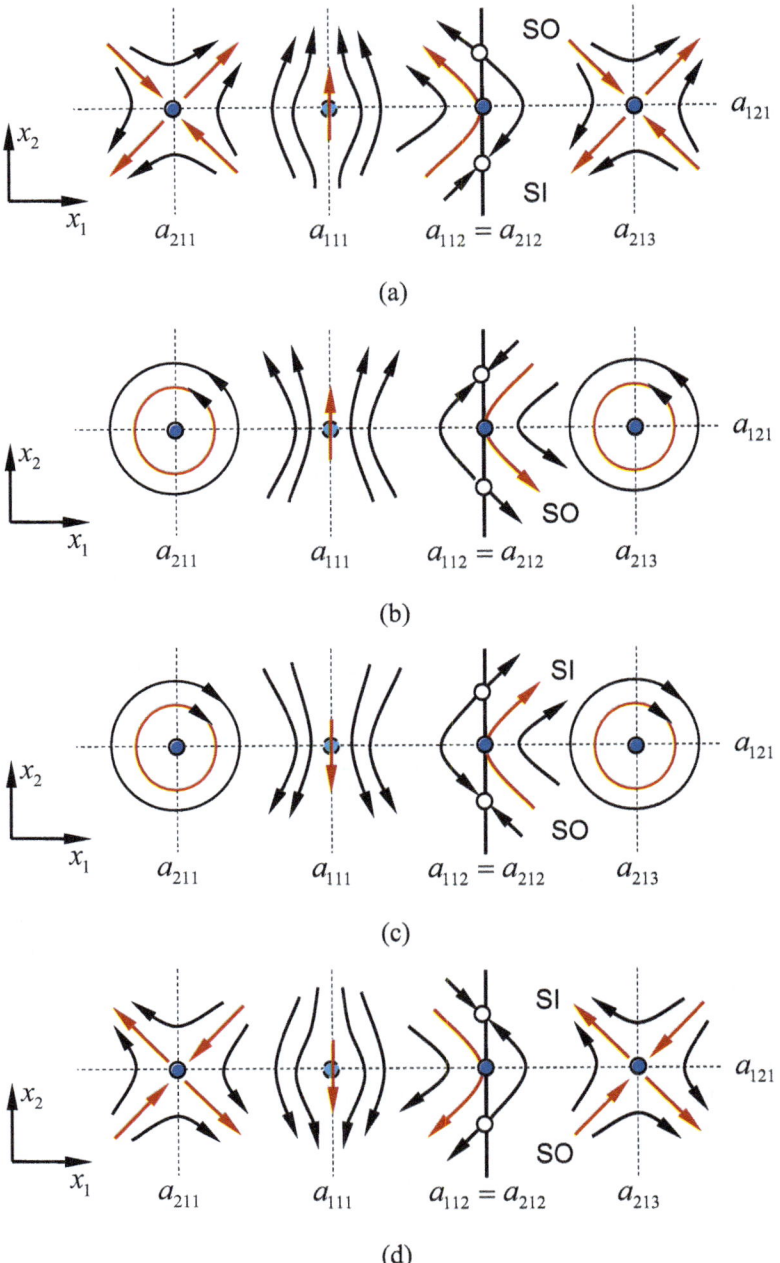

Fig. 5.19 Phase portraits ($a_{112} = a_{212}$, $a_{211} < a_{111}$) for two-dimensional systems on the x_1-direction with $x_1^* = a_{111}, a_{112}, a_{211}, a_{212}, a_{213}$ and on the x_2-direction with $x_2^* = a_{121}$. (**a**) ($a_{110} > 0$, $a_{220} > 0$), (**b**) ($a_{110} < 0, a_{220} > 0$), (**c**) ($a_{110} > 0, a_{220} < 0$), (**d**) ($a_{110} < 0, a_{220} < 0$)

$$\left.\begin{cases}(a_{121},a_{212}) & (a_{121},a_{213})\\ (a_{111},a_{121}) & (a_{121},a_{211})\end{cases}\right\} = \left\{\begin{array}{cc}\underbrace{(DP_-,DP_+)}_{CW\ center} & \underbrace{(_{II:DI}DP,_{SO:SI}LS)}_{down\text{-}parabola\ lower\text{-}saddle}\\[4pt] \underbrace{(UP:DP,nF)}_{hyperbolic\text{-}secant\ flow\ (-)} & \underbrace{(UP_+,UP_+)}_{positive\ saddle}\end{array}\right\} \quad (5.82)$$

for $a_{110}<0,a_{220}>0,$

$$\left.\begin{cases}(a_{121},a_{212}) & (a_{121},a_{213})\\ (a_{111},a_{121}) & (a_{121},a_{211})\end{cases}\right\} = \left\{\begin{array}{cc}\underbrace{(DP_+,DP_-)}_{CCW\ center} & \underbrace{(_{II:DI}DP,_{SI:SO}US)}_{down\text{-}parabola\ upper\text{-}saddle}\\[4pt] \underbrace{(UP:DP,pF)}_{hyperbolic\text{-}secant\ flow\ (+)} & \underbrace{(UP_-,UP_-)}_{negative\ saddle}\end{array}\right\} \quad (5.83)$$

for $a_{110}>0,a_{220}<0,$

$$\left.\begin{cases}(a_{121},a_{212}) & (a_{121},a_{213})\\ (a_{111},a_{121}) & (a_{121},a_{211})\end{cases}\right\} = \left\{\begin{array}{cc}\underbrace{(UP_+,UP_+)}_{positive\ saddle} & \underbrace{(_{DI:II}UP,_{SO:SI}LS)}_{up\text{-}parabola\ lower\text{-}saddle}\\[4pt] \underbrace{(DP:UP,pF)}_{hyperbolic\ flow\ (+)} & \underbrace{(DP_-,DP_+)}_{CW\ center}\end{array}\right\} \quad (5.84)$$

for $a_{110}<0,a_{220}<0;$

and the infinite-equilibrium of $x_1^* = a_{112} = a_{213}$ is summarized in Table 5.6.

The inflection-source and sink infinite-equilibriums of $x_1^* = a_{112} = a_{213}$ are presented in Fig. 5.20. The infinite-equilibriums are for the switching bifurcation for the equilibriums for $a_{112} \in (a_{212},a_{213})$ and $a_{213} < a_{112}$ with $a_{111} < a_{211}$. The simple saddle and center equilibriums are between an infinite-equilibrium and a hyperbolic or hyperbolic-secant flow.

(xi) For $a_{213} = a_{112} = a_{11l_1}$ with $a_{111} \in (a_{211},a_{212})$, the equilibriums of $x_1^* = a_{111},a_{112},$ a_{211},a_{212},a_{213} with $x_2^* = a_{121}$ have the following properties:

$$\left.\begin{cases}(a_{121},a_{212}) & (a_{121},a_{213})\\ (a_{121},a_{211}) & (a_{111},a_{121})\end{cases}\right\} = \left\{\begin{array}{cc}\underbrace{(UP_-,UP_-)}_{negative\ saddle} & \underbrace{(_{DI:II}UP,_{SI:SO}US)}_{up\text{-}parabola\ upper\text{-}saddle}\\[4pt] \underbrace{(UP_+,UP_+)}_{positive\ saddle} & \underbrace{(UP:DP,pF)}_{hyperbolic\text{-}secant\ flow\ (+)}\end{array}\right\} \quad (5.85)$$

for $a_{110}>0,a_{220}>0,$

Table 5.6 Infinite-equilibriums of $x_1^* = a_{213} = a_{112}$	$(a_{112},\bar{x}_2)$	$\bar{x}_2$	
	(a_{110},a_{220})	$(-\infty,a_{121})$	(a_{121},∞)
	$(+,+)$	(SI,DI)	(SO,II)
	$(-,+)$	(SO,II)	(SI,DI)
	$(+,-)$	(SI,II)	(SO,DI)
	$(-,-)$	(SO,DI)	(SI,II)

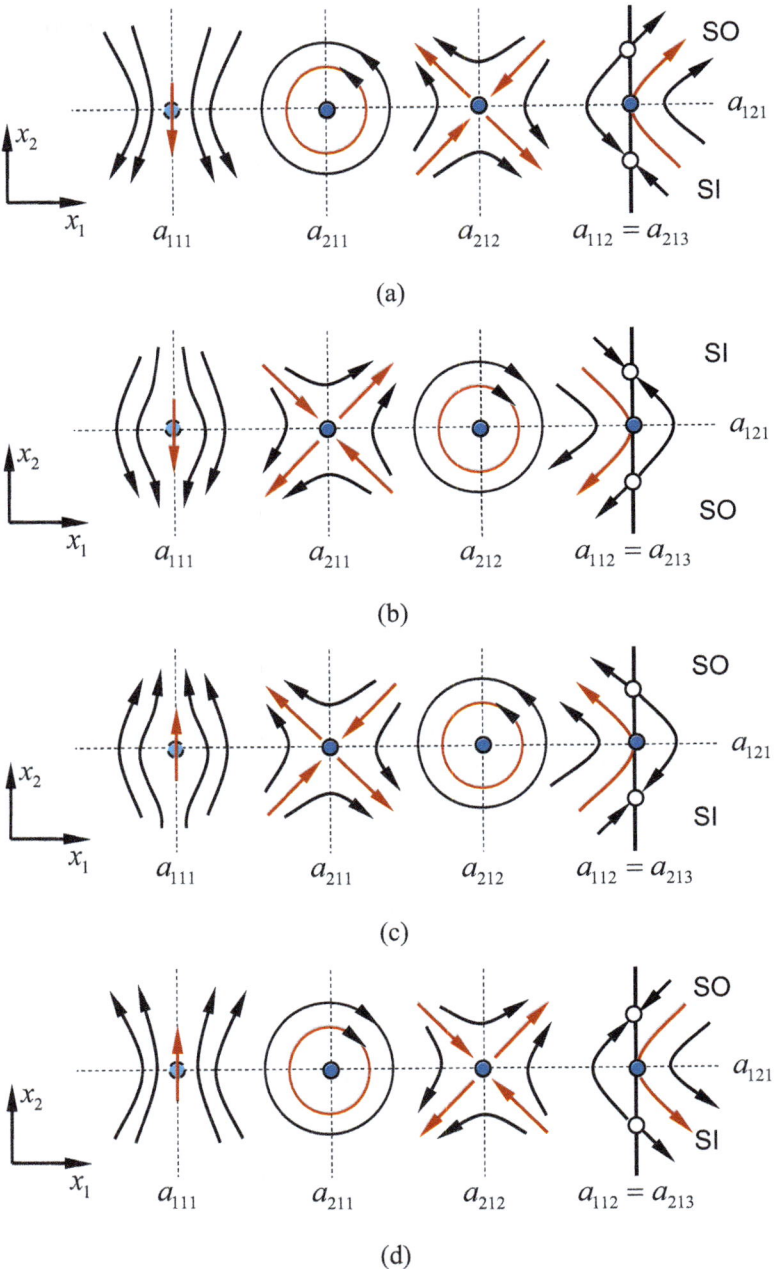

Fig. 5.20 Phase portraits ($a_{112} = a_{213}$, $a_{111} < a_{211}$) for two-dimensional systems on the x_1-direction with $x_1^* = a_{111}, a_{112}, a_{211}, a_{212}, a_{213}$ and on the x_2-direction with $x_2^* = a_{121}$. (**a**) ($a_{110} > 0$, $a_{220} > 0$), (**b**) ($a_{110} < 0, a_{220} > 0$), (**c**) ($a_{110} > 0, a_{220} < 0$), (**d**) ($a_{110} < 0, a_{220} < 0$)

$$\left\{ \begin{matrix} (a_{121}, a_{212}) & (a_{121}, a_{213}) \\ (a_{121}, a_{211}) & (a_{111}, a_{121}) \end{matrix} \right\} = \left\{ \begin{matrix} \underbrace{(\mathrm{DP}_-, \mathrm{DP}_+)}_{\text{CW center}} & \underbrace{\left(_{\mathrm{II:DI}}\mathrm{DP}, _{\mathrm{SO:SI}}\mathrm{LS}\right)}_{\text{down-parabola lower-saddle}} \\ \underbrace{(\mathrm{DP}_+, \mathrm{DP}_-)}_{\text{CCW center}} & \underbrace{(\mathrm{DP} : \mathrm{UP}, \mathrm{pF})}_{\text{hyperbolic flow } (+)} \end{matrix} \right\} \quad (5.86)$$

for $a_{110} < 0, a_{220} > 0$,

$$\left\{ \begin{matrix} (a_{121}, a_{212}) & (a_{121}, a_{213}) \\ (a_{121}, a_{211}) & (a_{111}, a_{121}) \end{matrix} \right\} = \left\{ \begin{matrix} \underbrace{(\mathrm{DP}_+, \mathrm{DP}_-)}_{\text{CCW center}} & \underbrace{\left(_{\mathrm{II:DI}}\mathrm{DP}, _{\mathrm{SI:SO}}\mathrm{US}\right)}_{\text{down-parabola upper-saddle}} \\ \underbrace{(\mathrm{DP}_-, \mathrm{DP}_+)}_{\text{CW center}} & \underbrace{(\mathrm{DP} : \mathrm{UP}, \mathrm{nF})}_{\text{hyperbolic flow } (-)} \end{matrix} \right\} \quad (5.87)$$

for $a_{110} > 0, a_{220} < 0$,

$$\left\{ \begin{matrix} (a_{121}, a_{212}) & (a_{121}, a_{213}) \\ (a_{121}, a_{211}) & (a_{111}, a_{121}) \end{matrix} \right\} = \left\{ \begin{matrix} \underbrace{(\mathrm{UP}_+, \mathrm{UP}_+)}_{\text{positive saddle}} & \underbrace{\left(_{\mathrm{DI:II}}\mathrm{UP}, _{\mathrm{SO:SI}}\mathrm{LS}\right)}_{\text{up-parabola lower-saddle}} \\ \underbrace{(\mathrm{UP}_-, \mathrm{UP}_-)}_{\text{negative saddle}} & \underbrace{(\mathrm{UP} : \mathrm{DP}, \mathrm{nF})}_{\text{hyperbolic-secant flow } (-)} \end{matrix} \right\} \quad (5.88)$$

for $a_{110} < 0, a_{220} < 0$;

and the infinite-equilibrium of $x_1^* = a_{112} = a_{213}$ is summarized as in Table 5.6. The inflection-source and sink infinite-equilibriums of $x_1^* = a_{112} = a_{213}$ are presented in Fig. 5.21. The infinite-equilibriums are for the switching bifurcation for the equilibriums for $a_{112} \in (a_{212}, a_{213})$ and $a_{213} < a_{112}$ with $a_{111} \in (a_{211}, a_{212})$. Two centers are separated by a hyperbolic flow, and two saddles are separated by a hyperbolic-secant flow. The infinite-equilibriums are on the right side.

(xii) For $a_{213} = a_{112} = a_{11l_1}$ with $a_{212} < a_{111}$, the equilibriums of $x_1^* = a_{111}, a_{112}$, a_{211}, a_{212}, a_{213} with $x_2^* = a_{121}$ have the following properties:

$$\left\{ \begin{matrix} (a_{111}, a_{121}) & (a_{121}, a_{213}) \\ (a_{121}, a_{211}) & (a_{121}, a_{212}) \end{matrix} \right\} = \left\{ \begin{matrix} \underbrace{(\mathrm{DP} : \mathrm{UP}, \mathrm{nF})}_{\text{hyperbolic flow } (-)} & \underbrace{\left(_{\mathrm{DI:II}}\mathrm{UP}, _{\mathrm{SI:SO}}\mathrm{US}\right)}_{\text{up-parabola upper-saddle}} \\ \underbrace{(\mathrm{UP}_+, \mathrm{UP}_+)}_{\text{positive saddle}} & \underbrace{(\mathrm{DP}_-, \mathrm{DP}_+)}_{\text{CW center}} \end{matrix} \right\} \quad (5.89)$$

for $a_{110} > 0, a_{220} > 0$,

$$\left\{ \begin{matrix} (a_{111}, a_{121}) & (a_{121}, a_{213}) \\ (a_{121}, a_{211}) & (a_{121}, a_{212}) \end{matrix} \right\} = \left\{ \begin{matrix} \underbrace{(\mathrm{UP} : \mathrm{DP}, \mathrm{nF})}_{\text{hyperbolic-secant flow } (-)} & \underbrace{\left(_{\mathrm{II:DI}}\mathrm{DP}, _{\mathrm{SO:SI}}\mathrm{LS}\right)}_{\text{down-parabola lower-saddle}} \\ \underbrace{(\mathrm{DP}_+, \mathrm{DP}_-)}_{\text{CCW center}} & \underbrace{(\mathrm{UP}_-, \mathrm{UP}_-)}_{\text{negative saddle}} \end{matrix} \right\}$$

for $a_{110} < 0, a_{220} > 0$,

$$(5.90)$$

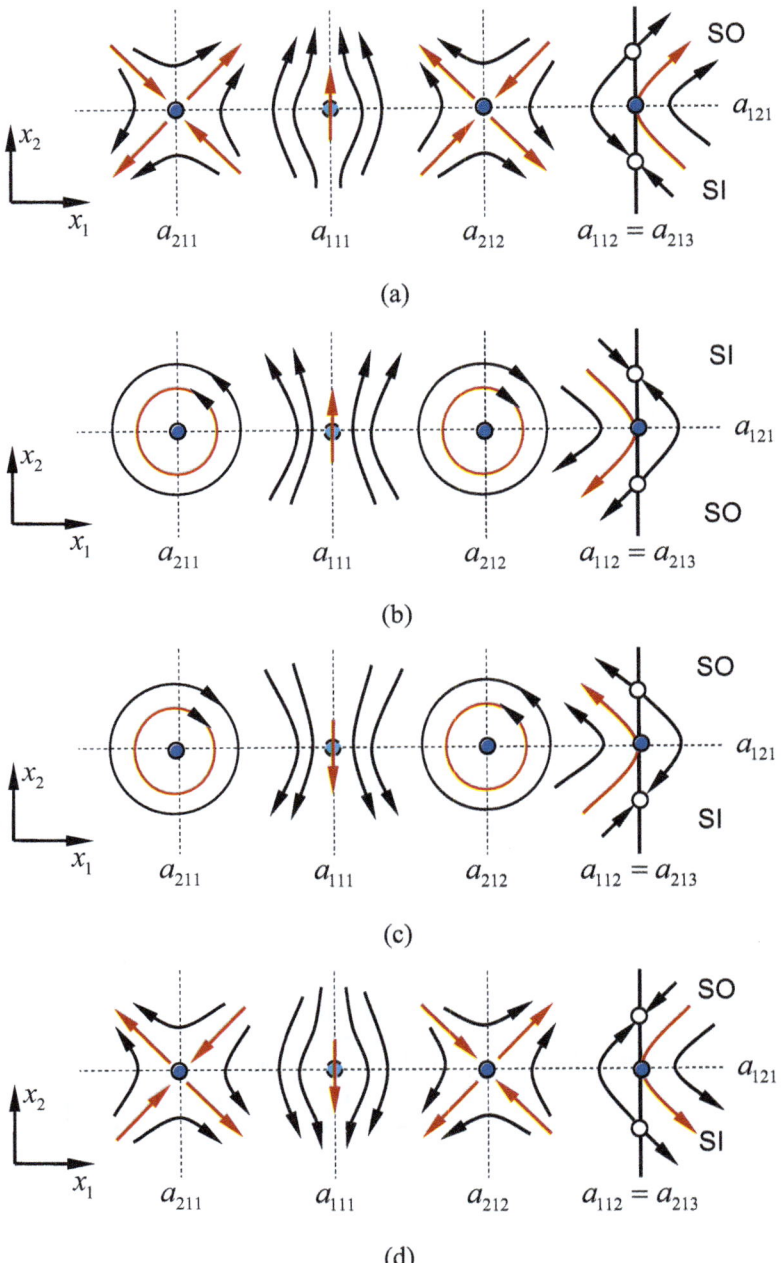

Fig. 5.21 Phase portraits ($a_{112} = a_{213}$, $a_{111} \in (a_{211}, a_{212})$) for two-dimensional systems on the x_1-direction with $x_1^* = a_{111}, a_{112}, a_{211}, a_{212}, a_{213}$ and on the x_2-direction with $x_2^* = a_{121}$. (**a**) ($a_{110} > 0$, $a_{220} > 0$), (**b**) ($a_{110} < 0, a_{220} > 0$), (**c**) ($a_{110} > 0, a_{220} < 0$), (**d**) ($a_{110} < 0, a_{220} < 0$)

$$
\left\{
\begin{array}{ll}
(a_{111}, a_{121}) & (a_{121}, a_{213}) \\
(a_{121}, a_{211}) & (a_{121}, a_{212})
\end{array}
\right\}
=
\left\{
\begin{array}{ll}
\underbrace{(\text{UP}:\text{DP},\text{pF})}_{\text{hyperbolic-secant flow }(+)} & \underbrace{(_{\text{II:DI}}\text{DP},\,_{\text{SI:SO}}\text{US})}_{\text{down-parabola upper-saddle}} \\
\underbrace{(\text{DP}_-, \text{DP}_+)}_{\text{CW center}} & \underbrace{(\text{UP}_+, \text{UP}_+)}_{\text{positive saddle}}
\end{array}
\right\} \quad (5.91)
$$

for $a_{110} > 0, a_{220} < 0,$

$$
\left\{
\begin{array}{ll}
(a_{111}, a_{121}) & (a_{121}, a_{213}) \\
(a_{121}, a_{211}) & (a_{121}, a_{212})
\end{array}
\right\}
=
\left\{
\begin{array}{ll}
\underbrace{(\text{DP}:\text{UP},\text{pF})}_{\text{hyperbolic flow }(+)} & \underbrace{(_{\text{DI:II}}\text{UP},\,_{\text{SO:SI}}\text{LS})}_{\text{up-parabola lower-saddle}} \\
\underbrace{(\text{UP}_-, \text{UP}_-)}_{\text{negative saddle}} & \underbrace{(\text{DP}_+, \text{DP}_-)}_{\text{CCW center}}
\end{array}
\right\} \quad (5.92)
$$

for $a_{110} < 0, a_{220} < 0$;

and the infinite-equilibrium of $x_1^* = a_{112} = a_{213}$ is summarized as in Table 5.6. The inflection-source and sink infinite-equilibriums of $x_1^* = a_{112} = a_{213}$ are presented in Fig. 5.22. The infinite-equilibriums are for the switching bifurcation for the equilibriums for $a_{112} \in (a_{212}, a_{213})$ and $a_{213} < a_{112}$ with $a_{212} < a_{111}$. Two centers are separated by a hyperbolic flow, and two saddles are separated by a hyperbolic-secant flow. The infinite-equilibriums are on the right side.

5.2.2 Double Inflection-Sink and Source Infinite-Equilibriums

Consider a dynamical system for $a_{11s_1} = a_{21l_1}$ and $a_{11s_2} = a_{21l_2}$ ($s_1, s_2 \in \{1,2,3\}$; $s_1 \neq s_2$; $l_1, l_2 \in \{1,2\}, l_1 \neq l_2$) with $s_1 < s_2$ and $l_1 < l_2$ as

$$
\begin{aligned}
\dot{x}_1 &= a_{110}(x_1 - a_{11l_1})(x_1 - a_{11l_2})(x_2 - a_{121}), \\
\dot{x}_2 &= a_{220}(x_1 - a_{21s_1})(x_1 - a_{21s_2})(x_1 - a_{21s_3}),
\end{aligned} \quad (5.93)
$$

and the corresponding first integral manifold is

$$
\frac{1}{2}\left[(x_1 - a_{21s_3})^2 - (x_{10} - a_{21s_3})^2\right] = \frac{1}{2}\frac{a_{110}}{a_{220}}\left[(x_2 - a_{121})^2 - (x_{20} - a_{121})^2\right]. \quad (5.94)
$$

(i) For $a_{211} = a_{111}$ with $a_{212} = a_{112}$, the equilibriums of $x_1^* = a_{111}, a_{112}, a_{211}, a_{212}, a_{213}$ with $x_2^* = a_{121}$ have the following properties:

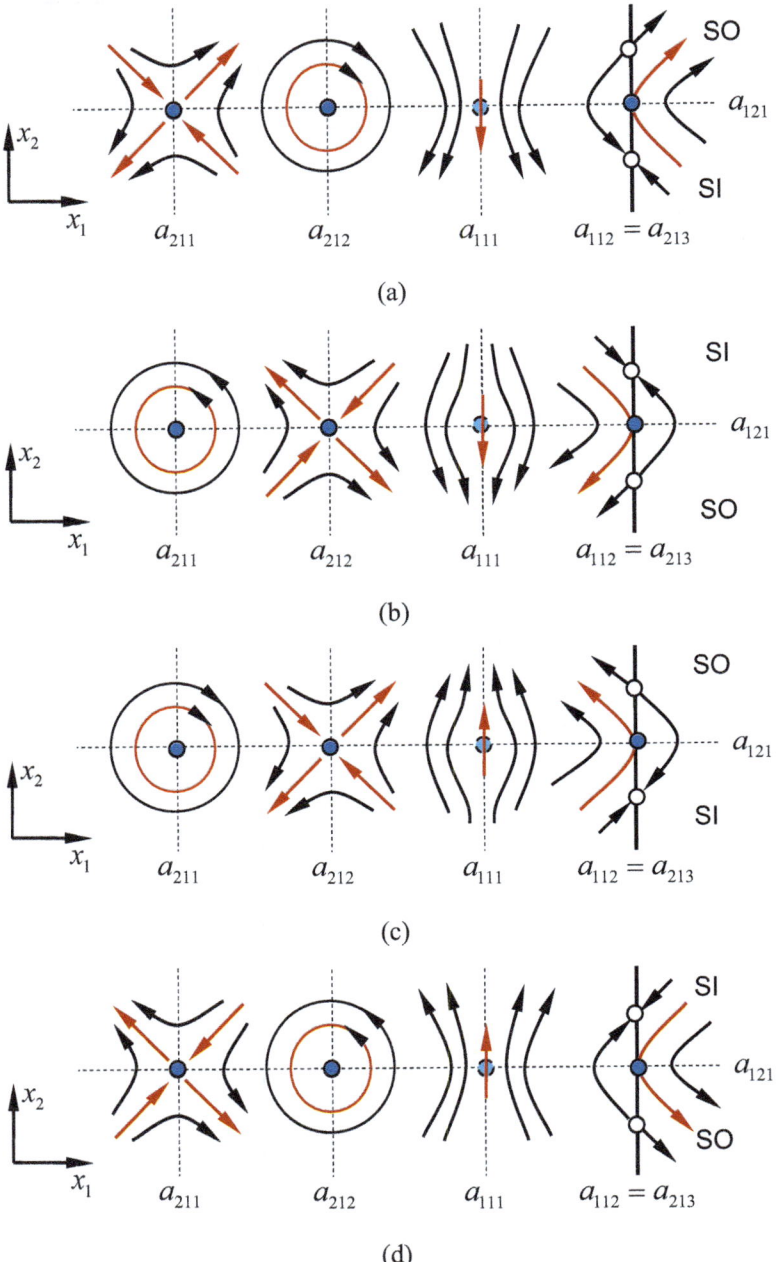

Fig. 5.22 Phase portraits ($a_{112} = a_{213}$, $a_{212} < a_{111}$) for two-dimensional systems on the x_1-direction with $x_1^* = a_{111}, a_{112}, a_{211}, a_{212}, a_{213}$ and on the x_2-direction with $x_2^* = a_{121}$. (**a**) ($a_{110} > 0$, $a_{220} > 0$), (**b**) ($a_{110} < 0, a_{220} > 0$), (**c**) ($a_{110} > 0, a_{220} < 0$), (**d**) ($a_{110} < 0, a_{220} < 0$)

$$\left.\begin{array}{l}(a_{121},a_{213})\\(a_{121},a_{212})\\(a_{121},a_{211})\end{array}\right\}=\left\{\begin{array}{c}\underbrace{(UP_+,UP_+)}_{\text{positive saddle}}\\\underbrace{\left(_{\text{II:DI}}DP,\,_{\text{SI:SO}}US\right)}_{\text{down-parabola upper-saddle}}\\\underbrace{\left(_{\text{II:DI}}DP,\,_{\text{SO:SI}}LS\right)}_{\text{down-parabola lower-saddle}}\end{array}\right\}\text{ for }a_{110}>0,a_{220}>0,\qquad(5.95)$$

$$\left.\begin{array}{l}(a_{121},a_{213})\\(a_{121},a_{212})\\(a_{121},a_{211})\end{array}\right\}=\left\{\begin{array}{c}\underbrace{(DP_+,DP_-)}_{\text{CCW center}}\\\underbrace{\left(_{\text{DI:II}}UP,\,_{\text{SO:SI}}LS\right)}_{\text{up-parabola lower-saddle}}\\\underbrace{\left(_{\text{DI:II}}UP,\,_{\text{SI:SO}}US\right)}_{\text{up-parabola upper-saddle}}\end{array}\right\}\text{ for }a_{110}<0,a_{220}>0,\qquad(5.96)$$

$$\left.\begin{array}{l}(a_{121},a_{213})\\(a_{121},a_{212})\\(a_{121},a_{211})\end{array}\right\}=\left\{\begin{array}{c}\underbrace{(DP_-,DP_+)}_{\text{CW center}}\\\underbrace{\left(_{\text{DI:II}}UP,\,_{\text{SI:SO}}US\right)}_{\text{up-parabola upper-saddle}}\\\underbrace{\left(_{\text{DI:II}}UP,\,_{\text{SO:SI}}LS\right)}_{\text{up-parabola lower-saddle}}\end{array}\right\}\text{ for }a_{110}>0,a_{220}<0,\qquad(5.97)$$

$$\left.\begin{array}{l}(a_{121},a_{213})\\(a_{121},a_{212})\\(a_{121},a_{211})\end{array}\right\}=\left\{\begin{array}{c}\underbrace{(UP_-,UP_-)}_{\text{negative saddle}}\\\underbrace{\left(_{\text{II:DI}}DP,\,_{\text{SO:SI}}LS\right)}_{\text{down-parabola lower-saddle}}\\\underbrace{\left(_{\text{II:DI}}DP,\,_{\text{SI:SO}}US\right)}_{\text{down-parabola upper-saddle}}\end{array}\right\}\text{ for }a_{110}<0,a_{220}<0;\qquad(5.98)$$

and the infinite-equilibriums of $x_1^*=a_{111}=a_{211}$ and $x_1^*=a_{112}=a_{212}$ are summarized in Table 5.7. The phase portrait of the inflection-source and sink infinite-equilibriums of $x_1^*=a_{111}=a_{211}$ and $x_1^*=a_{112}=a_{212}$ are presented in Fig. 5.23.

(ii) For $a_{211}=a_{111}$ with $a_{212}=a_{112}$, the equilibriums of $x_1^*=a_{111},a_{112},a_{211},$ a_{212},a_{213} with $x_2^*=a_{121}$ have the following properties:

Table 5.7 Infinite-equilibriums of $x_1^*=a_{111}=a_{211}$ and $x_1^*=a_{112}=a_{212}$

	$\bar{x}_2$ for $(a_{111},\bar{x}_2)$		$\bar{x}_2$ for $(a_{112},\bar{x}_2)$	
(a_{110},a_{220})	$(-\infty,a_{121})$	(a_{121},∞)	$(-\infty,a_{121})$	(a_{121},∞)
$(+,+)$	(SO,II)	(SI,DI)	(SI,II)	(SO,DI)
$(-,+)$	(SI,DI)	(SO,II)	(SO,DI)	(SI,II)
$(+,-)$	(SO,DI)	(SI,II)	(SI,DI)	(SO,II)
$(-,-)$	(SI,II)	(SO,DI)	(SO,II)	(SO,DI)

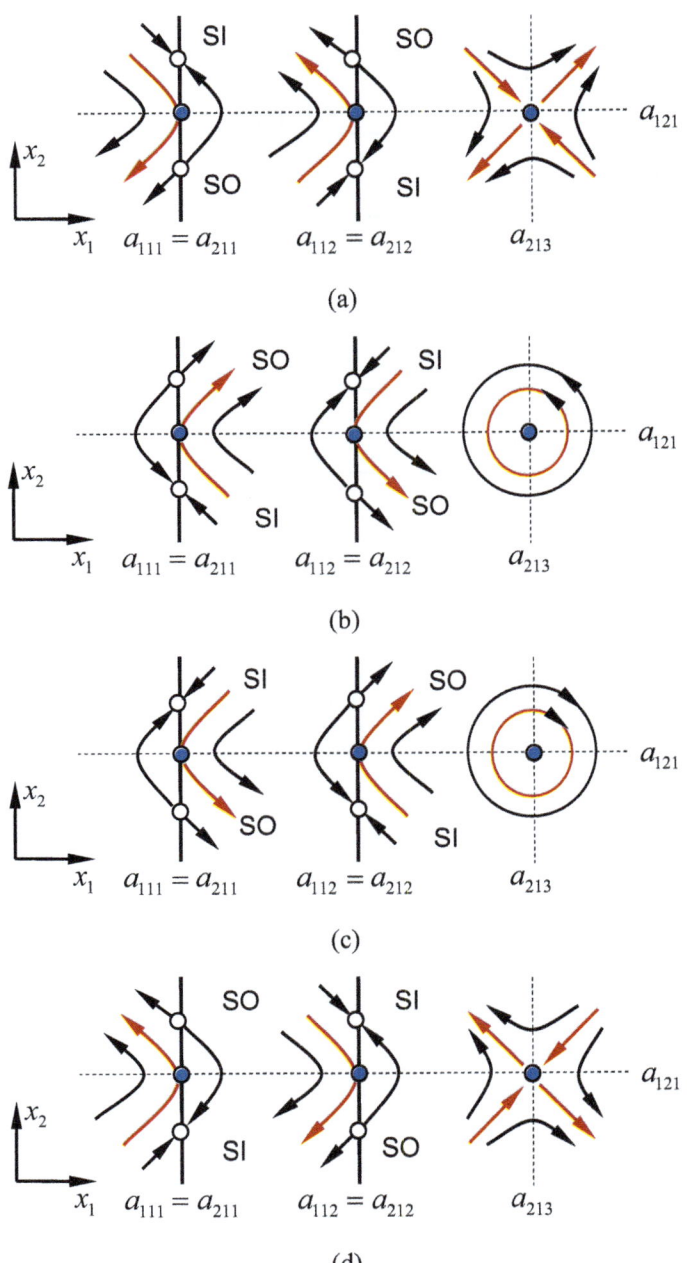

Fig. 5.23 Phase portraits ($a_{111} = a_{212}$, $a_{112} = a_{212}$) for two-dimensional systems on the x_1-direction with $x_1^* = a_{111}, a_{112}, a_{211}, a_{212}, a_{213}$ and on the x_2-direction with $x_2^* = a_{121}$. (**a**) ($a_{110} > 0$, $a_{220} > 0$), (**b**) ($a_{110} < 0, a_{220} > 0$), (**c**) ($a_{110} > 0, a_{220} < 0$), (**d**) ($a_{110} < 0, a_{220} < 0$)

$$
\left\{
\begin{array}{l}
(a_{121}, a_{213}) \\
(a_{121}, a_{212}) \\
(a_{121}, a_{211})
\end{array}
\right\}
=
\left\{
\begin{array}{l}
\underbrace{\left(_{\text{DI:II}}\text{UP}, _{\text{SI:SO}}\text{US}\right)}_{\text{up-parabola upper-saddle}} \\[4pt]
\underbrace{(\text{UP}_-, \text{UP}_-)}_{\text{negative saddle}} \\[4pt]
\underbrace{\left(_{\text{II:DI}}\text{DP}, _{\text{SO:SI}}\text{LS}\right)}_{\text{down-parabola lower-saddle}}
\end{array}
\right\}
\quad \text{for } a_{110} > 0,\, a_{220} > 0, \qquad (5.99)
$$

$$
\left\{
\begin{array}{l}
(a_{121}, a_{213}) \\
(a_{121}, a_{212}) \\
(a_{121}, a_{211})
\end{array}
\right\}
=
\left\{
\begin{array}{l}
\underbrace{\left(_{\text{II:DI}}\text{DP}, _{\text{SO:SI}}\text{LS}\right)}_{\text{down-parabola lower-saddle}} \\[4pt]
\underbrace{(\text{DP}_-, \text{DP}_+)}_{\text{CW center}} \\[4pt]
\underbrace{\left(_{\text{DI:II}}\text{UP}, _{\text{SI:SO}}\text{US}\right)}_{\text{up-parabola upper-saddle}}
\end{array}
\right\}
\quad \text{for } a_{110} < 0,\, a_{220} > 0, \qquad (5.100)
$$

$$
\left\{
\begin{array}{l}
(a_{121}, a_{213}) \\
(a_{121}, a_{212}) \\
(a_{121}, a_{211})
\end{array}
\right\}
=
\left\{
\begin{array}{l}
\underbrace{\left(_{\text{II:DI}}\text{DP}, _{\text{SI:SO}}\text{US}\right)}_{\text{down-parabola upper-saddle}} \\[4pt]
\underbrace{(\text{DP}_+, \text{DP}_-)}_{\text{CCW center}} \\[4pt]
\underbrace{\left(_{\text{DI:II}}\text{UP}, _{\text{SO:SI}}\text{LS}\right)}_{\text{up-parabola lower-saddle}}
\end{array}
\right\}
\quad \text{for } a_{110} > 0,\, a_{220} < 0, \qquad (5.101)
$$

$$
\left\{
\begin{array}{l}
(a_{121}, a_{213}) \\
(a_{121}, a_{212}) \\
(a_{121}, a_{211})
\end{array}
\right\}
=
\left\{
\begin{array}{l}
\underbrace{\left(_{\text{DI:II}}\text{UP}, _{\text{SO:SI}}\text{LS}\right)}_{\text{up-parabola lower-saddle}} \\[4pt]
\underbrace{(\text{UP}_+, \text{UP}_+)}_{\text{negative saddle}} \\[4pt]
\underbrace{\left(_{\text{II:DI}}\text{DP}, _{\text{SI:SO}}\text{US}\right)}_{\text{down-parabola upper-saddle}}
\end{array}
\right\}
\quad \text{for } a_{110} < 0,\, a_{220} < 0; \qquad (5.102)
$$

and the infinite-equilibriums of $x_1^* = a_{111} = a_{211}$ and $x_1^* = a_{112} = a_{213}$ are summarized in Table 5.8. The phase portrait of the inflection-source and sink infinite-equilibriums of $x_1^* = a_{111} = a_{211}$ and $x_1^* = a_{112} = a_{213}$ are presented in Fig. 5.24.

(iii) For $a_{212} = a_{111}$ with $a_{213} = a_{112}$, the equilibriums of $x_1^* = a_{111}, a_{112}, a_{211}, a_{212}, a_{213}$ with $x_2^* = a_{121}$ have the following properties:

Table 5.8 Infinite-equilibriums of $x_1^* = a_{111} = a_{211}$ and $x_1^* = a_{112} = a_{212}$

(a_{110}, a_{220})	$\bar{x}_2$ for $(a_{111}, \bar{x}_2)$		$\bar{x}_2$ for $(a_{112}, \bar{x}_2)$	
	$(-\infty, a_{121})$	(a_{121}, ∞)	$(-\infty, a_{121})$	(a_{121}, ∞)
(+,+)	(SO,II)	(SI,DI)	(SI,DI)	(SO,II)
(−,+)	(SI,DI)	(SO,II)	(SO,II)	(SI,DI)
(+,−)	(SO,DI)	(SI,II)	(SI,II)	(SO,DI)
(−,−)	(SI,II)	(SO,DI)	(SO,DI)	(SI,II)

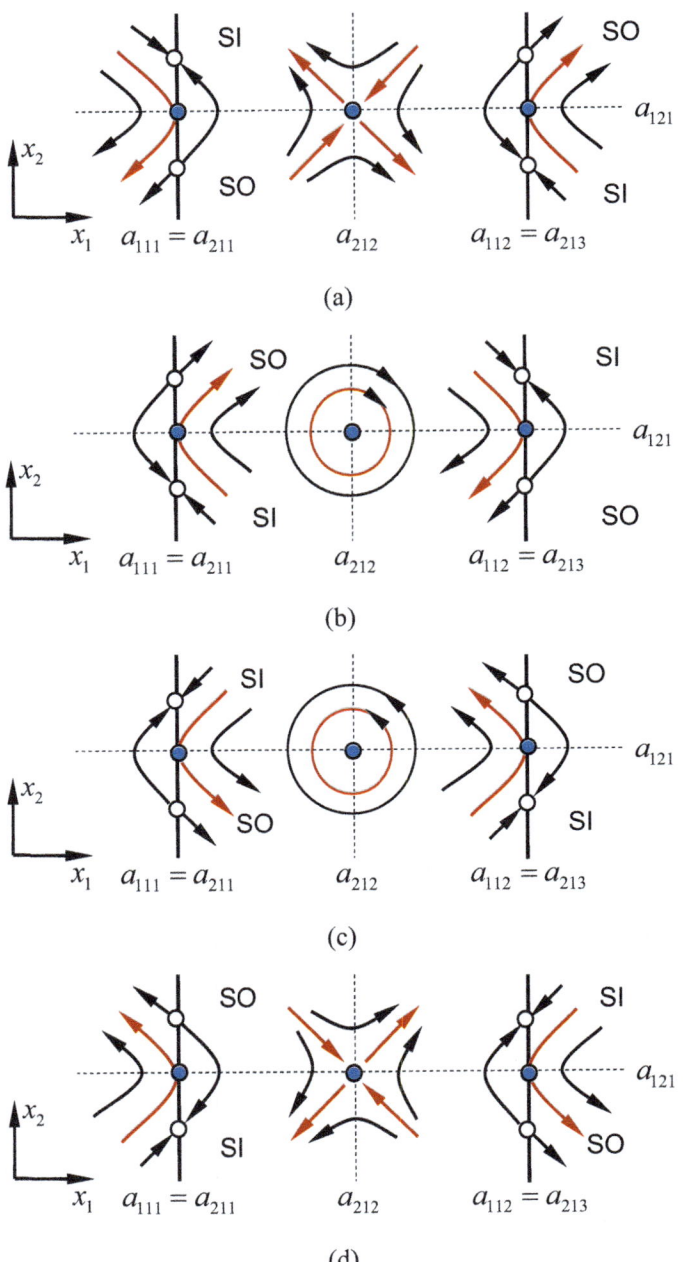

Fig. 5.24 Phase portraits ($a_{111} = a_{211}$, $a_{112} = a_{213}$) for two-dimensional systems on the x_1-direction with $x_1^* = a_{111}, a_{112}, a_{211}, a_{212}, a_{213}$ and on the x_2-direction with $x_2^* = a_{121}$. (**a**) ($a_{110} > 0$, $a_{220} > 0$), (**b**) ($a_{110} < 0, a_{220} > 0$), (**c**) ($a_{110} > 0, a_{220} < 0$), (**d**) ($a_{110} < 0, a_{220} < 0$)

$$\left\{\begin{matrix}(a_{121},a_{213})\\(a_{121},a_{212})\\(a_{121},a_{211})\end{matrix}\right\}=\left\{\begin{matrix}\underbrace{\left(_{\text{DI:II}}\text{UP},_{\text{SI:SO}}\text{US}\right)}_{\text{up-parabola upper-saddle}}\\\underbrace{\left(_{\text{DI:II}}\text{UP},_{\text{SO:SI}}\text{LS}\right)}_{\text{up-parabola lower-saddle}}\\\underbrace{(\text{UP}_+,\text{UP}_+)}_{\text{positive saddle}}\end{matrix}\right\}\text{ for }a_{110}>0,a_{220}>0,\qquad(5.103)$$

$$\left\{\begin{matrix}(a_{121},a_{213})\\(a_{121},a_{212})\\(a_{121},a_{211})\end{matrix}\right\}=\left\{\begin{matrix}\underbrace{\left(_{\text{II:DI}}\text{DP},_{\text{SO:SI}}\text{LS}\right)}_{\text{down-parabola lower-saddle}}\\\underbrace{\left(_{\text{II:DI}}\text{DP},_{\text{SI:SO}}\text{US}\right)}_{\text{down-parabola upper-saddle}}\\\underbrace{(\text{DP}_+,\text{DP}_-)}_{\text{CCW center}}\end{matrix}\right\}\text{ for }a_{110}<0,a_{220}>0,\qquad(5.104)$$

$$\left\{\begin{matrix}(a_{121},a_{213})\\(a_{121},a_{212})\\(a_{121},a_{211})\end{matrix}\right\}=\left\{\begin{matrix}\underbrace{\left(_{\text{II:DI}}\text{DP},_{\text{SI:SO}}\text{US}\right)}_{\text{down-parabola upper-saddle}}\\\underbrace{\left(_{\text{II:DI}}\text{DP},_{\text{SO:SI}}\text{LS}\right)}_{\text{down-parabola lower-saddle}}\\\underbrace{(\text{DP}_-,\text{DP}_+)}_{\text{CW center}}\end{matrix}\right\}\text{ for }a_{110}>0,a_{220}<0,\qquad(5.105)$$

$$\left\{\begin{matrix}(a_{121},a_{213})\\(a_{121},a_{212})\\(a_{121},a_{211})\end{matrix}\right\}=\left\{\begin{matrix}\underbrace{\left(_{\text{DI:II}}\text{UP},_{\text{SO:SI}}\text{LS}\right)}_{\text{up-parabola lower-saddle}}\\\underbrace{\left(_{\text{DI:II}}\text{UP},_{\text{SI:SO}}\text{US}\right)}_{\text{up-parabola upper-saddle}}\\\underbrace{(\text{UP}_-,\text{UP}_-)}_{\text{negative saddle}}\end{matrix}\right\}\text{ for }a_{110}<0,a_{220}<0;\qquad(5.106)$$

and the infinite-equilibriums of $x_1^*=a_{111}=a_{211}$ and $x_1^*=a_{112}=a_{213}$ are summarized in Table 5.9. The phase portrait of the inflection-source and sink infinite-equilibriums of $x_1^*=a_{111}=a_{211}$ and $x_1^*=a_{112}=a_{213}$ are presented in Fig. 5.25.

Table 5.9 Infinite-equilibriums of $x_1^*=a_{111}=a_{212}$ and $x_1^*=a_{112}=a_{213}$

	$\bar{x}_2$ for $(a_{111},\bar{x}_2)$		$\bar{x}_2$ for $(a_{112},\bar{x}_2)$	
(a_{110},a_{220})	$(-\infty,a_{121})$	(a_{121},∞)	$(-\infty,a_{121})$	(a_{121},∞)
(+,+)	(SO,DI)	(SI,II)	(SI,DI)	(SO,II)
(−,+)	(SI,II)	(SO,DI)	(SO,II)	(SI,DI)
(+,−)	(SO,II)	(SI,DI)	(SI,II)	(SO,DI)
(−,−)	(SI,DI)	(SO,II)	(SO,DI)	(SI,II)

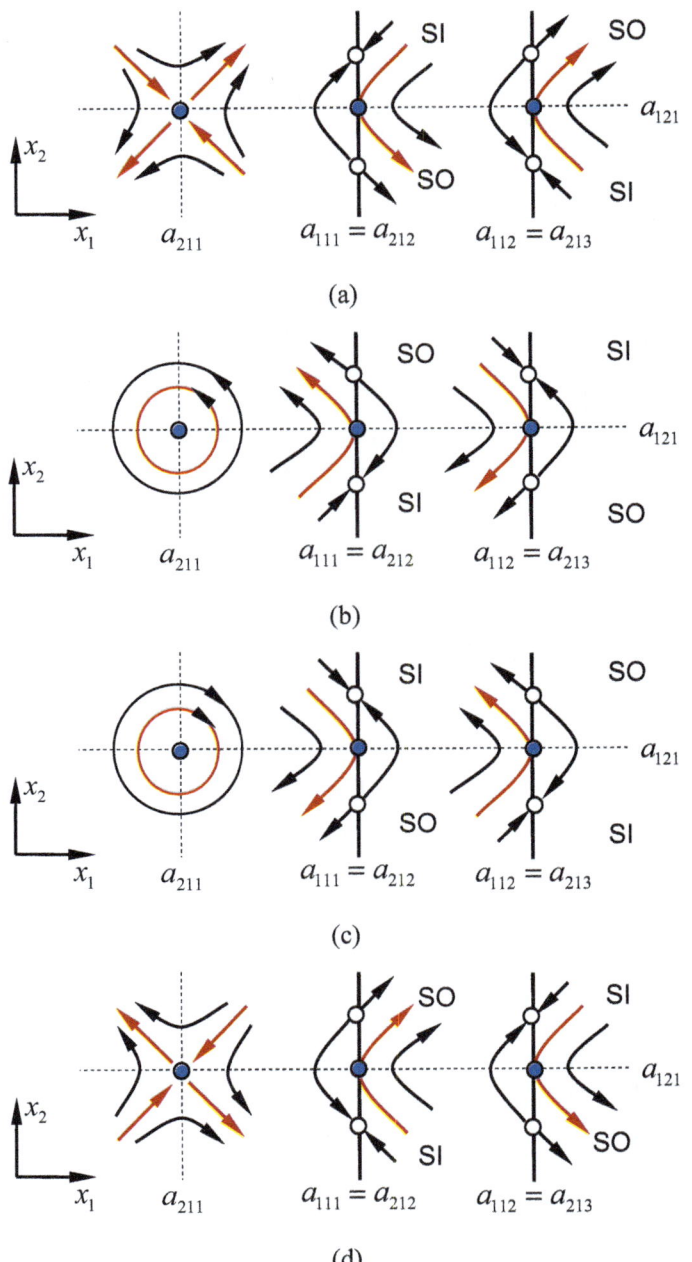

Fig. 5.25 Phase portraits ($a_{111} = a_{212}$, $a_{112} = a_{213}$) for two-dimensional systems on the x_1-direction with $x_1^* = a_{111}, a_{112}, a_{211}, a_{212}, a_{213}$ and on the x_2-direction with $x_2^* = a_{121}$. (**a**) ($a_{110} > 0$, $a_{220} > 0$), (**b**) ($a_{110} < 0, a_{220} > 0$), (**c**) ($a_{110} > 0, a_{220} < 0$), (**d**) ($a_{110} < 0, a_{220} < 0$)

Index

© The Editor(s) (if applicable) and The Author(s), under exclusive license to
Springer Nature Switzerland AG 2025
A. C. J. Luo, *Two-dimensional Crossing and Product Cubic Systems, Vol. II*,
https://doi.org/10.1007/978-3-031-57100-8

The manufacturer's authorised representative in the EU is Springer
Nature Customer Service Centre GmbH, Europaplatz 3, 69115 Heidelberg,
Germany. If you have any concerns regarding our products, please
contact ProductSafety@springernature.com

Printed and bound by CPI Group (UK) Ltd, Croydon, CR0 4YY

10/06/2025

01898343-0001